Lost Maps of the

Lost Maps of the Caliphs

DRAWING THE WORLD IN ELEVENTH-CENTURY CAIRO

Yossef Rapoport and
Emilie Savage-Smith

Bodleian Library
UNIVERSITY OF OXFORD

First published in the UK in 2018 by the Bodleian Library
Broad Street, Oxford OX1 3BG
www.bodleianshop.co.uk

ISBN: 978 1 85124 491 1

and in the US by
The University of Chicago Press, Chicago 60637
The University of Chicago Press, Ltd., London

Published 2018
Printed in the United States of America

Cover design by Dot Little at the Bodleian Library
Designed by Matt Avery and typeset by Graphic Composition, Inc., in Arno Pro (text) and Galliard ITC (display) fonts.
Printed and bound by Sheridan Books

British Library Catalogue in Publishing Data
A CIP record of this publication is available from the British Library

♾ This paper meets the requirements of ANSI/NISO Z39.48-1992 (Permanence of Paper).

Contents

Introduction

About a millennium ago, sometime between 1020 and 1050, in Cairo, a large illustrated book was completed. Its subject was nothing less than the entire universe, or what was then known of it. In the course of ten chapters, the author moved the reader from the outermost sphere of the stars through the spheres of the five planets visible to the naked eye down to the sublunary world of winds and comets. This was followed by twenty-five chapters on the Earth, beginning with calculation of the Earth's circumference, moving to overall views of the inhabited world, then to coastal lands and islands of the Indian Ocean and the Mediterranean, lakes of the world, the five major rivers, and ending with strange plants, animals, and birds inhabiting the Earth. In his treatise, the author assembled materials, including many large maps and diagrams, "encompassing the principles of the Raised-Up Roof [the Heavens] and the Laid-Down Bed [the Earth]"—a book, he tells his patron, "that will reveal to you their intricate and difficult aspects."[1] The rhyming title that he (and it was surely a "he") gave the book, *Kitāb Gharāʾib al-funūn wa-mulaḥ al-ʿuyūn,* loosely translates as "The book of curiosities of the sciences and marvels for the eye." For convenience, it will be referred to simply as the *Book of Curiosities.*

We do not know the name of the author nor that of the patron, and the treatise itself was unknown to modern scholars until a remarkable manuscript copy of it surfaced in the year 2000. This highly illustrated copy was probably made in Egypt around 1200, some hundred and fifty years after the original was completed, and some eight hundred years later, in June of 2002, it was acquired by the Bodleian Library at the University of Oxford. The story of the "discovery" of this highly illustrated manuscript of the *Book of Curiosities,* and the evidence for dating the copy as well as the composition of the treatise itself, will be discussed in the next chapter.

The manuscript evoked immense interest, particularly from historians of cartography, for it contains a remarkable series of early maps and astronomical diagrams, most of which are unparalleled in any Greek, Latin, or Arabic material. It was followed by the identification of later copies of the treatise (lacking most of the illustrations) that had lain unnoticed in various libraries among the thousands of unstudied Arabic manuscripts awaiting scholarly attention. In 2014 we published a critical edition along with a facsimile and annotated translation of the maps and text (using all available copies).[2] This had been preceded in 2007 by an electronic high-quality reproduction of the Bodleian manuscript and its illustrations, linked by mouse-overs to a modern Arabic edition (without full use of other copies) and a preliminary English translation. This preliminary edition and translation (no longer available at www.bodley.ox.ac.uk/bookofcuriosities) has now been superseded by the printed critical edition and translation of 2014. Images of all the folios of the manuscript are available through the Bodleian Library for digital images (https://digital.bodleian.ox.ac.uk/).

While the recent critical edition makes the treatise and its maps available for further study, it does not explain what they mean. In the 2014 publication, we limited ourselves to deciphering and translating the details of the text, and left out the analysis of the treatise as a whole, whether within the context of Fatimid Egyptian society and learning or within the traditions of astrology, astronomy, geography, and cartography as they had developed up to that time in the medieval Islamic world. In this book, we show what the *Book of Curiosities* can teach us about eleventh-century Egyptian views of the macrocosm and microcosm—that is, the celestial universe and the diminutive world of humankind. By doing so, we will open a window into the worldview of eleventh-century Islamic society and learning, its geographic horizons, and some of its scientific endeavors.

The image of the macrocosm and microcosm is essential for understanding the structure and composition of the *Book of Curiosities.* The macrocosm is the entire celestial universe, while the microcosm is the earthly realm of humans regarded as a miniature reflection of the universe. Astronomy, astrology, and star lore are the topics of the first half of the treatise, "On the Heavens." In the second chapter of the present volume, Emilie Savage-Smith shows that the *Book of Curiosities* sheds considerable light on the practice of astrology and the role of star lore in medieval Islam. The second half of the *Book of Curiosities,* "On the Earth," is discussed by Yossef Rapoport in chapters 3 through 10, which are devoted to the enigmatic maps, their geographical information, and the evidence for global trade and mission networks found in this remarkable treatise. The depictions of both the celestial sphere and of the terrestrial world reflect the Fatimid Egyptian intellectual milieu in which the author operated and the Ismaʿili missionary network of which he may have been a member.

Fatimid Cairo, where this unique treatise was composed, was an exceptional intellectual and political center. The Fatimids had built the city of Mahdia, in modern Tunisia, as their capital in 909–12, and indeed the *Book of Curiosities* has the earliest preserved depiction of the city (see fig. 2.4 and plate 9). From this North African base, the Fatimids came to occupy Sicily and to undertake naval operations against the Byzantines. In 969 the Fatimids entered Old Cairo (Fusṭāṭ) and built a new capital nearby—that of New Cairo, known as "The Victorious" (*al-Qāhirah*). Uniquely among Middle Eastern Islamic empires, the Fatimids made a strategic decision to rely on maritime power rather than land armies, and they were to become the most powerful Islamic empire in the Mediterranean at the time. From their capital in Cairo, the Fatimids ruled over Egypt and Syria, and faced the forces of the First Crusade in 1099. Greatly weakened in the twelfth century, they were eventually overthrown by Saladin in 1187.

Unlike the majority of Sunni Islamic states, the Fatimids were an Ismaʿili-Shiʿa dynasty that used a missionary network to propagate their sect throughout the Islamic world. By virtue of claiming descent from the Prophet's daughter Fāṭimah and his cousin ʿAlī, the Fatimid caliph imams saw themselves as the righteous leaders of the Muslims and the salvation guide for humankind. To fulfil this divine promise, the Fatimids relied on a clandestine missionary network, called *daʿwah*, established in the mid-ninth century. It continued to operate following the foundation of the Fatimid state in North Africa, and, after the conquest of Egypt, its headquarters were relocated to Cairo. The training of missionaries was sometimes associated with the House of Knowledge (Dār al-ʿIlm), founded by the caliph al-Ḥākim in 1005, which became a focal point for encyclopedic learning by missionaries as well as non-Ismaʿilis. As a result of caliphal patronage and the missionary network, Fatimid Cairo attracted some the most influential scientists, philosophers, and poets of the medieval Islamic world.

The chapters that follow go beyond the analysis of the work of one individual in one extraordinary capital of the Islamic world. We aim here to offer a reconsideration of the development of astronomy, astrology, geography, and cartography in the first four centuries of Islam. Through the comprehensive and structured discussion of the skies in the *Book of Curiosities*, this book outlines the medieval Islamic understanding of the basic structure of the cosmos and celestial phenomena, as they were understood outside the circles of practicing astronomers and astrologers. Moreover, the *Book of Curiosities* forcefully illustrates the pervasive assumptions about the effects on life on Earth of almost any visible celestial phenomenon—be it zodiacal or non-zodiacal constellations, the smaller star groups called lunar mansions, individual stars, planets, or comets. The amalgamation of Hellenistic, Coptic, Hindu, and other star lore was all channeled toward an astrological mindset, not limited to the construction of horoscopes or to the calculation of prayer times.

In the field of cartography, this book is the most authoritative reappraisal of the history of early, pre-Idrīsī, Islamic mapmaking since the publication of J. B. Harley and David Woodward's volume on *Cartography in the Traditional Islamic and South Asian Societies* in 1992. It reassesses the transmission of Late Antique geography to the Islamic world and covers areas that were scarcely touched upon in that volume, including the development of maritime diagrams before the era of the portolan charts, plans of towns and cities in the medieval Islamic period, and the depiction of the Nile. It points to Islamic parallels for recent work on the reception of Ptolemy's *Geography* in the medieval Christian world, and presents an Islamic angle on debates concerning the origins of the European portolan chart.[3]

Lost Maps of the Caliphs is also a contribution to the history of global communication networks at the turn of the previous millennium. We use the geographical materials of the *Book of Curiosities* to depict the Fatimid Empire as a global maritime power, with tentacles of military and religious authority in the Eastern Mediterranean, the Indus Valley, and along the East African coast. The discussion of the Fatimid Mediterranean adds an important dimension to recent scholarship on the Mediterranean horizons of contemporary Geniza merchants and on Mediterranean networks in general.[4] The extent of Fatimid knowledge of Byzantine coasts revises current scholarship on Fatimid-Byzantine relations and complements the findings of maritime archaeology. Surprisingly for a treatise written in Egypt, the material on East Asia sheds new light on Sino-Indian trade routes, and the treatise's familiarity with the East African coasts contributes to recent debates on the Islamization of the Swahili coast.[5]

Over the past decade, we, as well as other scholars, have published studies of individual maps in the treatise. This includes studies of the map of Sicily, a chapter on the names of Aegean islands that appear in the map of the Mediterranean, a study of the representation of East Asia, and a couple of articles on the rectangular world map.[6] Some of the maps have also been discussed in general surveys of the history of maps.[7] The maps of Mahdia and the Mediterranean feature on the cover jackets of three recent books on Fatimid history and Mediterranean trade as well as world history through maps, attesting to the enduring aesthetic appeal of the treatise.[8] This present volume draws on these earlier contributions, while often superseding and correcting them.

But *Lost Maps of the Caliphs* is also more than the sum of its parts. The overarching theme running throughout the *Book of Curiosities,* and reflected in the construction of our book, is the fine balance between the sky and the Earth, the neat procession from the outermost spheres to the Earth, its seas, islands, bays, lakes, and rivers, and the political and intellectual context of Fatimid Cairo. The Earth *together with* the Heavens formed the universe of eleventh-century Cairo. To a medieval person, whose night skies were not blanked out by city lights and pollution, the contents of the night sky—the "Raised-Up Roof" as our author, following the Qur'an, called it—revealed the workings of the universe and, if properly understood, heralded events on Earth.

CHAPTER ONE

A Discovery

A simple telephone call on a quiet autumn afternoon set in train a surprising chain of events that over the next decade disrupted the otherwise routine existence of a number of academics and injected some astonishing surprises into the field of medieval cartography. On the last Friday in September of 2000 the telephone rang, and a specialist in Islamic manuscripts at Christie's auction house in London asked if I—that is, Emilie Savage-Smith, who will be narrating the following story—had seen their catalogue for an upcoming sale of Islamic manuscripts to be held on the 10th of October. I had not, for I did not routinely follow the art market. I was then asked if I could come into London from Oxford and look at a manuscript that was up for sale on the 10th, for they thought they had "made a mistake." In response to this extraordinary request, I said that even if they had "made a mistake," there was nothing anyone could do about it at that point since the catalogue had been published some weeks earlier. Still, this somewhat bizarre request haunted me. As a result, I arranged to go round to the King Street office of Christie's the following Monday afternoon.

The manuscript that I was shown seemed at first glance a rather scruffy thing, bound in ill-fitting covers. The bird dropping visible on the cover (see fig. 1.1) suggested that its previous owners had stored it in a loft—or even a garden shed—and forgotten about it for some time.

Inside the covers were forty-eight sheets (or folios, as specialists call them) of paper, plus torn strips from two missing leaves. The sheets were larger than a standard page in size (32.4 x 24.5 cm) and made of a sturdy, relatively soft but lightly glossed biscuit-brown paper. The edges had at some point been trimmed for rebinding, with some writing in the margins cut off. The margins were soiled through use, and numerous amateurish repairs had been made on some paper tears. It looked old. My experience cataloguing Islamic medical manuscripts both for the Bodleian Library and

FIG. 1.1. Front cover of the *Book of Curiosities*, with bird dropping indicated by arrow. Oxford, Bodleian Library, MS Arab. c. 90.

for the National Library of Medicine in Bethesda, Maryland, told me that the papers and inks were consistent with an early manuscript.

My attention was transfixed, however, by the large maps and diagrams, as well as drawings of comets and groups of stars, that filled many of these leaves. Having for several years taught a seminar at Oxford's Museum of the History of Science on cartography in the medieval Islamic world, I recognized that nearly all of the maps or diagrams were unknown in either the published literature or in major manuscript collections. But what were they? I had only a half hour to look at them before I had to leave, knowing that they would be put up for sale in eight days' time, probably to go into a private collection and not to be seen again by historians for many years, if ever. I was anxiously asked by the specialist at Christie's, "Are they important?," and I am afraid I was almost incoherent in my reply. After saying that they probably were, I asked if they had a copy of any of the maps. They did not (except for the three published in their sale catalogue), but they offered me xeroxed copies.

Under ordinary circumstances I would never agree to placing a valuable, ancient manuscript on top a Xerox machine, for fear of the intense light harming the pages and the spine of the volume being damaged. But these were no ordinary circumstances. I asked if photocopies could be made of all the maps and diagrams plus most of the text, and this was immediately done for me in the office. The Xerox machine, of course, only produced a standard-size copy of each page, so the material outside that area was missing from what I was given. Still, it was surely better than nothing. My thinking was that at least scholars would have these black-and-white paper copies to work with, even if the manuscript itself disappeared from view after the sale. Armed with these Xeroxes, I headed for the National Theatre, where I had agreed to meet my husband. The tickets for the play we wished to see were sold out, however, but we had lots to talk about over wine and sandwiches as I showed him the Xeroxes of this remarkable manuscript.

Upon returning to Oxford, the problem to confront was whether there was anything that could be done to assure that the manuscript would end up where scholars could study it. A hasty message of its importance was sent the next day to David Khalili, for whom I had catalogued the scientific and magical manuscripts that form part of the important Nasser D. Khalili Collection of Islamic Art. With Christie's sale less than a week away, there seemed little else that could be done.

Meanwhile, I took the precious xeroxed copies to the Oriental Institute in Oxford to show colleagues over morning coffee. Jeremy Johns, an acknowledged authority on Sicily and its history, was among the group having coffee, and I showed him the curious large map of an oval island (see fig. 1.2 and plate 7), expecting him to be immediately elated by it. It bore no title, but was part of a chapter titled "The Twelfth Chapter Presenting a Brief Description of the Largest Islands in These Seas." The "seas" (*bihār*) in question, I surmised, must be the Mediterranean, and since Sicily is the largest island in that ocean, this, I thought, must surely be Sicily. Jeremy, however, informed me that it could not possibly be Sicily, since we all know that Sicily was always depicted as triangular in shape. A bit deflated, I left a copy of my Xerox of the map with Jeremy and returned to other matters needing my attention. About 10:00 p.m. that evening the telephone rang, and I heard a voice on the other end saying, "It *is* Sicily!" He had apparently spent the evening deciphering some of the numerous place-names, gate names, and names of markets written on the map in very tiny writing. From that point on, Jeremy was as adamant as I that this curious and old manuscript be available for study by scholars. But how that could be achieved we did not know.

A week after the sale, I received an unexpected telephone call from Sam

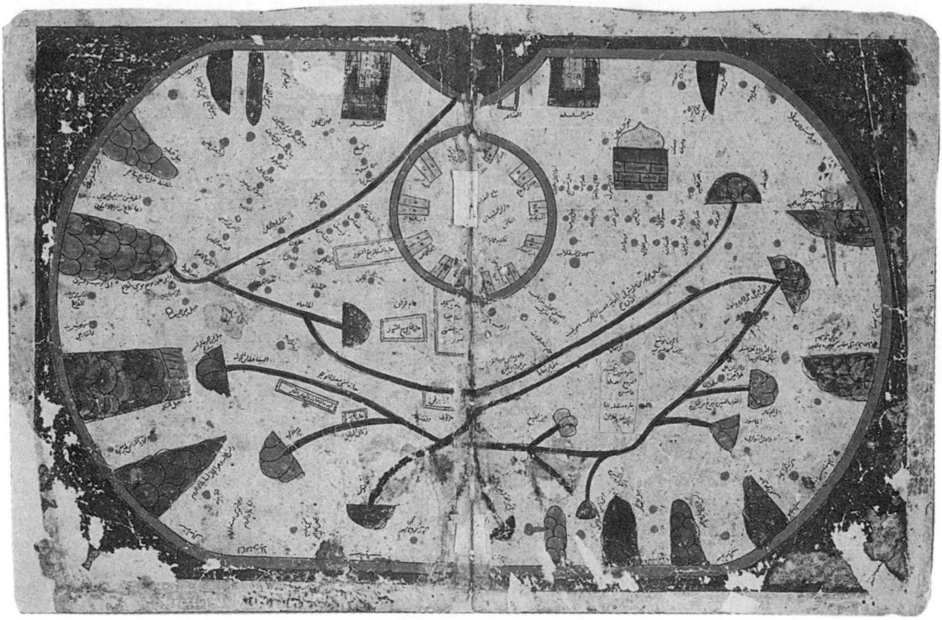

FIG. 1.2. The island of Sicily. Oxford, Bodleian Library, MS Arab. c. 90, fols. 32b–33a, copied ca. 1200.

Fogg, a highly regarded London dealer in rare (primarily Western) books and manuscripts. It turned out that he was the person who had purchased the manuscript at the auction on October 10th.[1] Although he is not an expert in Arabic materials, his professional instincts told him that this was a very important and rare item. The purpose of his call was to ask if I would prepare a write-up, perhaps even a small booklet, on this manuscript in order to sell it following its full description. Somewhat to his surprise, I suspect, I declined to do this, with the argument that the manuscript should not be put up for sale again and risk going into private hands and thereby become unavailable for study. It should be placed in a public collection. The Bodleian Library, I suggested, would be the perfect place.

At the end of the telephone conversation, Sam Fogg said he would call the keeper of the Oriental Collections at the Bodleian and offer the manuscript to the Bodleian for a reasonable price, but warned that if the Bodleian could not raise the funds, then the manuscript would go under the hammer again. Immediately upon hanging up the phone, I rang the newly appointed keeper of the Oriental Collections, Lesley Forbes. She was in fact so recently appointed that we had not yet had occasion to meet and she had no idea who I was. It must have been a most curious call for her to receive. Not only was it from someone she did not know, but the caller told her that she was soon to have a telephone call from a manuscript dealer who would offer the Bodleian—at probably a rather high price—what was possibly one of the most important Arabic manuscripts to come to light in the past hundred years. Why she listened to me and did not dismiss it as a prank, I do not know. She received this call, and the subsequent one from Sam Fogg, at a time of sharp financial restrictions and a pressing need for renovation of book storage areas in the Bodleian, so to anyone else it might have seemed the height of folly to undertake external fundraising for such an acquisition. Yet, if the purpose of the Bodleian is to preserve for posterity the major sources of our intellectual heritage, and to make them available to students and scholars worldwide, then wouldn't the acquisition of such an important historical and scientific manuscript remind people (and potential donors to the Bodleian) of the library's primary function and its role as one of the most important repositories of cultural artifacts? Moreover, its acquisition would complement the Bodleian's already important collection of medieval Islamic cartographic materials. Such thoughts may have been in her mind, for, when the manuscript and its potential importance were drawn to her attention, she agreed to set about trying to raise the funds for its purchase.

Not long thereafter an agreement was drafted whereby Sam Fogg would sell the manuscript to the library for four hundred thousand pounds, giving the Bodleian six months in which to raise the monies. The agreement also included the stipulation that if the manuscript were ever to be shown to be a forgery or a fake, the monies would be returned to the Bodleian—such was the confidence that Sam Fogg had in the legitimacy of this aged manuscript. The price set was well under what he thought to be the true market value. To put this sum in perspective, an Arabic manuscript copy dated 1229 of the *Canon of Medicine* by Ibn Sīnā (Avicenna to Europeans), who died in 1037, had sold at Sotheby's in London on October 12th of the same year for well over five hundred thousand pounds, following press stories highlighting its supposed importance. However, there are numerous copies of the *Canon* preserved today, the text has been fully printed several times, and no new chapters were contained in this copy, nor were there any illustrations. In contrast, "our" manuscript contained a treatise no one knew about, with fourteen completely unique maps plus other diagrams, and much new information on travel and communication between Byzantium and the Islamic world in the eleventh century (though that was not known at this stage). In the event, however, eighteen months rather than six months were required to raise the required funds and to verify various important details regarding the manuscript itself. Throughout this period Sam Fogg generously granted our repeated requests for an extension of time, while allowing the manuscript to be kept at the Bodleian for examination while we tried to raise the funds.

The ensuing eighteen months of fundraising were followed by more than a decade-long project that turned several of our lives upside down. Rather like giving birth to a baby, no one told us ahead of time how much effort and exhaustion was going to be involved.

The first hurdle was to obtain permission from certain committees of the university to undertake the fundraising. And therein we encountered our first major problem: prior to the sale at Christie's, no one knew that such a treatise existed. There was reluctance on the part of many to believe that the manuscript could possibly be as important as we suggested. How could an eleventh-century treatise containing numerous and large maps and diagrams have been completely unknown to modern scholars? Surely this was impossible. Such was the reaction of many who were puzzled, if not unbelieving, that an entirely unexpected text with massively important new material could still be discovered in the twenty-first century. And particularly in such a well-trodden field as early Arabic geographical writing. As the head of one of the most important committees put it: "If no one has known of this treatise, and no one has published or studied it, then surely it cannot be important." But the fact that it was "lost" to modern scholars for several hundred years does not mean that it was not important in its day nor that it cannot provide new insights into the intellectual life of the society for which it was composed, as well as its trade and travel.

With necessary committee support for the fundraising denied, it looked as if we would have to terminate our efforts before we even got started. Why not go then, I asked, to the vice chancellor, Sir Colin Lucas, who was a historian, and seek his support for fundraising to obtain this manuscript for the Bodleian? Much to my surprise, I was informed that it would be inappropriate for someone to go over the heads of the official committees to the vice chancellor, in addition to which I was not of sufficient rank to take up the vice chancellor's time. At this point my American background came into play, and I decided that I would on my own seek the help of the vice chancellor. As it happened, Professor Wadad Qadi of the Department of Near Eastern Languages and Civilizations at the University of Chicago was in town at this time, and since the vice chancellor had been professor of history at the University of Chicago before becoming master of Balliol College and subsequently vice chancellor at Oxford, I asked Professor Qadi to come with me to make the case for the manuscript. Making an appointment to see the vice chancellor was easy enough, after simply ringing his secretary and asking for one, and Professor Qadi and I were sufficiently convincing as to gain his support in principle for the project.

The provenance of the manuscript, of course, needed further investigation. Undated owners' notes and stamps occur on the title page of this manuscript, giving the names of three individuals who owned the manuscript sometime between the seventeenth and nineteenth centuries. Due diligence was expended in trying to determine as much as possible about more recent provenance, but little was revealed except that it was in a continental European collection for most of the twentieth century. All registers of stolen goods and all known catalogues of Arabic manuscripts in both public and private collections were thoroughly searched. The auction house had published four illustrations (three maps and a later owner's added drawing) from the manuscript in their sale catalogue, and these provoked no reaction from any previous owners. Various legal assurances were given by the vendor, Sam Fogg, and it was eventually concluded that we (that is, Lesley Forbes, Jeremy Johns, and myself) could proceed with fundraising.

Following on from that, the Friends of the Bodleian Library agreed to make a generous contribution of fifty thousand pounds. With the support and encouragement of the vice chancellor, we were then in a position to approach other agencies as well as individuals for funds. Although the manuscript was not beautifully illuminated, as some famous Western manuscripts are, its numerous colorful and curious maps and diagrams provided engaging material for presentations to potential donors.

Occasionally, however, when talking about our oval Mediterranean and Indian Ocean maps, or Sicily depicted as an oval and Cyprus as a square, someone would say—and some still do—"But are these really maps? Are they not just crude diagrams that should not be dignified with the name 'map'"? To which we reply, "They are indeed maps." Neither Euclidean geometry nor demarcations of longitude and latitude are required for a map to impart important spatial information. The definition of a map given by Samuel Johnson in his famous dictionary ("a geographical picture on which lands and seas are delineated according to the longitude and latitude")[2] is now long outdated. As the great modern historians of cartography, J. B. Harley and David Woodward, have said: "Maps are graphic representations that facilitate a spatial understanding of things."[3] And the "maps"—both geographic and celestial—in the *Book of Curiosities* do just that.

When making fundraising presentations, it was of course the rectangular world map in the volume that attracted most attention (see fig. 1.3 and plate 1). It is unlike any other recorded ancient or medieval map. At this time we thought it to be a copy made around 1200 of a map produced in Egypt toward the end of the eleventh century. In fact, the original map was probably made between 1020 and 1050, but we only discovered the evidence for that much later when we had time to study the manuscript in more detail.

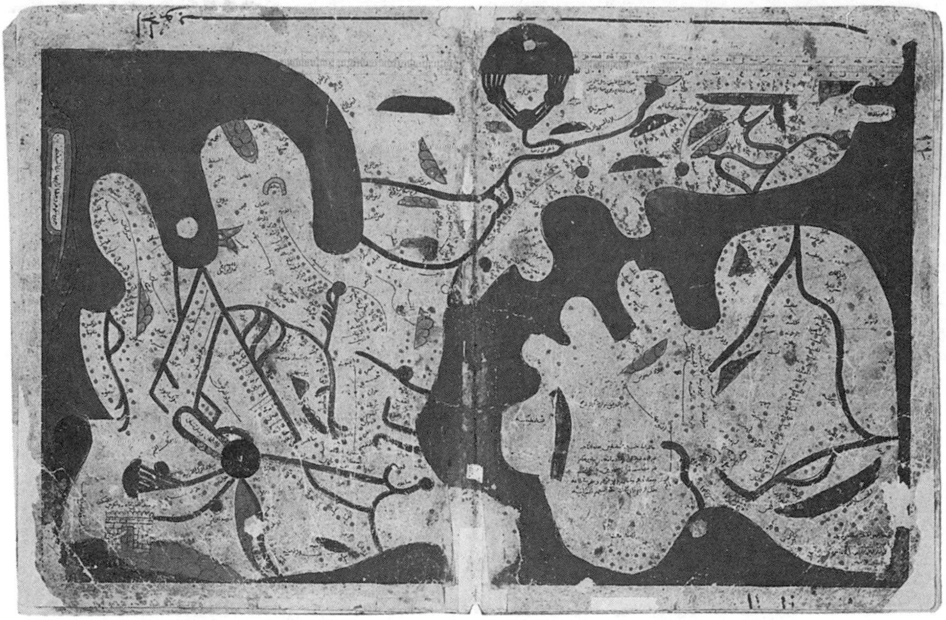

FIG. 1.3. The rectangular world map. Oxford, Bodleian Library, MS Arab. c. 90, fols. 23b–24a, copied ca. 1200.

The scale bar (technically called a graticule) at the top is unique testimony to the circulation and use during the medieval period of maps employing mathematical techniques. The map also differs from all earlier ones by being a stand-alone map—that is, a map that does not illustrate a historical or theological narrative and which conveys information independent of a related text. It is, moreover, the earliest world map to be annotated with names of cities (395 in number) rather than simply names of regions or countries. To a modern reader the layout of the landmasses appears strange and unfamiliar, but more will be said of this remarkable feature of the map in chapter 3.

The first major breakthrough in fundraising came when, following a presentation by Lesley Forbes to the National Art Collections Fund, the NACF granted us one hundred thousand pounds toward the purchase price. Curiously, however, while the panel was struck by the illustration of the Palaces of the Imams in the city map of Mahdia (see fig. 1.4 and plate 9), in modern Tunisia, they were perhaps even more interested in an illustration of the legendary *wāq wāq* tree, whose fruits are voluptuous women,[4] added to the manuscript by some reader in probably the fourteenth or maybe even fifteenth century (see fig. 1.5).

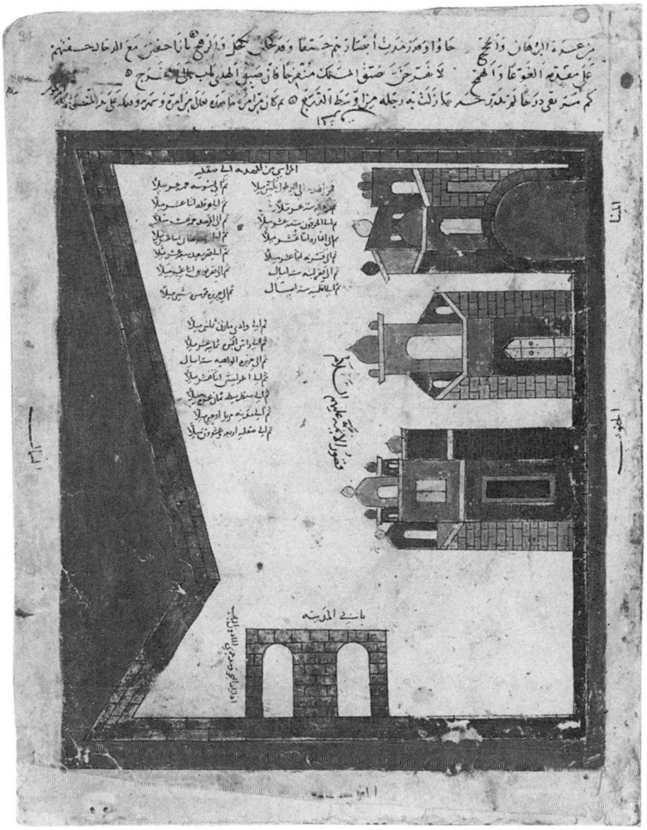

FIG. 1.4. The peninsular city of Mahdia, in modern Tunisia. Oxford, Bodleian Library, MS Arab. c. 90, fol. 34a, copied ca. 1200.

As a part of our application, the NACF required an inspection of the manuscript by two authorities of their choosing, one to report on its historical importance and one to comment on its market value. The results of these outside authorities' examination of the manuscript were very gratifying, for it confirmed in our minds the importance and validity of the manuscript and also made apparent to us that the price set by the seller was indeed well under the market value.

The oddness of the maps in the volume—not just the rectangular world map but maps of oceans and islands drawn as ovals and rectangles—contributed to the reluctance of many to accept its importance, for with its acceptance went a suspicion that much work would be required to accommodate something so completely different into the field of medieval cartography. Parallel with the reluctance to accept its importance, of course, ran the

FIG. 1.5. "Concerning the fruits of the *wāq wāq*," added by a reader in probably the fourteenth or fifteenth century. Oxford, Bodleian Library, MS Arab. c. 90, fol. 27a.

concern (and in some cases, a desire) that it would turn out to be a forgery. Once the manuscript had been deposited with the Bodleian we were able to have a number of other experts examine it. The leading authorities on Islamic paper and paleography were consulted, as well as experts in the history of medieval cartography, in order to confirm the treatise's importance and legitimacy. It was imperative that we not overlook something that might indicate a later date or—the greatest nightmare—a fake. We did not wish to find ourselves embroiled in an unsavory controversy such as that surrounding the now infamous Vinland Map at Yale University.[5] A number of specialists in Islamic science and cartography made special trips to the Bodleian to inspect it, some of them hoping, I suspect, to demonstrate it was a forgery. If so, then they were disappointed, for every scholar who personally examined the volume has been convinced of its genuineness and importance.

With the monies from the Friends of the Bodleian Library and the National Art Collections Fund behind us, we were then able to approach other possible sources of funding. A call was made to all the colleges comprising the university, and nine colleges responded with generous donations: All Souls College, Merton College, New College, Nuffield College, St. Antony's College, St. Cross College, St. Edmund's Hall, St. John's College, and Wadham College. Particularly gratifying were the donations from six individuals who wished to help us meet the final target.

Putting all these donations together, we were then in a position to proceed with an application to the Heritage Lottery Fund. Lesley Forbes and Jeremy Johns took the responsibility for writing the complicated application, which was submitted in February of 2002. The proposal to the Heritage Lottery Fund was to preserve the manuscript for the nation and make its content available to the widest possible audience, thereby increasing public awareness of the Islamic contribution to our common heritage. The HLF application was based on "partnership funding," and we could now offer the £182,000 that we had so far raised as our "partnership funding," requesting an additional £421,000 from the Heritage Lottery Fund. The funds requested would complete the purchase price, but also provide for (1) the conservation, pigment analysis, and digitization of the manuscript; (2) the exhibition of the manuscript for the general public; (3) the preparation of a schoolteacher's pack based on portions of the manuscript; (4) the creation of a website presenting the text and maps with translations; and (5) the initial work of transcribing, editing, and translating the text, maps, and diagrams. After we had completed the application, we received a somewhat unexpected donation from Saudi Aramco that allowed us to increase our "partnership funding" by some £22,000.

On June 13, 2002, the Bodleian received word that the Heritage Lottery Fund application had been successful. It was highly appropriate that this important manuscript should be acquired in 2002, for it helped mark the Bodleian's four-hundredth anniversary year.

Overnight we were suddenly committed to a very large project to make this remarkable and unique—or what we at this point thought to be unique—manuscript available to a large and diverse public.[6] A celebration was of course called for, and Lesley Forbes held a splendid garden party at her home, inviting all the people locally who had given so generously of their time and money in order to bring this about. The manuscript was duly assigned the shelfmark MS Arab. c. 90 and its acquisition by the Bodleian proudly announced to the media. We were particularly pleased with a lengthy piece (illustrated with the rectangular world map) by Hannah Hennessy titled "Arabic Atlas Offers Unique View of 11th-Century World" that appeared in *The Times* on July 20th.

Immediately upon the public announcement of the Bodleian's acquisition, the leading history of cartography journal, *Imago Mundi*, requested an article describing the manuscript. The resulting piece had by necessity to be a very preliminary description.[7] Since then, as we worked further and more carefully on the manuscript, we have amended a number of points. For example, we initially thought that the treatise was composed shortly after the Norman advance on western Sicily, but a more careful reading has revealed that it was written prior to 1050, *before* the Norman invasion. Nonetheless, it is this first publication, with its incorrect dating, that continues to be cited (much to our embarrassment) and has even been recently reprinted without our prior knowledge.[8] Later publications have tried to clarify the point.[9]

This original error in placing the composition of the treatise at the end of the eleventh century is a good example, as Jeremy Johns likes to say, of how much hangs on a dot. For it all hinges on where you place the dots on a single Arabic word that in the Arabic manuscript is written without any dots at all (typical of early copies). Unlike modern scripts, most words in the manuscript are undotted, with the result that many Arabic words could be read in several ways. In the *Imago Mundi* publication, the opening sentence of the twelfth chapter was rendered as

> The island of Sicily is the largest of the Islamic islands, and the most famous on account of the enemy—may God cast them down!—having reached **its western parts**, and the continuing struggle of its imams and governors against them.[10]

This sentence was interpreted as referring to the advance of the Normans on western Sicily, which took place from 1068 to 1071, for the Arabic word in question was read as *maghāribihā* (مغاربها), "its western parts." If, however, the dots are placed differently, you can read it as *maghāziyihā* (مغازيها), meaning "its military expeditions." Adopting this interpretation, the sentence then reads:[11]

> The island of Sicily is the largest of the Islamic islands and the most honourable on account of its continuous military expeditions against the enemy—may God forsake them!—and the perennial efforts of its people and governors in this respect.

By this interpretation, the Normans have not yet conquered the island of Sicily, and so, taken together with other internal evidence, the treatise was composed *before* the Norman invasion and not after it. There are three other references in the treatise that allow us to pinpoint its composition to between 1020 and 1050:

(1) The tribe of the Banū Qurrah is mentioned as still inhabiting the lowlands near Alexandria. Since Fatimid authorities waged several campaigns against them in 1050–51, eventually banishing them from the region in 1051–52, our author is writing before 1050.
(2) In the chapter on the city island of Tinnīs in the Nile Delta, it is stated that six large buildings for merchants were constructed in 1014–15, which means our author must have been working after 1015.
(3) Al-Ḥākim bi-Amr Allāh, the Fatimid ruler of Egypt and Syria from 996 to 1021, is referred to as if he were no longer reigning, meaning that our author is writing after the year 1020.

That so much can hang on a dot exemplifies the problems of working with a manuscript of this age.

Although undated, the Bodleian manuscript can be assigned, based on the characteristics of ink, paper, and script,[12] to the late twelfth century, some one hundred and fifty years after the original was completed. The Heritage Lottery Fund grant provided funds for the testing of pigments and inks in the manuscript. Two techniques were used: optical microscopy[13] and Raman spectroscopy.[14] Six pigments were identified in the illustrations: cinnabar (red), orpiment (yellow), lazurite (blue), indigo, carbon-based black, and basic lead carbonate (a "lead white"). No evidence of modern inks or pigments was revealed. The results of the scientific analyses were completely consistent with the suggested origin and age of the manuscript and, thankfully, supported our initial impressions.

When interpreting the text and maps, difficulties and ambiguities abound. For this reason, we were particularly pleased to have Yossef Rapoport join the project in February 2003. At that time he was a recent PhD from Princeton specializing in medieval Islamic social history. In the subsequent decade he has become a leading expert on medieval Islamic cartography (even if that was not his original intention) and a reader in Islamic history at Queen Mary, University of London. His knowledge of classical Arabic and medieval Islamic history has been invaluable, and without him we could not have prepared either the preliminary edition and translation that was released on the web in 2007,[15] or the fully critical and annotated printed edition and translation that appeared in 2014.[16]

Quite early on we discovered that our manuscript was not a unique copy of the text after all. There were in fact eight other copies of all or portions of the text preserved in manuscripts ranging in date from 1564 to 1741 that had lain unnoticed in other libraries.[17] And one of these even turned out to have been in the Bodleian since 1611, when it was given to the library by Paul Pindar, a merchant and Ambassador of King James I to the Ottoman court. On the one hand, their discovery was immensely comforting since it ruled out completely the idea that our treatise was a modern forgery, but, on the other, it meant a lot more work, for now we had to study as many of these copies as possible and incorporate differences into our own edition and translation. This we have done in excruciating detail with our printed edition of 2014.

The reason why these later versions had not been noticed was that they lacked most of the illustrations, and when they did have an illustration, they were distorted almost beyond recognition. As an example, compare the River Oxus as it is depicted in the Bodleian copy of about 1200 (plate 14 and fig. 1.6) with its depiction in a copy made in 1564 that is now in Damascus (fig. 1.7), or, the comet named "the bowl" (*qaṣʿah*) as drawn in the copy made about 1200 (fig. 1.8) with that in the copy made in 1571 that is also in the Bodleian (fig. 1.9).

All the large two-page maps—the two world maps, the Mediterranean Sea, Indian Ocean, Sicily, and Tinnīs—are missing entirely from the later copies, although in the case of the Mediterranean, the copyist working in 1564 drew rectilinear frames on consecutive pages and placed the title within the frames, leaving the rest of the area blank.[18] Lacking comprehensible maps and diagrams, it is no wonder that no one paid attention to the text as preserved in these later copies, especially given that there are tens of thousands of unstudied Arabic manuscripts resting in libraries around the world.

There appears to have been particular interest in the treatise during the sixteenth century and later among the Christians of Syria, for one copy—the one in the Bodleian Library since 1611—was made in 1571 by Manṣūr bi-ism Shammās (the ordained deacon) using Karshūnī, the Syriac script used by Christians of Syria and Mesopotamia to write Arabic.[19] Another copy, transcribed in Arabic script in 1741 by a physician of Aleppo named Muḥammad ibn ʿAbd al-ʿAzīz al-Shāfiʿī al-Ḥalabī, has an owner's note written by "the ordained deacon" (*al-shammās*) Ḥannā the physician, son of Shukrī Arūtīn the physician.[20]

The project to which we were now committed required not only that the manuscript be conserved and then analyzed by historians but also that it be presented to the public as soon as possible. With that in mind we laid the ground work for its presentation, along with comparative materials from the Christian and Islamic worlds, in an exhibition at the Bodleian Library

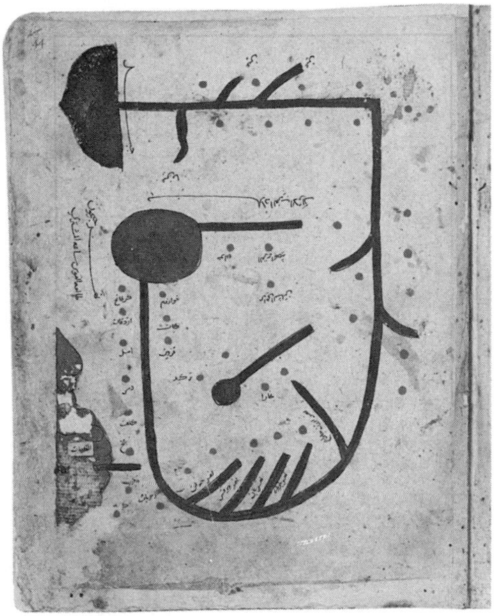

FIG. 1.6. The River Oxus. Oxford, Bodleian Library, MS Arab. c. 90, fol. 44a, copied ca. 1200.

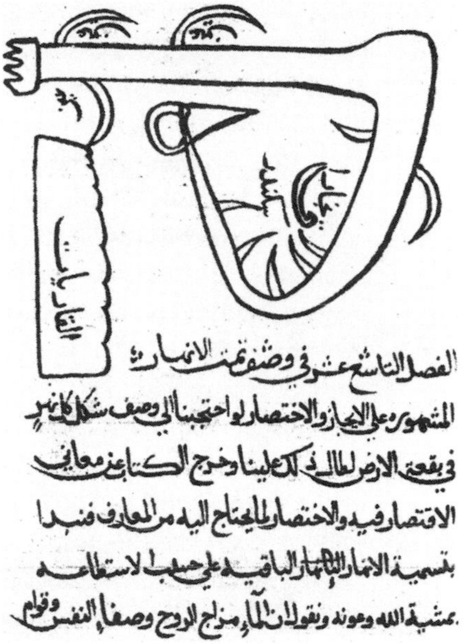

FIG. 1.7. The River Oxus. Damascus, Maktabat al-Assad al-Waṭanīyah, MS 16501, fol. 121a, copied 1564/972H.

FIG. 1.8. The comet called “the bowl” (*al-qaṣʿah*). Oxford, Bodleian Library, MS Arab. c. 90, fol. 14a, copied ca. 1200.

FIG. 1.9. The comet called “the bowl” (*al-qaṣʿah*). Oxford, Bodleian Library, MS Bodley Or. 68, fol. 131a, copied 1571 and written in a combination of Arabic and Karshūnī script.

titled "Medieval Views of the Cosmos: Mapping Earth and Sky at the time of the *Book of Curiosities*." Fortunately, by this time we had realized the correct date of the composition to have been between 1020 and 1050 rather than *after* the Norman invasion of Sicily! The manuscript was disbound so as to allow as many illustrated pages to be displayed as possible. It remains disbound because the binding in which it came was not original and of little value and because it can be more easily displayed and studied in the future. We were very fortunate in being able to have the exhibition opened by the noted Egyptian novelist Dr. Ahdaf Soueif and the Oxford-trained historian and member of the Monty Python comedy team, Mr. Terry Jones. The exhibition, which ran from June through October of 2004, attracted what was at that time a record attendance for a Bodleian exhibition, and it received some outstanding media coverage, including a marvelous piece in *The Times* (June 23, 2004) by Michael Binyon, "Mapping the Marvels: Muslims Explored the World before Europe Left the Dark Ages."

We also talked Terry Jones into writing a foreword to a small but highly illustrated book that was published to coincide with the exhibition, titled, not surprisingly, "Medieval Views of the Cosmos."[21] Professor Evelyn Edson, an acknowledged authority on medieval European cartography, joined me in preparing this book and never complained even once about the pressurized deadline that we had to meet in order to have the book out before the exhibition was over. We have been gratified by the continued interest in its sales (it has had three different covers!) and by the publication of a Korean translation as well as a German version augmented by the authoritative comments of the German historian of medieval cartography, Anna-Dorothee von den Brincken.[22]

At the same time that interest in the *Book of Curiosities* was increasing through the exhibition and media exposure, our work analyzing the details of the treatise continued on. We began to appreciate more fully the distinctive Egyptian perspective on the eleventh century that this treatise displays, for it reflects the exceptional intellectual and political context of Fatimid Cairo. For example, the extraordinary firsthand knowledge of Byzantine waters and ports that was discovered in this treatise reflects the naval operations and ambitions of the Fatimids. The Isma^cili interest in esoteric thought and analogies between the celestial and terrestrial worlds no doubt accounts for the wealth of star lore and the inclusion of arcane works attributed to the legendary Egyptian-Greek sage known as Hermes Trismegistus, including a previously unknown illustrated text on comets. More will be said of these features of the *Book of Curiosities* in the chapters that follow.

As we worked on this complicated text, new details as well as occasional surprises continued to emerge. Some were rather curious discoveries, such as the earliest recorded use on a map of the word "Angle-Terre" (*Inqilṭirrah*) to designate England.[23] Or what appears to be the earliest Arabic description of the kapok fibre used for filling mattresses and pillows, which our author observed in the home of a local elder who had traveled extensively in Africa.[24] We have to admit that this "discovery" alarmed us slightly, for kapok is a product of trees of the genus Bombax, usually said to be indigenous to Southeast Asia or the New World. However, much to our relief, it turns out that there are two species indigenous to Africa, one of them, *Ceiba pentandra,* being the tallest tree in Africa. Another "first" is the earliest mention in Arabic texts of the island of Zanzibar, here called Unjuwa, a corruption of the Swahili Unguja, referring to Unguja Ukuu, a site on the southern coasts of Zanzibar where excavations have uncovered evidence of extensive medieval trade with the Mediterranean.[25]

There are, in fact, so many unique or significant features in the text and maps of the *Book of Curiosities* that a person hardly knows how to select the single most important element. There were, of course, certain discoveries whose importance was immediately apparent and which attracted wide attention. We found, for example, the map of Sicily to be the earliest preserved of this island. The same turned out to be true of the diagrammatic map of the harbors of Cyprus. The map of the commercial town of Tinnīs in the Nile Delta is not only the earliest map, but the only one, for the town was completely destroyed in 1227 during the crusades. The map of Mahdia is the only known representation of the city earlier than the European engravings published to celebrate its capture by the emperor Charles V in 1550.

And if that is not enough, four of the five river maps are unique to the *Book of Curiosities,* with only that of the River Nile found elsewhere. Those of the Rivers Tigris, Euphrates, Oxus, and Indus have no known parallels. The large maps of the Indian Ocean and the Mediterranean, while not unique in subject matter, are certainly unique in conception and design. It is not their uniqueness, however, that makes them most important, but rather it is the information they contain and the perspective they provide the modern reader on the early medieval world.

The accommodation of so much previously unknown material requires a revision of many aspects of medieval Islamic history, such as the history of cartography, trade routes to India and China, knowledge of the source and course of the Nile, perceptions of the night sky and its portents, contacts with Byzantium, and many other topics. The chapters that follow in this book are our attempt to begin the process of reassessment.

Many puzzles still remain. For example, we still do not know the name of the person who compiled this treatise, nor the name of the patron for whom it was composed. All we know is that our author wrote an earlier composition called *al-Kitāb al-Muḥīṭ*, or "The comprehensive book," that he mentions in the course of the *Book of Curiosities.* It is not known to have been preserved today, but, like the *Book of Curiosities*, it appears to have also concerned both geography and star lore, for it is said to have included descriptions of islands in the Mediterranean as well as the astrological significance of zodiacal signs.[26]

There will no doubt continue to be new discoveries, new interpretations, and reevaluations of the treatise and its relationship to similar material. And there will continue to be points on which people disagree. In fact, there is one point on which Yossi Rapoport and I, after a decade of work on this manuscript, continue (amicably) to disagree—and that is the circular world map.

The manuscript contains among its many maps a circular world map of a well-known type usually associated with the geographical work of al-Idrīsī, composed in 1154 for Roger II, the Norman king of Sicily. Al-Idrīsī is the most familiar to Europeans of all Arab geographers, for an abridgement of his work, published in Rome in 1592, was one of the first Arabic books printed in Europe, and a Latin translation followed twenty-seven years later. Born in Morocco in 1100 and educated in Córdoba, al-Idrīsī traveled widely during his lifetime and in 1138 was invited to Roger II's court in Palermo. It is for his illustrated geography—happily titled *Entertainment for Someone Who Longs to Travel the World* (*Nuzhat al-mushtāq fī ikhtirāq al-āfāq*)—that he is now remembered. Six (and only six) of the preserved manuscript copies—two of them in the Bodleian Library—open with a circular map of the world. See fig. 1.10 and plate 4 for an example. Crucially, for our purposes, al-Idrīsī does not ever refer to this circular world map in the course of his treatise. While the map is frequently reproduced today (most often using the copy shown in fig. 1.10 and plate 4), and routinely referred to as the "Idrīsī World Map," it does not represent al-Idrīsī's approach to cartography as displayed in his treatise. His treatise is based on the idea of dividing the inhabited world into seven horizontal strips or "climes," each of which was divided into ten sections representing 18° in longitudinal width. In other words, his treatise is made up of seventy large regional rectangular maps covering what was then thought to be the inhabited world. What role al-Idrīsī intended the circular world map to play is unclear, if indeed he intended it to have any role at all.

The circular world map found in the *Book of Curiosities* is virtually identical (see fig. 1.11 and plate 3). Though undated, it was probably copied around the year 1200. Of the six other known copies of the "Idrīsī World Map," none are dated earlier than 1348, although one undated version has been

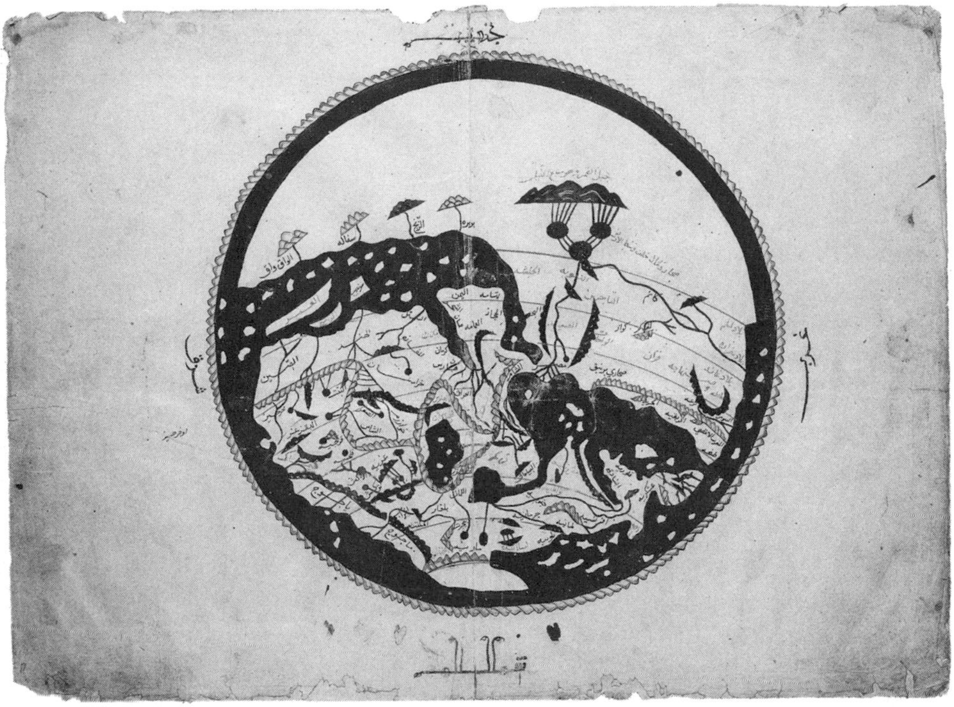

FIG. 1.10. Circular world map, from al-Idrīsī. Oxford, Bodleian Library, MS Pococke 375, fols. 3b–4a. Copy completed July 25, 1553/13, Shaʿbān 960H, by ʿAlī ibn Ḥasan al-Hūfī al-Qāsimī.

assigned to about 1300.[27] So the question becomes: Was this circular world map first designed by al-Idrīsī for his treatise in 1154, or was it part of the *Book of Curiosities* compiled between 1020 and 1050? If the former is the case, then the map in fig. 1.11 was added when our copy of the *Book of Curiosities* was made about 1200, but if the latter is correct, then the map was designed a hundred years before al-Idrīsī wrote his treatise—in which case it should no longer be designated the "Idrīsī World Map." In other words, is it a Fatimid map or a Norman one?

In favor of it not being part of the original *Book of Curiosities*, Yossi Rapoport argues that the circular world map was not originally part of this manuscript copy now in the Bodleian, and hence, by extension, not part of the original treatise of 1020–50. The evidence put forward is that this particular map in our manuscript employs a slightly different color palette than the other maps in the volume, and it is the only map in the volume to employ copper greens.[28] This, he argues, indicates that it was not designed or executed by the same person who made the other maps. He suggests

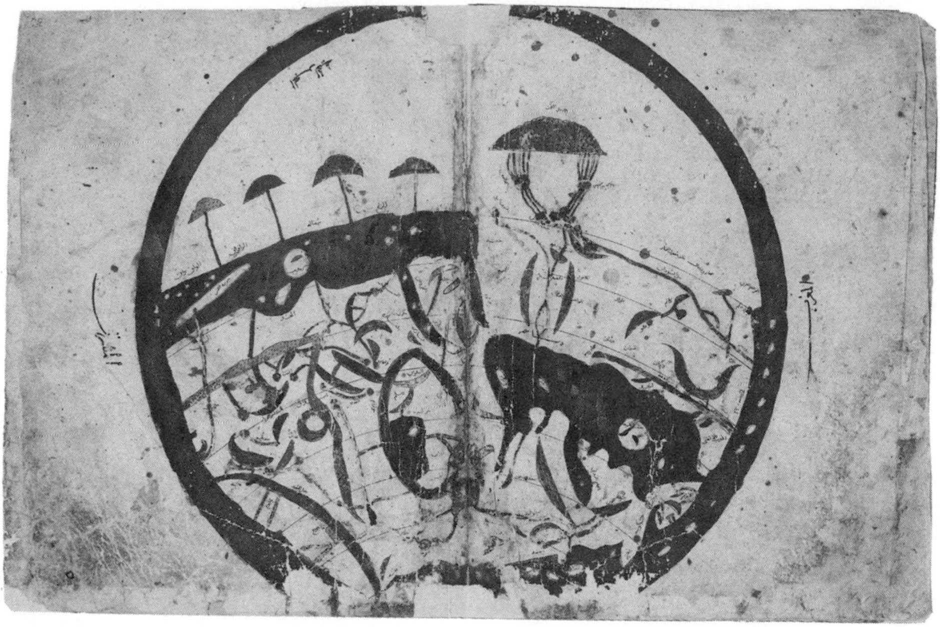

FIG. 1.11. Circular world map from the *Book of Curiosities*. Oxford, Bodleian Library, MS Arab. c. 90, fols. 27b–28a, copied ca. 1200.

that this map had its own separate history going back to al-Idrīsī's day and was easily available to the copyist of our manuscript, who, some fifty years later, either made the copy himself or gave the folio to another copyist well known for making copies of this particular map. The map occurs at the end of the chapter "On the Cities (*amṣār*) of the Remoter Regions," which is concerned with the four climatic extremes where certain types of people reside. No individual city is actually named in the chapter, but only regions such as "cities of the extreme South" or "the land of the Turks." The map, he argues, has no particular relevance to the chapter and appears to have been an afterthought on the part of the copyist.

I counter this by reminding him that the paper on which the map is drawn is identical in every way to the paper used in the rest of the volume, and that it is worn, soiled, and thumbed in the same way as the other folios in the volume.[29] Furthermore, I feel the content of the map coincides with that of the chapter in that no cities or towns are named on the map either, just regions or countries, such as "the land of the Lamlam" or "the Bulghars" or *Inqilṭirrah* (Angle-Terre). Curiously, the word *Inqilṭirrah* occurs only on the Bodleian map, and not on any of the six copies associated with al-Idrīsī, though it is hard to see how that bit of information helps solve the puzzle. The Bodleian map also has the White Sand Dunes indicated as the source for a western tributary of the Nile, and that is a feature unique to the text of the *Book of Curiosities* and not to be found in al-Idrīsī's treatise or on the other six Idrīsī-type circular maps. On the other hand, many of the other names on the map, such as "the Stinking Land," are not mentioned elsewhere in the *Book of Curiosities* but are in al-Idrīsī's treatise. I agree that the map had its own separate history, but one going back not just to al-Idrīsī's day but at least as far back as 1020–50, and that by 1200 a workshop in Cairo had become well known for grinding out copies of it. Our copyist simply took the bi-folio from the volume, had the map drawn by the map-maker specializing in this design (hence the slightly different palette and the copper greens) and put it back in its place in the manuscript.

Given the fact that al-Idrīsī does not himself refer to such a map in his text, and given that no copies predate the one in the Bodleian copy of the *Book of Curiosities*, the only reason I can see for not accepting the map as an integral part of the *Book of Curiosities* is that to make such an assertion would be to argue for a major revision of the history of Islamic cartography. Al-Idrīsī's treatise and the "Idrīsī World Map" are icons of cartographic history. To displace them requires considerable courage and more evidence than presently available. These are, of course, the kinds of arguments that keep academics going for years, but at this point we simply cannot say whose interpretation is correct.

The *Book of Curiosities* is a complicated treatise with much material packed into terse paragraphs or densely labeled maps. Its remarkable series of early maps and astronomical diagrams are for the most part unparalleled in any Greek, Latin, or Arabic material. The apparent simplicity of its geometric presentations of geographical space has attracted considerable attention, even from book cover designers.[30] Yet the significance of these maps and diagrams is neither simple nor readily apparent to the modern beholder. It is the purpose of *Lost Maps of the Caliphs* to draw out the most important features of the manuscript and to suggest where some of this material—both textual and graphic—fits into the history of ideas, to highlight the new or unusual perspectives on travel and exchange of commodities, and to point out for the reader the non-European perspective of the world as presented by our anonymous Egyptian author viewing the Mediterranean from its southeastern corner.

CHAPTER TWO

Macrocosm to Microcosm

READING THE SKIES AND STARS IN FATIMID EGYPT

Urban dwellers today pay little attention to the night skies, unless the media tells them of some unusual event such as a lunar eclipse or Halley's comet. They might be able to identify a major constellation such as Ursa Major (or the Big Dipper) or Orion when it rises over the horizon, or possibly recognize the planet Venus, which is dazzlingly bright in the west around sunset or in the east at sunrise. They can probably do little more. In modern cities only a few dozen stars are often visible because of the pollution and the city lights.

In contrast, for a person living in the eleventh century—and particularly in a nearly cloudless region such as Egypt—the night skies would be brilliant, filled with over three thousand stars visible to the naked eye. An equal number would be hidden below the horizon, waiting to be seen later in the night or at a different time of the year. An observer then could easily identify the five planets visible to the naked eye (Venus, Mercury, Mars, Jupiter, Saturn) and notice that they appear never to stray far (8° at most) from the path of the Sun through the stars. It would have been noticed that the stars visible on the eastern horizon just before sunrise appear to shift, giving the appearance of the Sun "moving" through the stars on an orbit that took approximately a year to complete.

Virtually every society has devised its own distinctive way of "mapping" this rotating sphere of stars with its moving celestial objects.[1] The Greeks divided the skies into forty-eight areas or constellations, which we call the classical constellations. The Greek system formed the basis of Arabic scientific astronomy and astrology, and Arabic astronomers added some star names (still used today).

What helps make the *Book of Curiosities* so important is that the author was as interested in other ways of mapping the sky and using the stars as he was in the dominant classical Greek system. Moreover, through the *Book of Curiosities* we have the fullest picture of what an ordinary educated Egyptian during the first half of the eleventh century would have thought about the structure and contents of the night skies. By "ordinary educated Egyptian," we mean someone other than a professional astronomer or even a professional astrologer—perhaps an upwardly mobile Egyptian seeking advancement at court or in the bureaucracy of the day. We need to keep in mind that in the early eleventh century a much larger percentage of the population in Egypt could read and write than in Europe, for literacy was not restricted to a small religious scholarly community. In addition, the use of paper (rather than parchment, as in Europe) in making books meant that many more could afford to own and compose books, while the easy availability of paper may also have encouraged people to undertake graphic depictions of maps and diagrams.[2]

In the first of the two books comprising the *Book of Curiosities*, the author devoted ten chapters to the Heavens and their influence on the Earth, with illustrative diagrams and drawings. From the outermost limit of the universe, the author works his way downward through the various layers of celestial phenomena until finally, in the last chapter, reaching the Earth, where earthquakes and winds reflect the intersection of celestial and earthly events. His progress is very methodical: one chapter on basic principles, four on stars (his favorite topic), two on comets, one on the "wandering stars" or planets, one on "lunar mansions," and the final one concerned with earthly winds and earthquakes. Throughout, there is a careful avoidance of mathematical astrology or astronomy, being for the most part star lore requiring little or no knowledge of trigonometry or mathematical astronomy.

These chapters did not form a technical discourse on astrology as a form of divination. Rather, they present in simple and relatively accessible terms what our author could learn of the structure of the universe and those occurrences in the skies that might foretell events on Earth. To achieve this, our author drew on a range of sources, some uniquely preserved today in this work.

For virtually all writers of the day, the study of astronomy was a way of glorifying the divine creation by examining the complex and sophisticated way in which the universe was structured, while at the same time gaining the practical skills of timekeeping and prediction of earthly events. This is reflected in an aphorism occasionally encountered in manuscripts:[3]

> Whoever does not know astronomy and anatomy[4]
> is deficient in the knowledge of God

One can also imagine the material proving quite useful to the chattering classes of the day on occasions when it was advantageous to display a bit of knowledge about celestial or esoteric matters.

In the medieval period, virtually everyone—even those who were not outspoken advocates of astrological practices—adhered in some way to an astrological worldview. According to that viewpoint, events of the microcosm mirror those in the macrocosm. An assumption that celestial events influence the terrestrial world was shared by all peoples, regardless of their particular religious or sectarian differences. Such a view of the cosmos had permeated, to one degree or another, all learned thought well before the advent of Islam and continued to do so for centuries. Stars, planets, and comets were all seen as indicative of future events on Earth.

The practice of astrology was at one end of a spectrum extending from the complex mathematical models of mathematical astronomy and timekeeping to the prediction of events from the simple observation of a bright star or comet.[5] The historian Richard Lemay has put it this way:[6]

> In the mind of medieval Arab writers there is but one science of the sky with the moving bodies set in it. It was called "the science of the stars" (*ʿilm al-nujūm*), and it consisted of two distinct treatments of the subject matter of the heavens: a purely mathematical one (or *ʿilm al-falak*) corresponding to our astronomy, and a humanistic but rather conjectural one which aimed at deducing from the celestial motions their probable significance for the evolution of human affairs, more directly what we now call astrology. The name for this latter discipline was "the science of the judgements of the stars" (*ʿilm aḥkām al nujūm*). . . . The two methods of treatment were indissolubly linked in the overall picture, and it must be further stressed that the dominating interest of medieval Arabic civilization was the "science of the judgements of the stars."

When the *Book of Curiosities* was composed in Egypt between 1020 and 1050, the distinction between the "science of the stars" (*ʿilm al-nujūm*) and mathematical astronomy (*ʿilm al-falak or ʿilm al-hay'ah* in Arabic) was blurred.[7] Only after the eleventh century did the distance between the two begin to widen as the profession of the mathematically trained timekeeper emerged. The office of an "official" mathematically skilled mosque timekeeper (*muwaqqit*) is first attested in thirteenth-century Egypt.[8] But, even then, the *muwaqqit* would have continued to adhere to the astrological conviction that celestial events influence the terrestrial world.

Egypt was at the time of the *Book of Curiosities'* composition under Fatimid rule, an Ismaʿili-Shiʿa dynasty. The very structure of the book reflects the Ismaʿili emphasis on analogies between the celestial and terrestrial worlds.[9] The spheres of the planets played a major role in Fatimid Ismaʿili cosmology, and astrology was a subject of considerable interest to Fatimid rulers and thinkers.

Astrology was not, however, held in high regard by all Ismaʿilis at all times, and Ismaʿili imams occasionally pronounced verdicts against it.[10] The caliph al-Ḥākim bi-Amr Allāh, the Fatimid ruler of Egypt from 996 to 1021, was inconsistent and unpredictable in his support for astrology (and in other matters as well). On the one hand, there are a number of reports that al-Ḥākim would go to the Muqattam mountain near Cairo to study the stars himself and even try to draw down the planetary influence of Saturn.[11] On the other hand, in the year 1013 al-Ḥākim issued an edict against astrology and astrologers. According to the fifteenth-century historian al-Maqrīzī, al-Ḥākim[12]

> forbade idle talk about the stars. Several astrologers thereupon emigrated, but some of them stayed behind. These were banished, and the population was warned against hiding any of them. Then some of the astrologers showed remorse and were forgiven, and they swore that they would never again look at the stars.

The fact that al-Ḥākim felt the need to issue such an edict suggests that belief in celestial influences was so widespread as to be potentially dangerous. As for the impact of the edict upon the activities of astrologers and astronomers, there is little available evidence. In any case, in the *Book of Curiosities,* the author refers to al-Ḥākim as if he were no longer reigning, and it is probable the book was composed after 1021, when the environment was presumably more conducive to astrological discourses.

The Fatimid caliphate at the height of its power was committed to the promotion of encyclopedic, multidisciplinary knowledge, not limited to the religious sciences. The House of Knowledge, established by the caliph al-Ḥākim in 1005, was a state-sponsored institution, open to all, in which Islamic law and Qur'anic readings were taught alongside astrology and medicine. It was not, however, the only institution of its kind in the Islamic world, for al-Ḥākim was following the recent example of the Buyid vizier Sābūr ibn Ardashīr, who established his own House of Knowledge in Baghdad in the 990s. But the Cairo institution was larger, and the personal involvement of al-Ḥākim in scholarly debates reinforced the learned image of the Fatimid caliph, a key facet of his public persona.[13] The *Book of Curiosities* conveys the same high regard for scholarship, and particularly for those who study the sky and stars. Within the House of Knowledge, the *munajjimūn*—a term that referred to both astronomers and astrologers—were accorded the highest respect.[14]

With support from the Fatimid caliphs, eleventh-century Cairo boasted some of the best-known figures in the history of Islamic science. Ibn Yūnus was a highly original, observational astronomer, who worked in the House of Knowledge since its inception. He prepared a famous corpus of astronomical tables for al-Ḥākim, known as *al-zīj al-Ḥākimī*. Ibn al-Haytham, who died circa 1040, is renowned today for his work on optics, but he also composed treatises on mathematical astronomy, including a summary of astronomical principles in Ptolemy's *Almagest*. The self-taught Ibn Riḍwān, who died in 1061, was appointed by al-Ḥākim as chief physician. He wrote books on the difference between charlatans and true astrologers, a commentary on Ptolemy's *Tetrabiblos*, and a treatise on the effect of Egypt's climate and environment on the health of its population. Even the amir al-Musabbiḥī (d. 1029) tried his hand at two astrological manuals, though now lost. The *Book of Curiosities* was written in an intellectual milieu where the power of the stars was assumed and their study highly valued.

Throughout the *Book of Curiosities*, astronomy and astrology are confined to nonmathematical, nontechnical aspects, based on the observation and interpretation of constellations, star groups, individual bright stars, planets, and comets, as well as winds and other geophysical events. There are no horoscopes, with their houses precisely calculated. To calculate a horoscope required considerable mathematical skill, and, as will become evident in what follows, our author displays no interest in such technical details and may have lacked the mathematical skills. Nonetheless, horoscopes were part of the Egyptian intellectual landscape in which he was functioning, so he referred to them in the course of his writing and assumed that his reader would understand the references to them. See the appendix for further information about horoscopes and what our author had to say about zodiacal signs, which play a fundamental role in the production of horoscopes.

The basic cosmological scheme presented is Aristotelian and Ptolemaic rather than Qur'anic[15]—that is, the cosmos described in the *Book of Curiosities* reflects concepts developed first by Aristotle in the fourth century BC and later applied rigorously to astronomy by the great Alexandrian astronomer of the second century AD, Ptolemy. There is an immobile spherical Earth at the center of a finite universe. The Earth is surrounded by nine spheres, the ninth sphere being starless and planetless, marking the physical limit of the cosmos. This outermost sphere provided the westward motion of the whole. It enclosed eight lower spheres that supported (in descending order) the fixed stars, then Saturn, Jupiter, Mars, the Sun, Mercury, Venus, and the Moon.

It should be kept in mind that at this time there was only "naked-eye astronomy," for there were no telescopes or any other means of magnifying the view of the skies. Medieval astronomers, following earlier Greek traditions, mapped the night sky by dividing it into forty-eight areas containing groups of stars known as constellations. These consisted of twenty northern constellations (including Ursa Major, Ursa Minor, Andromeda, etc.), twelve zodiacal constellations (Aries, Taurus, Gemini, Cancer, Leo, Virgo, Libra, Scorpio, Sagittarius, Capricorn, Aquarius, and Pisces) and fifteen southern ones (including Orion, Cetus, and so forth). They calculated the total number of stars in these forty-eight constellations to be 1,022,[16] which they grouped into six categories of magnitude or size. The first category comprised 15 stars, each said to be "107 times the size of Earth." The other five categories had 45, 208, 474, 217, and 63 stars, respectively, with the stars of sixth magnitude said to be a mere 18 times the size of Earth.[17]

As will be seen, however, other sources were available that described bright stars not included in the 1022 comprising the classical constellations, and these other lists of stars generated their own star lore and prognostications separate from those arising from classical Greek astronomy. The author of the *Book of Curiosities* described the prominent stars in the classical forty-eight constellations, but he spends more time on the stars recognized in other traditions, particularly those reflecting pre-Islamic Bedouin customs and those derived from a Late Antique tradition attributed to the legendary Egyptian-Greek sage known in Arabic as "Hermes the Wise."[18]

So besotted by stars was our author that he even included, at the conclusion of his discourse on the classical Greek constellations, the stars supposedly seen by the prophet Joseph in his dream:[19]

> The stars seen by the prophet Joseph (Yūsuf), may the peace and the blessings of God be upon him, in his dream. These are *jryan, al-ṭāriq, al-riʾāl, qābis, ʿmwran, fnlq, al-muṣbah, dhū al-farʿ, ryab, dhū al-nakafatayn, alṣwdḥ,* the Sun and the Moon.

This tradition concerning the stars seen by Joseph is related by several other sources, all of them nonastronomical. In particular they are found in commentaries on Qur'an 12:4, for there it is said that Joseph saw in a dream eleven stars plus the Sun and Moon. In the commentaries, the eleven stars are given names, but none can be aligned with any star recognized today. Our author makes no further comments about them other than to list them, thereby displaying his knowledge of them.

The Cosmos

The *Book of Curiosities* opens with a large circular diagram of the skies, occupying two facing pages. It is labeled "The Illustration of the Encompassing Sphere and the Manner in which It Embraces All Existence, and Its Extent."[20] See fig. 2.1 and plate 19. Around the outermost circle the twelve zodiacal signs are named, in counterclockwise sequence, beginning with Aries at the top. The zodiacal signs are the twelve equal parts into which the zodiac is divided, with the Sun passing through one of them in each month. They are named after the twelve zodiacal constellations (Aries, Taurus, Gemini, etc.) which at a much earlier time—some two thousand years ago—they contained.

The zodiac is the broad pathway or belt (some 18° in width) among the stars through which the Sun, Moon, and planets appear to move when viewed against the background of the bright stars. Today we say that this apparent movement is due to the movement of the Earth around the Sun, though in antiquity and the Middle Ages it was attributed to the East-to-West movement of the largest sphere with its fixed stars. In practical terms it makes no difference whether the sphere of stars rotates *westward* around the Earth or whether the Earth itself has an *eastward* rotation, for it will appear the same to an observer anchored on the Earth. The median line of the band is designated the path of the Sun, or ecliptic, so called because eclipses can happen only when the Moon is very near this line and hence very close to the Sun.

The diagram reproduced in fig. 2.1 is not a map of the skies as they appeared in the eleventh century, but an abstract map of the zodiacal signs and their approximate relationship to the classical Greek constellations as well as star groups used by Bedouins to tell the passage of time. Large spaces beneath the name of each zodiacal sign have been left blank in the diagram, and it is likely that the original treatise, of which this is a copy made some 150 years later, contained illustrations of each zodiacal sign. These illustrations were probably emblematic depictions with no attempt to represent the stars themselves. An idea of how they might have looked can be gathered by looking at an elaborate manuscript painting from the fourteenth century (see fig. 2.2) showing the Heavens as they were on April 25, 1384 (3 Rabīᶜ I 786), the birthdate of Iskandar Sultan, the grandson of Tamerlane.[21] Near the center there is a ring of twelve medallions, each with a zodiacal sign, running in counterclockwise sequence beginning with Aries at 9 o'clock. At the 3 o'clock position, Libra can be seen as a seated male with a pan balance held above his head, while the roundel below has Virgo as a male using a scythe to harvest wheat.

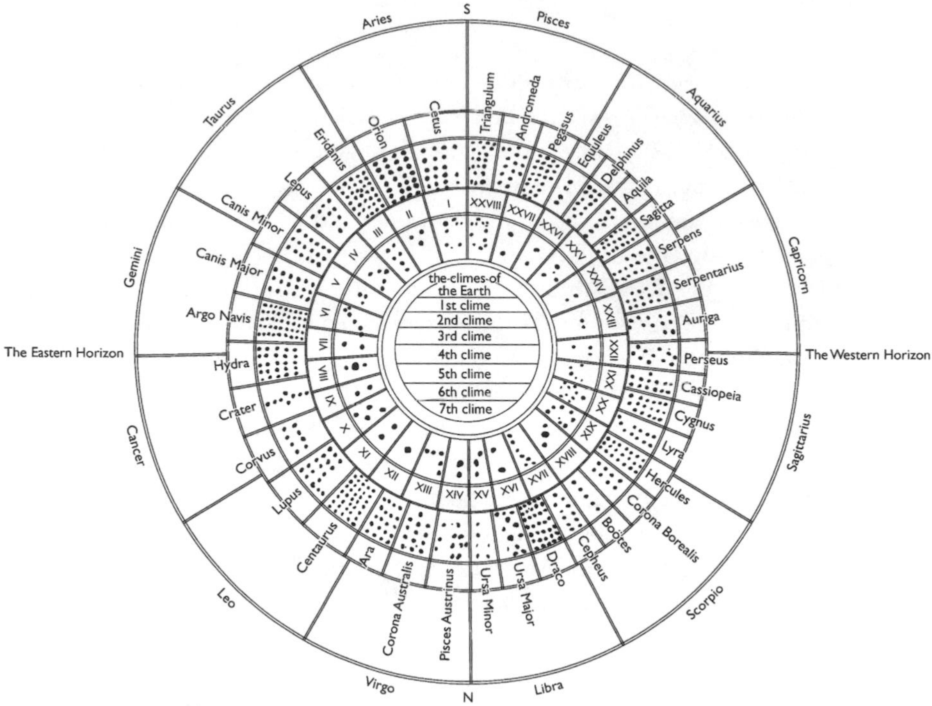

FIG. 2.1. The opening diagram of the *Book of Curiosities*, titled "The Illustration of the Encompassing Sphere and the Manner in Which It Embraces All Existence, and Its Extent," interpreted in a schematic diagram.

Returning to the diagram from the *Book of Curiosities* in fig. 2.1 and plate 19, the next innermost ring contains the names of thirty-six classical constellations—the twenty-one northern constellations and the fifteen southern ones—with each depicted in the ring beneath by a pattern of dots representing stars. The subsequent inner two rings contain the names of the twenty-eight Bedouin "lunar mansions"—a favorite topic of the author—accompanied by representations of their star groups.

At the center of the diagram is the Earth, with its seven "climes" or zones of the inhabited world. At the point marked "eastern horizon" at the left-hand side of the diagram, there is a label written vertically reading:

> This horizon is very dry because, when the Sun rises there, it absorbs the dampness and it dries and expels the nocturnal moisture. The dry wind coming from this direction is called *ṣabā* (east wind).

The western horizon is labeled:

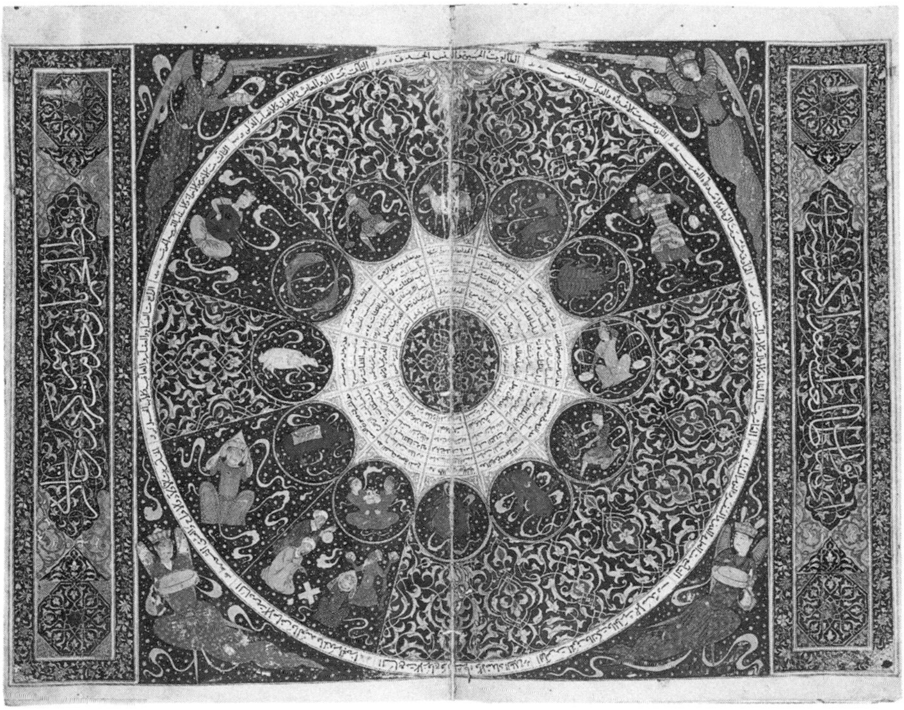

FIG. 2.2. The horoscope of Iskandar Sultan (grandson of Tīmūr, also known as Tamerlane), calculated for 3 Rabīʿ I 786/April 25, 1384. In the ring of twelve medallions, the zodiacal signs are represented by allegorical forms arranged in a counterclockwise sequence, beginning with Aries at 9 o'clock. London, Wellcome Library, MS Persian 474, fols. 18b–19a.

> This horizon is associated with moisture, because the Sun, as it moves towards it, distributes the dew and moisture that had accumulated during the day as a result of the Sun's absorption and removal of the moisture. The wind coming from this direction is called *dabūr* (west wind) and is wet and damp.

The northern horizon, at the very bottom of the diagram, has a label that in this early copy is nearly obliterated, but is preserved in some of the later copies of the treatise in which the contents of this diagram are treated as text but not in a diagrammatic form. Its label reads:

> This northern horizon is very cold due to its distance from the orbit of the Sun and its heat. It comes close to the pole of the Earth. The wind that blows from that direction is called the north wind (*shimāl*).[22]

The diagram is annotated on either side with statements regarding the dimensions of the largest sphere—that is, the sphere that contains all the fixed stars—said to be 130,715,000 miles in diameter, with each of the 360 degrees equal to 1,100,160 miles.[23] This data, he states, he took from the twenty-first chapter of the *Book of Chapters* (*Kitāb al-Fuṣūl*) by al-Farghānī, a widely influential writer on nonmathematical astronomy active in Baghdad between 833 and 861.[24] Farghānī's text was also known as *A Compendium of Astronomy and the Principles of Celestial Motion* (*Jawāmiʿ ʿilm al-nujūm wa-uṣūl al-ḥarakāt al-samāwīyah*), and he is one of the few authorities specifically named by our author.

With this diagram, our author introduced the major topics that interested him—namely, the zodiacal signs, star groups from the Greek tradition, stars of the Bedouin tradition, and the winds. Omitted from this diagram are comets (another particular interest of his) and stars attributed to the legendary Egyptian-Greek sage Hermes, both of which have chapters later in the treatise devoted to the topics.

Following this diagram, the text itself begins modestly enough:[25]

> The knowledge of the celestial sphere, and the characteristics of its revolutions and movements, is a knowledge that eludes humans, who are unable to ascertain its nature and verify its size.

Despite these reservations, our author continues on shortly thereafter to say:[26]

> People are in complete agreement that the celestial sphere rotates and is spherical, bringing about through its revolving around the Earth the phenomena of longitude and latitude. The Earth is placed in its midst like a mid-point of a circle, surrounded by the revolving spheres, which in turn are encompassed by the one largest sphere. It (the largest sphere) rotates from East to West around two opposite poles, one southern and one northern, revolving on these two poles in a natural and continuous movement, according to the will of its Creator Who set it forth.

After providing a few more details about the basic characteristics of the universe with its nested spheres and planets, the author begins a very long discourse, extending over two chapters, about zodiacal signs. This appears also to have been a topic of great interest to our author—and presumably to his unnamed patron—and indeed they today play a fundamental role in modern practices of astrology.

Because the topic of zodiacal signs involves terminology unfamiliar to the modern reader and employs many technical details developed in the Greek system of horoscopic astrology, we have placed a discussion of them in the appendix. Anyone who enjoys considering how many different ways you can align and realign the same groups of items can then turn to the appendix for details. Those who are not attracted to such technicalities and mental gymnastics, or are not modern astrologers, can simply avoid the appendix altogether.

India and the Origins of Astronomy

In the midst of discussing the general principles of the universe, our author takes a rather abrupt digression into the history of astronomy in India. This account appears to be unique to this treatise, or at least it has not yet been found in any other preserved book. The story places the origins of astronomy and astronomical tables in the Indian city of Kannauj. That city lies today in the Farrukhābād district of Uttar Pradesh, and it is also indicated on the map of the Indus River in the second book of the *Book of Curiosities* (see fig. 8.5). Clearly it played a large role in our eleventh-century author's view of world history.

The narrative concerns an unidentified person named H-b-w-d (or H-n-w-d) and one "Nābaṭah, king of India." Nābaṭah is probably referring to either Nāgabhaṭṭa the First (reg. 750–80), founder of the Gurjara-Pratihāra dynasty of Ujjain and Kannauj, or Nāgabhaṭṭa II who ruled from 805 to 833.[27] The city Shawilābāṭṭ mentioned in the story stands for the Indian city of Kapilavastu, the birthplace of Gautama Buddha.[28] The king Aṭqā, said to have ruled six hundred years before the Prophet, is no doubt intended to be King Asoka or Ashoka (reg. 270–32 BC), the third king of the Mauryan dynasty and, according to Buddhist literature, a cruel and ruthless king who converted to Buddhism and thereafter established an exemplary reign of virtue.[29]

This remarkable account of the compilation of astronomical computations merits a full quotation:[30]

> As for the sages of India, it is unanimously agreed that the one devoted to the extraction of the science of the stars from the ancient books was *H-b-w-d*, king of Kannauj, which is one of the largest and most glorious cities in India, located near the equator, three degrees north of the tropic (of Cancer).[31] It is also a city of wise men and a centre of the learned in India. *H-b-w-d* had been observing the stars in this city four hundred years before the time of the Prophet, and he exerted great efforts and gained much knowledge in studying astronomy.

Others have said that Nābaṭah, king of India, ordered his scholars to observe the planets, and to calculate their mean motions[32] and their planetary functions.[33] Whenever their results conformed to the observations of his forefathers and to the knowledge passed down from the king *H-b-w-d*, they completed the calculations and placed them on the idol of the exalted Brahmins. He ordered that the observations should be repeated each day and the date recorded, and that they should be written in gold-water[34] on ivory plates. Nābaṭah lived to be 113 years old and had, since his childhood, spent his life observing the stars, his thoughts completely absorbed with them, relying on the assistance of the scholars of his age.

Towards the end of his life there appeared before him a very learned scholar from the edges of his country, from a city known as Shawilābāṭṭ, which is the land of the elephants. This scholar provided the king with mean motions and planetary equations that he claimed to have found buried amongst the treasures of the king Aṭqā, who reigned six hundred years before the time of the Prophet. Nābaṭah accepted these calculations, as he found them to be correct and conforming to what he had already determined and observed. He kept these calculations in his House of Learning and made them into the rule of law that should be followed in his kingdom.

To this day, the philosophers in the city of Kannauj prepare planetary equations related to the time of the Buddha, the great sage.[35] They claim that if one studies the planets using these planetary equations and mean motions, they reveal the obscure truths. It is said that these are preserved in the great temple of the idol, and no one can see them except the Brahmin worshippers.

Every day, the Brahmin keeper of the temple of the idol takes out a book containing the course of the seven planets, and the beneficial and malfeasant attributes of the sphere that result from the conjunctions of the planets. They copy it, and hang the copy they have made on the exterior wall of the House of Learning, so that anyone who wants to benefit from it on that day may look at it. It is left there until midday, and then it is hidden again in the Treasury of Knowledge until the keeper appears the next morning with another book to explain the condition [of the celestial sphere].

The tale merges the biography of Gautama Buddha with the origins of the great Indian manual of astronomy that was translated in the eighth century into Arabic and known as the *Sindhind*.[36] While India often served as a topos for astrological and astronomical authority, this particular story may well reflect (in a somewhat garbled form) an account of the origins of the art that circulated along the Fatimid commercial trade route to China passing through northern India and Tibet. This trade route will be discussed in chapter 8. At the very least the tale provides a notion of what was circulating in Egypt at that time regarding India and its astronomical observations.

Thirty Bright Stars of the Hermetic Tradition

Divination by seeing groups of bright stars has a long history. The *Book of Curiosities* is important in providing us with an additional source for this once popular topic. In Late Antiquity, a special list of Thirty Bright Stars was constructed and eventually transmitted to the Arabic-speaking world, where they were known as *bābānīyah* stars and associated with the legendary Egyptian-Greek sage known as Hermes the Wise or Hermes Trismegistus (Thrice-Great). The *Book of Curiosities* devotes an entire chapter to these stars, presented in a tabular format.[37]

Prior to presenting this table, our author states that the Persians and Indians assert that there are thirty stars having hidden properties related to the five planets (Mercury, Venus, Mars, Jupiter, Saturn). When any one of these stars are found to occur in particular places on horoscopes, they indicate the favorable or unfavorable consequences associated with the temperament (*mizāj*) of the planets assigned to them. In other words, even if a planet is not found in a particular sign on a horoscope, should one of these thirty stars be seen in the skies in that area, the astrologer can then base the prediction on the properties associated with its planet, even though the planet is not actually in that location.

The association of the "temperaments" of a planet with certain fixed stars was expounded in the second century AD by the Alexandrian astronomer Ptolemy in his defense of astrology known as the *Tetrabiblos* ("the four books"),[38] and it may well have had an even longer history going back to Babylonian practices. The oldest preserved text presenting Thirty Bright Stars is a fragment by an anonymous Egyptian astrologer writing in Rome in the year 379. The next oldest is that of an Alexandrian astrologer active in 601 known as Rhetorius Aegyptius. The attribution to Hermes or Hermes the Wise (*Hirmis raʾs al-ḥukamāʾ*) occurs only in the Arabic version of the Thirty Bright Stars, which, in the early thirteenth-century, was translated into Latin at Toledo by Salio of Padua.[39] There are a considerable number of Arabic treatises preserved today on astrology or alchemy and related topics that are attributed to Hermes, but this treatise on the Thirty Bright Stars is one of only two for which there is surviving evidence of an actual Greek original.[40]

Our author provides a list of these bright stars and the pair of planets associated with each one. A typical entry in the table is the following:[41]

> *al-Dabarān* (the follower)—that is, the eye of the bull
> In the zodiacal sign of Gemini
> Persian name: *s-k-d-w-l*
> Temperament: Mars and Venus

The star in this case is *α Tauri*, whose "modern" name is Aldebaran. It is the thirteenth brightest star in the Heavens and often given the common alternative name "the eye of the bull."

Only a few fragments of comparable material have been otherwise preserved from the Greek, Arabic, and Hebrew traditions of Thirty Bright Stars. In these related fragments, the stars are accompanied by their celestial latitudes and longitudes. In the *Book of Curiosities*, however, the latitudes and longitudes are omitted, with only the name of the zodiacal sign given. Our author also provides for every star a name that he calls "Persian" (*bi-l -fārisīyah*).[42]

A table relating the stars given in the *Book of Curiosities* to other recorded fragments of the Thirty Bright Stars is given in fig. 2.3.[43] Column I provides modern identifications of the stars, while Column II gives the sequence of stars as listed in the *Book of Curiosities.* In Column II the thirtieth place has been left blank because in the *Book of Curiosities* that field in the table was left open, without a name. Stars not named in the *Book of Curiosities'* list but given in other versions appear beneath the thirtieth row.

Column III gives the sequence of these same stars as presented in the Greek treatise by Rhetorius Aegyptius. Column IV lists the stars in the sequence found in a copy of an anonymous Arabic version preserved today in a manuscript in Dublin (unfortunately incomplete).[44] The Latin translation of the Arabic made by Salio of Padua is indicated in Column V.[45] A treatise titled *The Judgments of Nativities* (*Kitāb Aḥkām al-mawālīd*) and attributed to Abū Maʿshar al-Balkhī (d. 886), the famous astrologer of Baghdad, incorporates this star list within it,[46] and the sequence of its stars comprises Column VI. A second Arabic author, possibly Māshāʾallāh (d. ca. 810), apparently included a version of the treatise on Thirty Bright Stars in his astrological tract, for the latter was translated into Latin by Hugo of Santalla, probably between 1141 and 1151. While the Arabic of Māshāʾallāh does not appear to be preserved today, the Latin has been edited and published,[47] and its stars are indicated in Column VII.

Two additional Arabic authors also incorporated within their astrological treatises a chapter on the Thirty Bright Stars: Abū Naṣr Ḥassan ibn ʿAlī

I. Modern Nomenclature	II. *Book of Curiosities*	III. Rhetorius Aegyptius [Greek]	IV. Chester Beatty Arabic MS. 5399	V. Latin trans. by Salio of Padua	VI. Abū Ma'shar	VII. Latin trans. of Hugo of Santalla	VIII. al-Qummī	IX. Kūshyār ibn Labbān
θ *Eridani*	1	26	—	25	23	23	1	—
β *Persei*	2	15	14	14	16	14	2	—
α *Tauri*	3	30	—	30	27	27	3	2
β *Orionis*	4	11	12	12	12	12	4	9
γ *Orionis*	5	22	—	18	18	18	6	7
α *Aurigae*	6	16	16	16	15	--	5	10
ε *Orionis*	7	12	—	—	13	16	7	8
α *Orionis*	8	20	--	21	21 [?]	20	8	6
β *Aurigae*	9	13	10	10	10	10	9	11
α *Canis Majoris*	10	10	9	9	9	9	11	12
α *Geminorum*	11	19	5	5	5	6	12	14
β *Geminorum*	12	17	—	17	17	17	13	15
α *Canis Minoris*	13	21	—	19	—	19	14	13
α *Hydrae*	14	29	—	28	25	24	15	—
α *Leonis*	15	6	7	7	7	7	16	18
δ *Leonis*	16	28	—	27	24	25	17	—
β *Leonis*	17	27	—	29	26	26	26	—
α *Virginis*	18	1	1	1	1	1	18	22
α *Coronae Borealis*	19	5	4	4	4	4	21	21
β *Librae*	20	18	6	6	6	5	—	24
α *Centauri*	21	25	—	—	22	22	20	—
α *Scorpionis*	22	9	8	8	8	8	22	23
$\beta^{1,2}$ *Sagittarii*	23	14	15	15	14 (α *Sgr*)	13	25	—
α *Lyrae*	24	2	2	2	2	2	24	26
α *Aquilae*	25	8	11	11	—	11	27	27
α *Piscis Austrini*	26	3	—	—	—	—	28	28
α *Cygni*	27	4	3	3	3	3	29	29
β *Pegasi*	28	24	—	24	—	—	30	30
α *Andromedae*	29	23	—	20	20	21	—	—
	30	—	—	—	—	—	—	—
β *Virgini*	—	—	—	22	—	—	—	—
α *Boötis*	—	7	—	26	—	—	19	20
α *Ophiuchi*	—	—	—	—	—	—	23	—
α *Carinae*	—	—	—	—	—	—	10	—
β *Cassiopeiae*	—	—	—	—	—	—	—	1
M44 Praesepe	—	—	—	—	—	—	—	16
$\nu^{1,2}$ *Sagittari*	—	—	—	—	—	—	—	25

FIG. 2.3. Table comparing the Thirty Bright Stars as listed in the *Book of Curiosities* with other preserved sources.

al-Qummī, working in Baghdad around 976 and the Iranian astronomer Kūshyār ibn Labbān, active around 1000. The stars as given by al-Qummī and Kūshyār are in Columns VIII and IX.[48] In addition, there is also a Hebrew version made from the Arabic, preserved in a unique but defective manuscript in Paris,[49] but in the interests of conserving space, it has not been included on the table below.

A look at fig. 2.3 reveals that the author of the *Book of Curiosities* had before him a text very like that used by al-Qummī. However, al-Qummī does not specify his source for the Thirty Bright Stars, and so it is difficult to know what strand of Hermetic literature was contained in the Arabic literature used in compiling his list. From this comparative table it is evident that the list of Thirty Bright Stars was, nonetheless, remarkably uniform throughout the Greek, Arabic, and Hebrew traditions.

The placement of prominent stars on a horoscope, as advocated by the author of the *Book of Curosities,* is unusual, and its practice at this time probably reflects the wide circulation in Egypt of Arabic texts attributed to Hermes. Two of these Thirty Bright Stars occur on a horoscope made in Old Cairo for our author's contemporary, Ibn Riḍwān (d. ca. 1061), an astrologer and physician (see the appendix for details).

Star Groups of Bedouin Origin

Two hundred and twenty-nine stars or star groups that were not from the classical Greek tradition, nor from the later Greek tradition attributed to Hermes, also receive attention from our author. Their names derive from the Bedouin pre-Islamic tradition of mapping the sky. Bedouins used the risings and settings of prominent star groups for simple timekeeping and to predict rainfall and seasonal changes prompting the migration of herds.[50] In the pre-Islamic agricultural astronomy, the year was divided into intervals of thirteen days each plus one of fourteen days, defined by the setting in the west just before sunrise of one of the star groups and the simultaneous rising at the eastern horizon of the opposite group.

The star lore of pre-Islamic times featured prominently in early Arabic poetry, with lexicographers and grammarians recording the material in collections known as Books of Anwāʾ (*kutub al-anwāʾ*), of which over twenty are known to have been compiled in the ninth and tenth centuries, though only four survive today.[51] *Anwāʾ* is the plural of *nawʾ*, whose meaning is uncertain and consequently usually left untranslated, but seems to refer to the fact that when one star rises in the east another sets in the west. Three treatises on this topic were employed by ʿAbd al-Raḥmān al-Ṣūfī, an astronomer in Shiraz. His illustrated *Book of the Constellations of the Fixed Stars* (*Kitāb Ṣuwar al-kawākib al-thābitah*), composed in the year 964, formed the basis of most of the later celestial iconography in the Islamic world.[52] For this reason al-Ṣūfī's *Book of the Constellations of the Fixed Stars* is an important source of information on Bedouin asterisms as well as Greek-Ptolemaic constellations.[53]

In his book on the constellations, al-Ṣūfī compared the Bedouin star groups with the Greek-Ptolemaic constellations, occasionally illustrating

FIG. 2.4. The Bedouin constellation of a camel superimposed over the classical constellation of Cassiopeia, from the *Book of the Fixed Stars* (*Kitāb Ṣuwar al-kawākib al-thābitah*) by ʿAbd al-Raḥmān al-Ṣūfī. Oxford, Bodleian Library, MS Huntington 212, fol. 40b, copied in 1170/566H and possibly dedicated, in a now partially illegible dedication, to Sayf al-Dīn Ghāzī, at that time ruler in Mosul.

the Bedouin constellation image alongside the classical forms, though these composite images are rarely preserved today. For an example, see fig. 2.4 where the Bedouin constellation of a camel is superimposed over that of the classical Greek constellation of Cassiopeia. It should be noted that the illustrations in al-Ṣufi's treatise always show the constellation upright, even when it might not be seen that way in the sky, and the four cardinal directions are never given. The illustrations are not plotted, for there is no grid or point of reference. In other words, they are not "scientific" diagrams in any measured or quantitative sense, though they are often excellent visual representations of the relative positions of the stars.[54]

There is no concrete evidence anywhere in the *Book of Curiosities*, however, that its author used al-Ṣūfī's treatise, if indeed he had access to a copy. Moreover, it is improbable that our author would not have tried to incorporate into the *Book of Curiosities* some of the constellation illustrations forming such a prominent feature of al-Ṣūfī's treatise, since our author was intent on compiling a highly illustrated guide to both the Heavens and the Earth.[55]

The *Book of Curiosities* presents in tabular format the names and arrangement of 229 clusters of stars from this Bedouin tradition.[56] Each cluster is named and then illustrated with a small diagram beneath, occasionally accompanied by additional information on the location of the stars. They are divided into two groups, those north of the ecliptic and those south, though this distinction is not consistently followed. An example from those of the northern skies:

> The dyed hand (*al-kaff al-khaḍīb*): white stars in the Milky Way [illustrated with two stars]

The "dyed hand" refers to the well-known W-shaped asterism in the constellation of Cassiopeia and reflects the Bedouin image of a woman whose right hand was visualized as extending toward Cassiopeia, with the fingers represented by the asterism (βαγδε *Cassiopeiae*). In the illustration from al-Ṣūfī's treatise (fig. 2.4), however, this star is labeled: "The dyed hand (*al-kaff al-khaḍīb*)—that is, the hump of the camel." The latter information (that the star is also the "hump of the camel") is not in the *Book of Curiosities*, again suggesting that al-Ṣūfī's treatise was not a direct source for our author.

The source used by our author is unidentified. He appears not to have used al-Ṣūfī, and he seems to have had little interest in the poetry and proverbs that were a typical feature of the Books of Anwāʾ. In the table in the *Book of Curiosities*, the Bedouin star names have no astrological attributes assigned to them and there are no directions as to their astrological use. The focus is solely on recording the names of these 229 Bedouin star groups and depicting the number of stars in each.

Lunar Mansions

Another form of mapping the skies came to be known as "lunar mansions." It originated not with Bedouin Arabs nor with Greek astronomers, but in Asia. Focusing upon the orbit of the Moon, early observers of the sky in China, Central Asia, and India noted prominent groups of stars near the path of the Moon's orbit. A series of twenty-eight star groups near this path were singled out as a means of demarcating areas of the sky. For a month, the Moon stays, or "resides," in one of these twenty-eight "lunar mansions" each night. Sometime after the rise of Islam, knowledge of these star groups merged with the pre-Islamic Bedouin agricultural astronomy described in the previous section.[57] Early astronomers tried to align the two systems, meshing the twenty-eight divisions of the Asian lunar zodiac (the "lunar mansions") with the twelve signs of the Greek-Mesopotamian solar zodiac.[58]

Completely separate from the chapter given over to Bedouin star groups, the author of the *Book of Curiosities* devotes a lengthy and illustrated chapter to the topic of lunar mansions.[59] In this chapter, our author provides small maps or diagrams for each of the twenty-eight lunar mansions—maps on which he also indicated nearby stars from the Bedouin tradition and occasionally from the Greek tradition, apparently to identify the locations of these imported "lunar mansions." Most of the Arabic names given these "lunar mansions" are not Ptolemaic nor do they reflect classical Greek delineations of the sky into constellations. Many of the star names used in the miniature "maps" given by our author are of Bedouin origin, but there is not a large overlap with the Bedouin star names given in the chapter discussed above devoted to Bedouin-*anwāʾ* stars. Granted that the chapter on Bedouin star names covered a larger area of the sky since it was not restricted to the area near the Moon's path, it is nonetheless significant that the chapter on Bedouin stars contains 171 star names that are *not* found in the chapter on lunar mansions, while the latter chapter has 155 star names that do not occur in the former chapter. Moreover, the chapter devoted to Bedouin stellar names has seventy-three star names that are unidentified and not recorded in any other known source, while the chapter on lunar mansions has only eleven of that category. They share only fifty-eight star names. Clearly, different textual sources were used when composing these two chapters.

Even though this chapter in the *Book of Curiosities* is titled "On the Lunar Mansions, Their Attributes and Occult Influences," our author in fact presents only a guide to the location of the stars forming the lunar mansions, employing small diagrams or "maps" as aids for locating them in the skies (see plate 20 and fig. 2.5). He fails to give any prognostications or to specify how an astrologer or timekeeper might use them.

Although our author does not mention it, muezzins—the criers who, from a minaret, call the faithful to prayer five times a day—were often expected to recognize the risings of lunar mansions, as is evident from a guide composed about 1300 in Egypt for the regulation of the markets and public functions. In this manual, Ibn al-Ukhuwwah states the requirements of a muezzin:[60]

> The muezzin must know the lunar mansions and the arrangement of the stars in them, in order to know the passage of the hours at night. There are twenty-eight mansions. . . . Daybreak occurs in each of these man-

sions for thirteen days and then progresses to the next mansion. If the muezzin knows which mansion is rising [at the local eastern horizon] and looks at the mansion visible at the meridian, he will know which is rising and which is setting and what is the interval between that moment and daybreak. This involves knowledge (*ʿilm*) and calculation (*ḥisāb*) whose explanation would be very lengthy.

One of the twenty-eight small celestial "maps" in the *Book of Curiosities* is illustrated in fig. 2.5—in this case the third Lunar Mansion centered around the Pleiades, the famous open star cluster in the constellation of Taurus. The text above the "map" reads as follows:[61]

> *Al-thurayyā* (the Pleiades): Thereafter rise the Pleiades (Lunar Mansion III), a cluster of six semi-nebulous stars, in the form of an isosceles triangle. They rise laterally from their position, disappear from it, and then rise to the north-west of Lunar Mansion I and Lunar Mansion II. Rising with them in the north is Capella (*α Aurigae*), which is a luminous star of the first magnitude on the left shoulder of the constellation Auriga. Then there are two small stars known as The Two Feet of Capella. Also rising with it is a not very bright star called The Shoulder-Blade, close to the Pleiades. Toward the south, a cluster of stars known as The Cattle rises.[62] . . . Because of the ugliness and the magnitude of the stars in this asterism, Valens called it "A devil carrying lanterns." This is what this lunar mansion and its "indicator"-stars look like:

On the left-hand side of the "map," diagonally alongside the word for north, there is the statement that the Pleiades rise on the nineteenth of *Bashnas* (the ninth Coptic month), which is the fourteenth of the month of May (*Ayyār*).

The quotation from the second-century Greek astrologer Valens has not been found in any other preserved source. Vettius Valens, a contemporary of Ptolemy, worked from about 152 to 162 in Antioch and later in Alexandria. His Greek astrological treatise called the *Anthology* was translated and commented upon in Middle Persian by Buzurjmihr, a sixth-century Sasanid minister. It was then translated into Arabic, where Valens was known as *Wālīs*. Though full Arabic versions of his writings are now lost, both Valens and his Sasanian commentator were quoted by many early Arabic writers,[63] particularly the tenth-century Baghdadi astrologer Ibn Hibintā, with whom our author shares many sources. In Egypt there may have been particular interest in Valens, for all the extant copies of a small Arabic table for com-

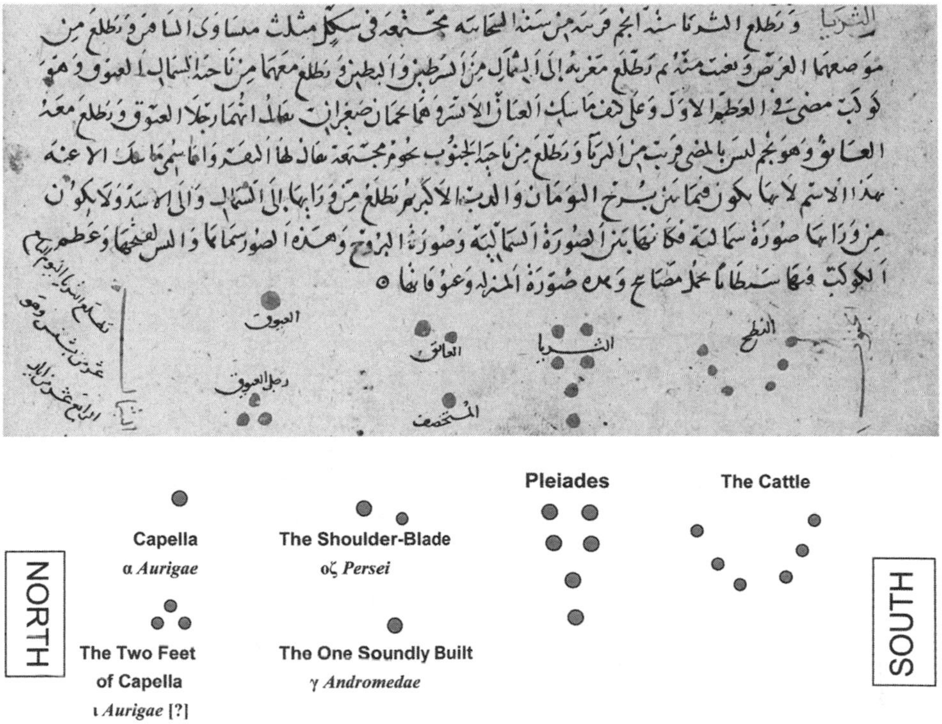

FIG. 2.5. The "map" of the third lunar mansion as drawn in the *Book of Curiosities*, with a diagram beneath providing translations and modern identifications of star-groups, when possible. Oxford, Bodleian Library, MS Arab. c. 90, fol. 18b.

puting the length of a person's life and attributed to Valens are of Egyptian provenance.[64]

Numerous Arabic similes for the Pleiades were composed, and more than four hundred have been published,[65] yet none compare the Pleiades to a devil (*shayṭān*) nor consider it as ugly. To the Arabs the Pleiades always had very auspicious connotations, and comparisons with lanterns were common. In this quotation the "devil" possibly refers to another nearby star group containing the star β *Persei* (Algol) which was called "the head of the ghoul" (*raʾs al-ghūl*)—with the lanterns being the Pleiades.

We do not know if our author originated the idea of drawing these twenty-eight small maps of the lunar mansions. It is not impossible. What we do know is that they continued to circulate in Egypt, for they are to be found in a thirteenth-century manuscript now in Dublin (see fig. 2.6). In that manuscript, however, the accompanying short texts are very different from those in the *Book of Curiosities*.[66] The Dublin manuscript is cataloged

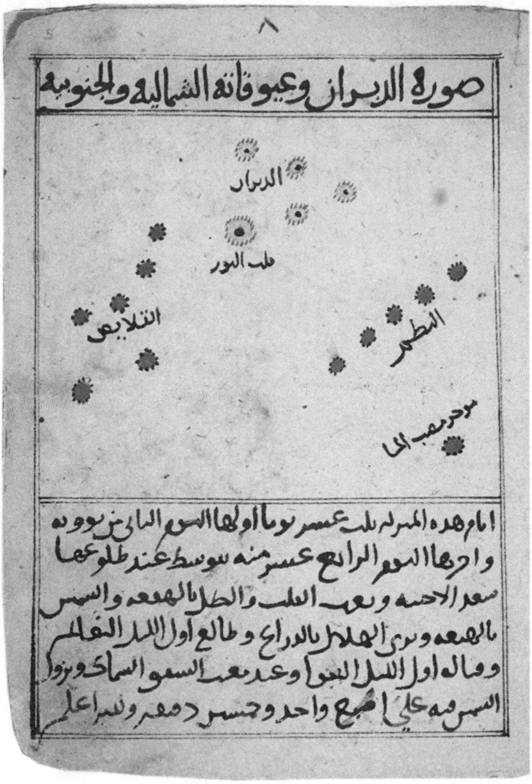

FIG. 2.6. A "map" of the fourth lunar mansion, as illustrated by an anonymous author, possibly the Egyptian astronomer Ibn Hārūn al-Ṣiqillī. The five stars at the top are labeled "Aldebaran" (*al-dabarān, α Tauri*), with the lowest of the five made more prominent and labeled "The Heart of the Bull" (*qalb al-thawr*, another name for *α Tauri*). The six red stars to the left are labeled "The Young Camels" (*al-qalāʾiṣ*) and represent the Hyades. Chester Beatty Library, MS Arabic 4538, fol. 5a, undated, ca. thirteenth century. © The Trustees of the Chester Beatty Library, Dublin.

as anonymous, though Professor David King reads an inscription on the defaced title page as providing an author's name of Ibn Hārūn al-Ṣiqillī, presumably from a Sicilian family. Professor King suggests a possible identification with one Abū al-Qāsim ibn ʿAbd Allāh ibn ʿAbd al-Raḥmān ibn Ḥasan al-Qurashī al-Ṣiqillī, recently identified by Professor Julio Samsó as an astronomer active in Cairo and Alexandria but at unknown dates.[67]

The occurrence of these sets of twenty-eight diagrams or maps in Egyptian treatises—and their omission from preserved material originating in other regions—suggests that making drawings of sections of the sky containing the lunar mansions may have been a distinctly Egyptian preoccupation.

On maps in the *Book of Curiosities* and in the treatise possibly by al-Ṣiqillī, stars designated as *ʿayyūq* or *ʿayyūqāt* are mentioned. Translated here as "indicator stars," these are groups of bright stars that serve as important visual indicators or signals of nearby lunar mansions. ʿAbd al-ʿAzīz ibn Aḥmad ibn Saʿīd al-Dīrīnī, an Egyptian Ṣūfī who lived from about 1215 to 1297, included in his treatise *Gems in the Art of Timekeeping* (*Kitāb al-Yawāqīt fī ʿilm al-mawāqīt*) a poem titled "A Discourse on the Indicator-Stars of the Lunar Mansions" (*Bāb dhikr al-ʿayyūqāt liʾl-manāzil*).[68] The margins of the treatise contain small illustrations of these groups of bright stars.

The *ʿayyūqāt* or "indicator stars" of lunar mansions were also discussed roughly a half-century later by another Egyptian scholar, Abū al-ʿAbbās ibn Abī ʿAbd Allāh Muḥammad ibn Aḥmad al-Miṣrī. His *Book of Pearls and Sapphires on Astronomical Observation and Timekeeping* (*Kitāb al-Durar wa-al-yawāqīt fī ʿilm al-raṣad wa-al-mawāqīt*), written in 1334, is preserved in a mixed volume known for its illustrated late fourteenth-century astrological treatise with the untranslatable title *Kitāb al-Bulhān*.[69] The *Book of Pearls and Sapphires* is a manual on the determination of approximate prayer times by means of lunar mansions and opens with a set of prayer tables for Cairo, in which the lunar mansion rising at daybreak is given for each day of the Coptic year and the corresponding date in the Syrian-Greek calendar. There follow tables for the lunar mansions rising, culminating and setting at nightfall, and at one-third, one-half, two-thirds, and three-quarters of the night—all computed for the Coptic calendar. Other tables display the thirteen divisions of the lunar mansions rising at daybreak, and yet others provide the shadow length at midday for each day in the Copic year, the duration of daylight and the solar altitude at the beginning of the afternoon prayer and when the Sun is in the direction of Mecca. When compared with the *Book of Curiosities*, this manual is much more technical (but still not really mathematical).

Like the small star maps themselves, the use of "indicator stars" may have been introduced into folk astronomy only by Egyptian or Cairene astrologers.[70] In any case their continued use for simple timekeeping is evident today in Oman.[71]

The topic of lunar mansions occurs again in the second half of the *Book of Curiosities*, where the influence of the macrocosm on the microcosm is evident through talismans employing lunar mansion imagery. In the chapter concerned with marvelous aquatic creatures, our author devotes a subsection to talismanic designs associated with each of the twenty-eight lunar mansions. Fabulous creatures, often semi-human in form, are aligned with the shapes and names of the twenty-eight lunar mansions. For example, it describes one creature as[72]

> a beast called *q-r-s,* having the form of a woman sitting cross-legged, with a crown on her head, tresses hanging down, and holding a fig leaf in her hand. It is associated with the name of *al-thurayyā* (the Pleiades, Lunar Mansion III).

A tradition of describing fantastical semi-human talismanic designs that were to be drawn when the Moon was in a given lunar mansion appears to have been fairly widespread. The magical use of lunar mansions is evident by the tenth century in al-Andalus. This is shown by the tenth-century treatise *The Goal of the Sage* (*Ghāyat al-ḥakīm*), known in Europe as the *Picatrix,* probably composed by the Andalusian occultist Abū al-Qāsim Maslamah ibn Qāsim al-Qurṭubī.[73] His contemporary and compatriate Ibn al-Ḥātim composed a treatise giving the designs of talismans for the lunar mansions, but the designs differ radically from the creatures described by our author.[74] A short treatise attributed to Aristotle titled *Kitāb al-Istamākhīs*[75] also provided astrological uses of lunar mansions in talismans to be made when the Moon "lodges" in a mansion.

Talismanic designs associated with lunar mansions are also occasionally illustrated. They can be seen, for example, on a large astrolabe made in Egypt in 1227 (625H) by ʿAbd al-Karīm al-Miṣrī.[76] Another example can be seen in fig. 2.7 and plate 21 from a late fourteenth-century astrological miscellany produced in Iraq.[77]

Thus the lunar mansions were not only a crude method of timekeeping through observing their dawn risings, but they also became associated with magical powers. These powers could be elicited by drawing—when the Moon was visible in one of the mansions—talismanic designs. These designs, however, were far removed from the configuration of the lunar mansions so carefully mapped by our author.

"Stars with Tails": Comets, Meteors, and Fireballs

"Stars with wisps of tails"—that is, comets and related phenomena—were universally considered to be omens of future events.[78] The phrase "stars having wisps of tails" (*al-kawākib dhawāt al-dhawāʾib*) reflects the earlier Greek designation of comets as *komētai* (having long hair).[79]

This category included not just comets, but also meteors and fireballs.

FIG. 2.7. Talismanic designs of fourteen lunar mansions; third down in the right-hand column is the design for Lunar Mansion III (the Pleiades). From the illustrated astrological treatise *Kitāb al-Bulhān* compiled for the Mongol ruler of Baghdad, Sulṭān Aḥmad, who ruled 1382–1410. Oxford, Bodleian Library, MS Bodl. Or. 133 fol. 27b.

A comet, with its vaporous tail pointing away from the Sun, moves about the Sun in an eccentric orbit. A streak of light also emanates from a meteor, giving the impression of a tail. The brightest meteors are called in English "fireballs" and can take half a minute to cross the sky, leaving trails (called "trains") visible for several minutes.[80] When accompanied by noises, they are called "bolides," and indeed such an occurrence is mentioned by our author, as are meteor showers, called "a storm of stars (*inqiḍāḍ al-kawākib*)."

Comets, meteors, and related phenomena are seldom mentioned in astronomical treatises, presumably because they did not lend themselves to the mathematical modeling and prediction of their occurence. They were, however, of interest to chroniclers, historians, and astrologers. Comets, as well as meteors, were generally considered to be atmospheric phenomena of the sublunar sphere. In the sequence of chapters in the *Book of Curiosities,* however, this topic occurs between the fixed stars and the planets. Curiously, the usual term for meteors (*shihāb*, pl. *shuhub*) does not occur in the treatise, nor does the term *nayzak,* which was used by others for a variety of sublunar phenomena, including shooting stars and supernovae. [81]

The first of two chapters in the *Book of Curiosities* devoted to this subject opens with what one might call "generic" comets, for they bear no particular name and are not associated with any particular place or time. For example:[82]

> If the star with a wisp of a tail appears in the east, then [the misfortunes will occur] in the east; if it appears in the west, then in the west; or if it appears in the north, then in the north. If, however, the comet appears in the south, the tribulations will be felt worldwide, though mostly in the middle of the region. When it appears to the east of the Sun, the events it indicates will happen promptly; if the comet appears to the west of the Sun, the events it indicates will be delayed. If it fades quickly, then the events it indicates will be minor, but if its ascent is prolonged, the events it indicates will be long lasting. The region towards which the tail is inclined will be the one most affected by its evil influence. Only God knows His mysteries.

The significance of a comet's appearance in each of the zodiacal signs occupies a subchapter. For example:[83]

> If it appears in the sign of Gemini, hot sandstorms will scorch fruits and produce, while birds will perish from the excessive heat. Epidemics will strike, killing children and causing pregnant women to abort. Meteor showers will be frequent. A great thud will be heard in the sky, together with terrifying sights, such as thunder and glowing lights and strong flashes of lightning.

Comets Ascribed to Ptolemy

After the discourse on comets appearing in various zodiacal signs, the subject moves to eleven "named" comets whose occurrence at any point in the Heavens was usually (but not always) considered to portend much disorder and many unfortunate consequences. Though the Greek astronomer Ptolemy is cited in this section as the authoritative source, no precise parallel has been found in his *Tetrabiblos*.[84] Each celestial object was illustrated and provided a name and full description (see fig. 2.8 and plate 22).

As an example, the seventh comet is described as follows:[85]

> Concerning the appearance of the comet known as *ṭayfūr*: It has an ugly appearance. It is round, black on the inside and red on the outside. It contains an image and flares, and it is likened to the devil. It possesses no beauty, and it travels slowly in the sky, with a mane behind it. It descends after the Sun along a northern course. Its appearance indicates a widespread evil, soaring prices, rotting of fruits, and the destruction of lowly people, robbers, slanderers, poisonous animals, and other animals harmful to men, such as wild beasts and crocodiles. Moreover, it will better any malice in human hearts, and cause men and women to fall in love. It will cause the demise of cattle. Minerals and medicaments, such as the myrobalans[86] and the like, will become expensive. The kings of the east and the west will perish. This is what it looks like. [see no. 7 in fig.2.8 and plate 22]

The meaning of the name *ṭayfūr* is obscure. It possibly corresponds to the Greek word for typhoon (τυφῶν), which is also one of the ten comet names given in Late Antique lists.

Of the eleven comets given names by our author, three names also occur in another Arabic treatise purporting to present Ptolemy's knowledge of comets. This treatise is attributed (probably falsely) to the ninth-century Baghdadi translator Ḥunayn ibn Isḥāq and preserved today in at least three copies.[87] It presents general views of Ptolemy on comets, followed by the implications of their appearance in each zodiacal sign. Several are given names, and eleven are illustrated in small circular marginal diagrams. Only three names, however, correspond fully to ones in our treatise: *al-miṣbāḥ, al-liḥyānī, ṭayfūr*, which are equivalent to nos. 3, 4, and 7 in fig. 2.8.[88] Ptolemy is also cited in another small Arabic treatise as an authority on comets, but in this instance the topic is the appearance of an unnamed comet (*nayrūz*) on specified days of the week in which the Coptic New Year occurs.[89]

The early Arabic tradition of named "comets" derived in part from late Greek literature. A number of Greek manuscripts preserve a small text attributing a list of comets to Aristotle, with Greek names assigned to the comets.[90] Of the eleven comet names given in the *Book of Curiosities* (see fig. 2.8 and plate 22), seven have clear parallels with these Late Antique

1. *ʿurf al-faras* (the mane of the horse)

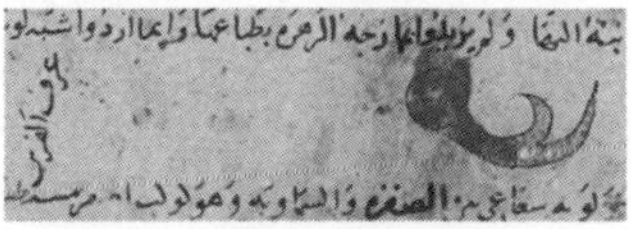

2. *al-ḥarbah* (the lance)

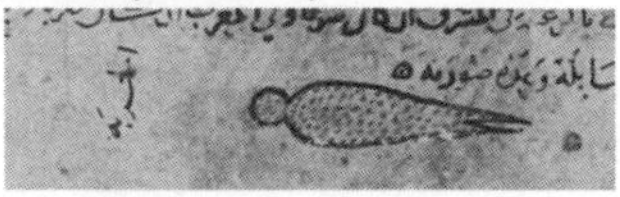

3. *al-miṣbah* (the lamp)

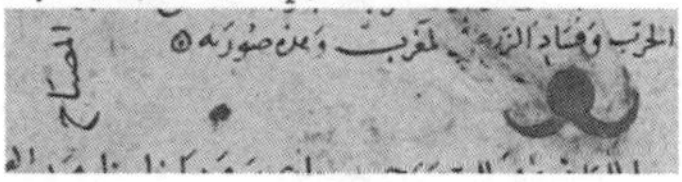

4. *al-liḥyānī* (the long-bearded)

5. *al-qaṣʿah* (the bowl)

6. *al-muwarrad* (the rosy one)

7. *ṭayfūr*

8. *al-ḥabashī* (the Ethiopian)

9. *al-saffūd* (the skewer)

10. *al-khābiyah* (the cask)

11. *al-kayd* (the deception)

FIG. 2.8. Eleven comets in the *Book of Curiosities* said to have been described by Ptolemy. Oxford, Bodleian Library, MS Arab. c. 90, fols. 13b, 14a, and 14b.

Greek names (nos. 1–5, 7, and 9). The same Late Antique Greek comet names and their Arabic equivalents are also reflected in many of the comet names found in medieval Latin treatises.[91] The eighth comet in our list, the Ethiopian (*al-ḥabashī*), has no identifiable Late Antique equivalent nor does the tenth one, "the cask" (*al-khābiyah*).

It is evident that our author employed treatises on this topic that were widely available a century or two earlier, for almost the entire discussion of these eleven supposedly Ptolemaic comets is a slightly expanded version of a chapter found in *The Ultimate Book on Astrology* (*al-Mughnī fī aḥkām al-nujūm*)[92] composed by a Christian astrologer named Ibn Hibintā, who was still active in Baghdad in 929.[93] Ibn Hibintā is not, however, cited as a source for this material. The fact that (with the exception of the eleventh comet, *al-kayd*) the discourse on each comet is more extensive in the *Book of Curiosities* than that found in Ibn Hibintā, together with the fact that two comets (nos. 9 and 10, "the skewer" and "the cask") are not mentioned at all by Ibn Hibintā, suggests that there was not direct employment of Ibn Hibintā's *Mughnī*.

The eleventh comet (*al-kayd*, "the deception") played a special role in Arabic astrological writings,[94] and there appears to be no Late Antique equivalent for this comet name nor a comparable Latin one. Ibn Hibintā seems to have given the first account of this fanciful comet. In the course of detailing its reappearance every one hundred years and its retrograde movement through one zodiacal sign every twelve years, he allotted much more space to the topic than did our author.

It has been suggested that the comet *al-kayd* derived from the Indian tradition of Ketu, which was the tail of the Indian demon Rāhu, whose head was severed from his body with both parts forming the eclipse monster identified with the two lunar nodes where the Moon crosses the ecliptic. In another Indian tradition Ketu took the form of a comet (*dhūmaketu*, "smoke-ketu") appearing at irregular intervals to threaten mankind.[95] And indeed the "comet" *al kayd* is always associated with India and its implications for humanity are always catastrophic.

Of the fearsome *al-kayd*, the author of the *Book of Curiosities* has only the following to say:[96]

> As for the red, round star surrounded by a dark blackness, the Indians say that it is seen in their lands. They call it "the deception" (*al-kayd*) and it ascends in their lands like a huge sack. It is one of the most inauspicious and ill-omened stars, and the most disruptive of essential needs. The ancient nations that have perished, such as ʿĀd and Thamūd, the

> people of Madyan, and the generation of Noah, all perished when this star appeared. The Indians believe that no other star brings destruction as this one, and that it is more ominous than the conjunction of Mars with Saturn. This is what it looks like: [see no. 11 in plate 22 and fig. 2.8].

The ʿĀd were an ancient tribe, mentioned in the Qur'an and said to have lived immediately after the time of Noah, while the Thamūd were an old Arabian tribe that disappeared before the rise of Islam.[97] In the Qur'an, following Old Testament accounts, the people of Madyan are said to have been punished for not believing their prophet Shuʿayb.[98]

Ibn Hibintā provided a long list of dire consequences if *al-kayd* is configured with certain planets in particular zodiacal signs, and he attributed all this to the legendary Egyptian-Greek sage Hermes. The fact that our author displayed great interest in star lore attributed to Hermes and yet did not make use of the prognostications following the appearance of *al-kayd* that are attributed by Ibn Hibintā to Hermes, suggests that this particular Hermetic treatise was not circulating in Egypt in the first half of the eleventh century and confirms the conjecture that our author had no access to Ibn Hibintā's treatise.

The first eight out of the eleven comet names given in the *Book of Curiosities* (see fig. 2.8) are also found in some later Arabic or Persian lists of comets.[99] The *Book of Curiosities*, or a common source, was evidently employed some seven centuries later in an illustrated treatise on the astrological significance of various kinds of comets and pseudo-comets written in Cairo around 1675 by an Egyptian astronomer named ʿAbd Allāh ibn Aḥmad al-Maqdisī al-Ḥanbalī.[100] It is unknown at this point just how much of our text is repeated in this seventeenth-century treatise, for only two folios have been published, but those correspond to the end of the section on comets according to Ptolemy in the *Book of Curiosities.*

"Stars with Faint Lances": Comets from the Hermetic Tradition

A completely separate chapter in the *Book of Curiosities* concerns yet more comets. Here, twenty-eight comets are attributed to the legendary Egyptian-Greek sage Hermes, but they do not include the fearsome comet *al-kayd* that Ibn Hibintā attributed to Hermes. This chapter is titled "On the Obscure Stars Having Faint Lances in the Ninth Sphere, which Have Immense Favourable and Malevolent Influences."[101] At the end it states that the chapter presented all twenty-eight comets mentioned by Hermes "as having a profound effect at times of birth, sometimes undermining actions without being noticed and sometimes bringing success without being detected."[102]

In the preserved text, however, only twenty-seven such "stars" are actually described and illustrated.

Some of the descriptions suggest auroral phenomena rather than comets. An unexplained curiosity is the statement made in the chapter title that these twenty-eight comets or stars are "in the ninth sphere," for the ninth sphere was in principle considered starless. On the other hand, the twentieth comet in the list is said to be a "yellow star under the orbit of the Moon," which contradicts the chapter title and suggests the usual placement of comets in the sublunary realm.

The descriptions of the first twenty-two comets in this section are accompanied by small groups of dots. As an example, the eighteenth (illustrated by one large star surrounded by six smaller ones) is described as follows:[103]

> A luminous large[104] star which is encircled by seven stars in the form of a necklace. Its rays, when they glow, almost obscure its surroundings. Its path follows that of the star Procyon,[105] but deviates slightly. The comet is known as The Upright (*al-qāʾim*). It traverses its orbit every one hundred years. Whenever it enters the sign of Scorpio, it causes in that year discord, bloodshed, wars, wide-spread killing, calamities, destruction, ruin, and death on land and sea through thirst, hunger, and the sword. Hermes called it The Ripper (*al-hattāk*). This is the star that appeared in the days of the khārijite Abū Rakwah, when, following its ascent, one hundred thousand men died by the sword, by drowning, by starvation, and by thirst.

The mention of the "khārijite" Abū Rakwah in connection with this comet clearly places our author in an early eleventh-century Fatimid Egyptian context. Walīd ibn Hishām Abū Rakwah was a leader of a rebellion against the Fatimid caliph al-Ḥākim that lasted from 1005 until his execution in Cairo in 1007.[106] He was not a khārijite in the sense of being a member of the religous sect of that name, but rather this was a common derogatory term employed by the Fatimids and other dynasties against their political enemies.

At the end of this chapter, the author employs the more common designation of comets or meteors when he speaks of twenty-eight "stars with wisps of tails." It is unclear what distinction the author wished to make between "stars with wisps of tails" (*dhawāt al-dhawāʾib*) and stars with faint lances (*dhawāt al-ḥirāb al-marsūmah*). Both terms may have been used interchangeably for all comets with at least two distinct tails—what today we know to be a dust tail and a gas tail. On the other hand, the use of two different terms may simply reflect two different sources used to construct the chapter, one Hermetic and the other Aristotelian.

While the first twenty-two comets were illustrated only by small groups of dots, the last five were depicted by larger images having sharp pointed tails, as shown in fig. 2.9 and plate 23. The celestial object called "the commander" (*al-qāʾid*) or "the archer" (*al-rāmī*)—fig. 2.9 no. 1—is described as[107]

> a star located on the same course as Saturn in terms of its altitude and course. Its saffron-like hue resembles that of Canopus (*Suhayl*). It is known as The Bridle (*al-zimām*). Hermes called it The Commander (*al-qāʾid*) and also The Archer (*al-rāmī*). When it appears, it casts light and large rays over the horizon. Its tail extends behind it for a length of about one hundred cubits. It appears every one hundred years. It is auspicious, bringing about happiness and joy. This is what it looks like: [see fig. 2.9, no. 1]

The common feature of all the comets attributed to the legendary Hermes is that they are recurrent or returning comets. The stars comprising the comet are said to reappear at specified intervals of time. However, comets bright enough to be visible to the naked eye are usually of such a large orbit that their reoccurrence is unpredictable and undetectable within a single life span.[108] The notion of periodicity may have arisen from the fact that at certain predictable times of the year the Earth's orbit crosses a stream of particles (in most cases probably from a comet) producing meteor showers. For example, the Perseids around August 12th or the Geminids near December 14th.[109] Such predictable meteor showers may have prompted medieval observers to associate periodic returns with meteors or comets.

No other account attributing twenty-eight comets to the legendary Egyptian-Greek sage Hermes is known to be preserved today, making this a unique remnant of Hermetic literature.

Astro-Meteorology

Completing his journey from the outmost sphere downward to Earth, the author devoted his last chapter to astro-meteorology, a subfield of divination and astrology. In astrological meteorology the focus is upon the sublunary and the earthly phenomena of winds, lightning, thunder, and earthquakes, and what they might portend.[110]

The final chapter on the Heavens in the *Book of Curiosities* opens with a circular diagram of winds (plate 24 and figs. 2.10 and 2.11). It reflects classical Greek theories of winds, with many of the wind names simply

1. *al-qāʾid* (the commander) / *al-rāmī* (the archer)

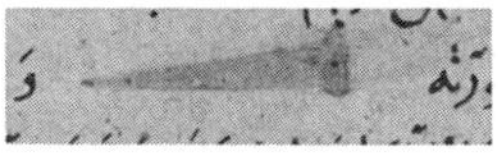

2. *kawkab al-dhanab* (the star of the tail)

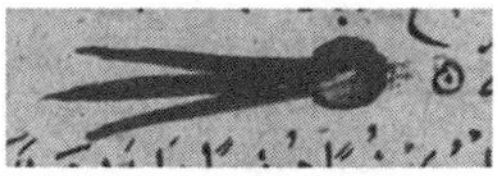

3. *al-waqqād* (the stoker)

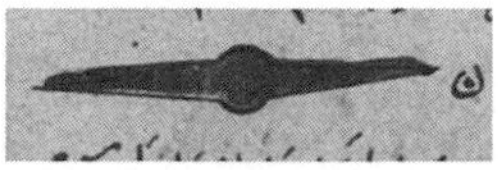

4. *al-muʿtaniqayn* (the embracing couple) / *alyat al-ḥamal* (the lamb's fat-tail)

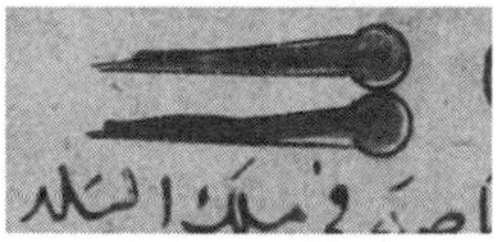

5. *al-rāmiḥ* (the lancer)

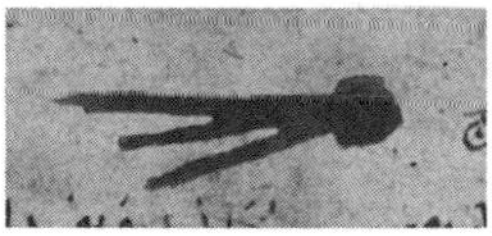

FIG 2.9. The last five of the twenty-eight "stars with faint lances," attributed to Hermes in the *Book of Curiosities*. Oxford, Bodleian Library, MS Arab. c. 90, fols. 15b and 16a.

transliterated Greek wind names. For example, Zephyros (ξέφυρος) meaning "west wind" is rendered in Arabic as *rīḥ y-f-w-r-w-s*, or Boreas (βορέας), "north wind," as *rīḥ būriyās.*[111]

In the diagram, however, the winds are not directly associated with the four cardinal directions, but rather with the zodiacal signs. In other words, it illustrates graphically the relationship between the macrocosm and the microcosm. Aries (linked with Zepyros, the west wind) is placed at the top, in the same position it held in the diagram opening the treatise (see fig. 2.1). Thus while the earthly dimension is specified by the placement of the Earth in the small circle at the center of the diagram, it is the temporal and seasonal dimensions of the cosmos that are illustrated rather than local, terrestrial occurrences of winds.

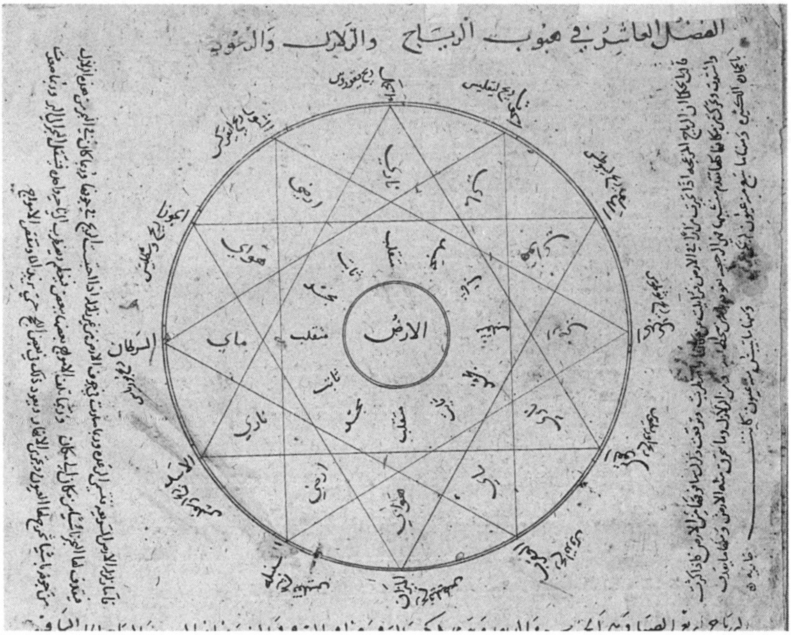

FIG. 2.10. A diagram of the winds in the *Book of Curiosities*. Oxford, Bodleian Library, MS Arab. c. 90, fol. 21b.

At first glance, this diagram appears to follow the tradition of "wind-roses," which were diagrams indicating the relative force of winds from various compass points at some given location. On Greek and early medieval wind diagrams, the outside circle represented the horizon of the observer on Earth, while on the *Book of Curiosities*' wind diagram it is the zodiac with the zodiacal signs. The Aristotelian tradition of wind diagrams presented the eight winds aligned with the four cardinal directions in a clockwise sequence, which in turn were aligned with the (local) solstices and specific localities, such as India or Ethiopia.[112]

An example of a Late Antique wind-rose can be seen in a circular Greek map of Egypt enclosed in a wind-rose (figs. 2.12 and 2.13).[113] This curious map originating from Egypt is preserved today in twelve copies.[114] It is in effect a composite map representing the celestial sphere, the inhabited world of Egypt, and the underworld, all set within a wind-rose. Though it is centered in Egypt, it shows neither the Mediterranean nor Alexandria, and the most recent studies suggest its prototype was made in Egypt during the second or first part of the third century AD. It is an example of some of the astrological and geographical material circulating

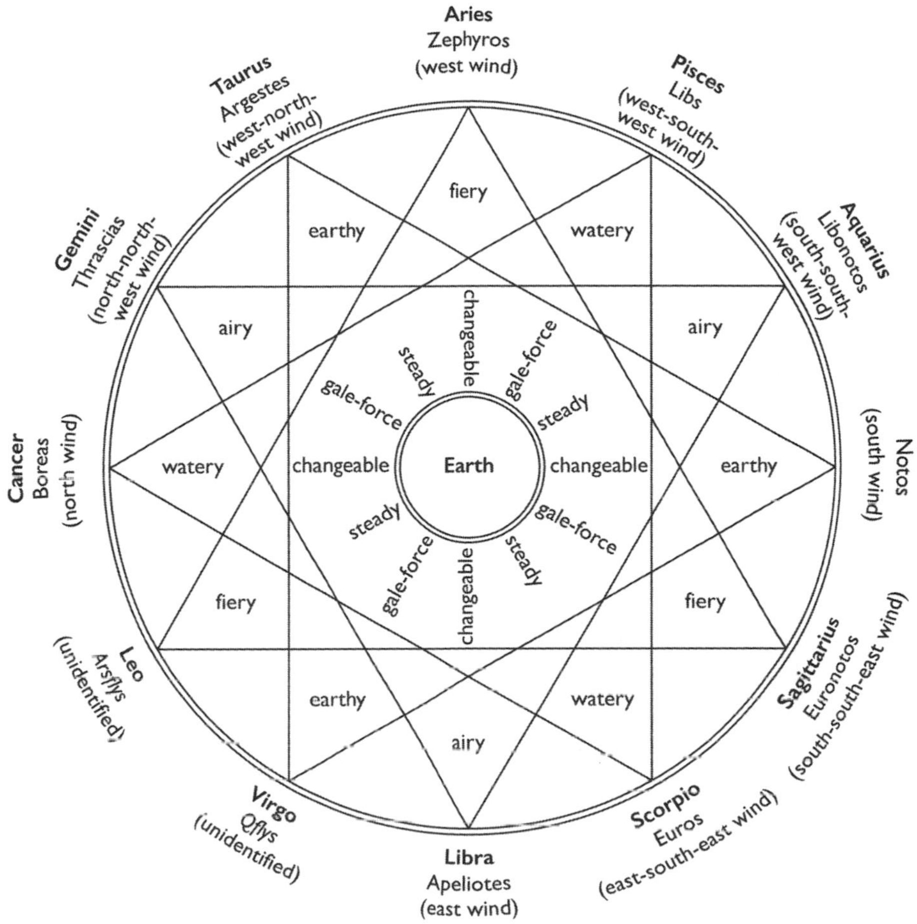

FIG. 2.11. The diagram of the winds from the *Book of Curiosities*, interpreted in a schematic diagram.

in Egypt prior to the time of our author. In the Greek map there are ten wind names arranged clockwise, with north at the top, around a highly schematic map of Egypt, thus very roughly aligning winds and compass directions.

The diagram in the *Book of Curiosities*, in contrast, has twelve wind names arranged counterclockwise with the west wind at the top, and they are aligned with the zodiacal signs and the four elements, with only a passing reference to the Earth, which occupies the center. Even though it transliterates Greek wind names, it does not directly reproduce a Greek diagram of a wind-rose. Rather, it is an amalgam of Greek wind diagrams with Is-

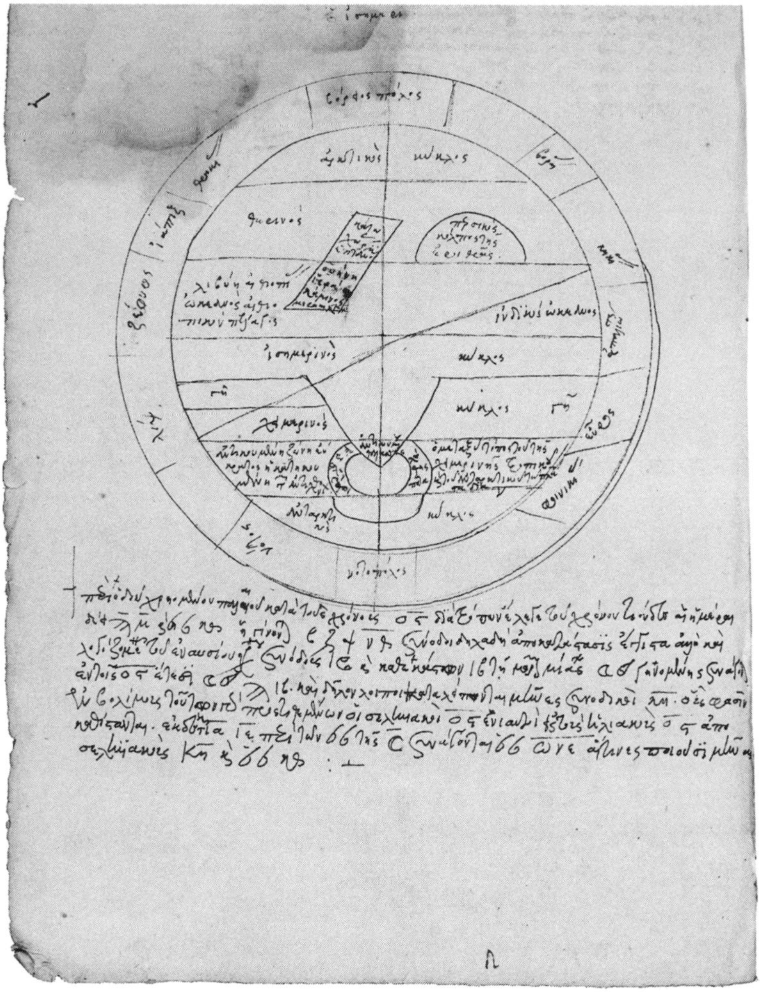

FIG. 2.12. Map from an anonymous astronomical miscellany. Oxford, Bodleian Library, MS Barocci 94, fol. 118v, undated, copied by Andreas Donas, presumably in Messina, Sicily, end of fifteenth or early sixteenth century.

lamic (or perhaps Late Antique Coptic) astrological and astronomical traditions in order to illustrate the central theme of the *Book of Curiosities*—the inextricable relationship between macrocosm and microcosm.

The circular wind diagram (fig. 2.10) is framed by two vertical side panels concerned with earthquakes. Here the author reproduces the classical Greek theories attributing earthquakes to subterranean winds or escaping gases:[115]

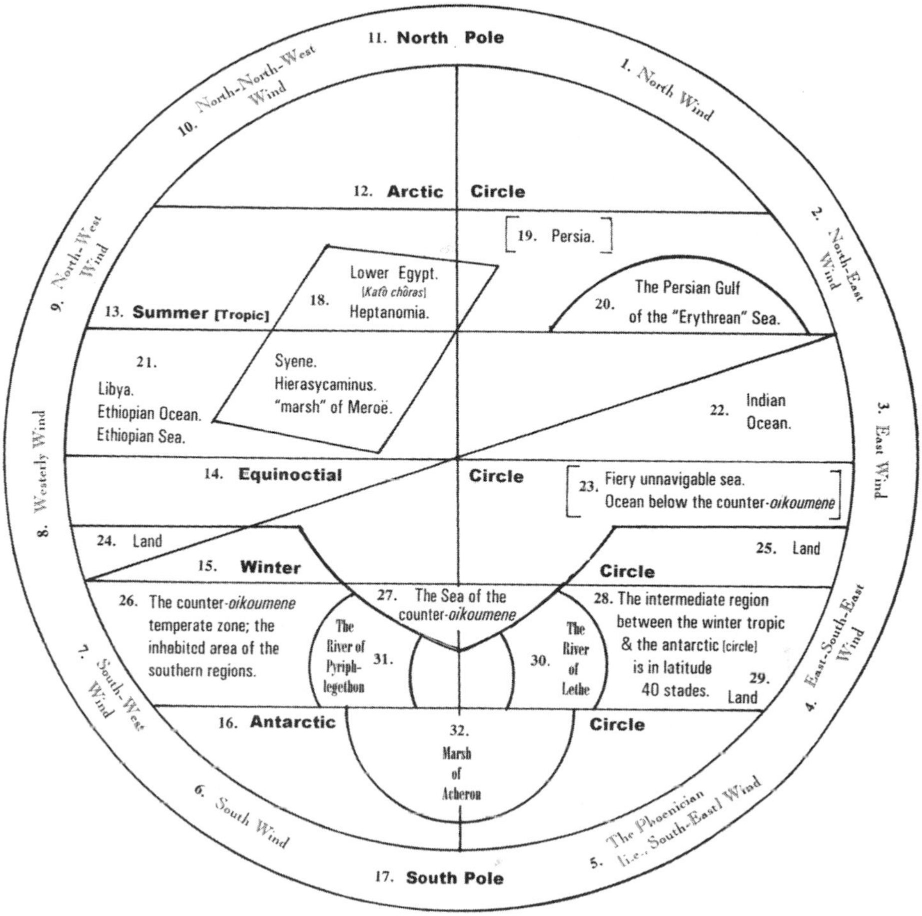

FIG. 2.13. A composite version of the Greek terrestrial/celestial map set in a wind-rose, based on the twelve known copies. From "An Astrologer's Map: A Relict of Late Antiquity," *Imago Mundi* 52 (2000), p. 12, fig. 5. Reproduced with permission from Imago Mundi Ltd.

The sages said: When the disruptive winds over time grow very strong inside the Earth, and they break out from their [trapped] position and tremble and move about, they shake the Earth above them. When these winds are abundant and forceful, and they leave their place so that all of them rise at the same time, by the will of their Creator, then they are called "the reaction" (*al-raj*ʿ*īyah*).[116] May God protect us from His wrath. Some earthquakes cause fires. Others fling out huge stones. Other cause springs to gush forth that were previously dry, while others desiccate springs that were flowing.

> As for the quick earthquake, it is called a tremor (*raʿdah*). Sometimes it is subterranean, but without being a true earthquake (*zalzalah*), such as occurs when the wind gets blocked in the bowels of the Earth. Sometimes these earthquakes occur under the sea, in which case they cause the sea to cast things from one place to another. Other times the waves roll up and on top of each other, creating a huge wave that dashes together into one spot, so the sea is transported onto the land. Yet other times it lifts things from the bowels of the Earth, causing springs to appear and rivers to flow. This occurs repeatedly in the depths of the sea until the water [sea level] swells and the waves dwindle.

This theory was developed in the fourth century BC by Aristotle in his *Meteorology* and commented upon by later authors, among them Ibn al-Biṭrīq (d. ca. 830) and Ibn Sīnā (d. 1037).[117] Note that our author only describes earthquakes and does not provide any prognostications based on them.

However, immediately following the circular wind diagram, our author gives prognostications based on the nature of the wind as it occurs on specific days (the 7th to 11th) of the Coptic month of Tūbeh, which is the fifth month in the Coptic calendar.[118] Here there is use of Coptic sources rather than Arabic ones—evident not only from the use of the Coptic calendar but also from some preserved fragments of a ninth-century Coptic treatise. Though not identical, the Coptic fragments and the text in the *Book of Curiosities* differ only slightly.

An example from the *Book of Curiosities* reads:[119]

> If the wind is easterly on the seventh day of *Ṭūbeh*, the Nile will flow in abundance. Domestic animals will survive, crops and fruits will be good, feverish shivering will increase, and honey will become rare.

The earlier Coptic fragment states:[120]

> If the east wind comes forth on that day (the 7th of *Ṭūbeh*), the water is good and it will cover the entire earth; the cattle will live, the crops will increase, the gardens will blossom, the honey will diminish, and the last of the crops of the field will perish.

The Coptic month of Ṭūbeh was especially important in divinatory material written in Egypt. Coptic Christians employed a calendar (also called the Alexandrian calendar) that begins its reckoning from August 29, 284, of the Julian calendar, the year Diocletian became Roman Emperor. The calendar is thus permanently synchronized with the Julian calendar, with the month of Ṭūbeh forming the first month in which the new year of the Julian calendar begins, for January 1st in the Julian calendar corresponds to the 6th of Ṭūbeh in the Coptic calendar.

Continuing in the Coptic tradition of divination,[121] the final portion of book 1 in the *Book of Curiosities* is devoted to predictions based on the nature of the winds on the 6th day of Ṭūbeh—that is, on New Year's Day. It provided forecasts for when New Year's Day occurs on each day of the week. This subchapter is titled "What the Sage Dīqūs Said Regarding the Days of the Week that Fall on the Sixth of Ṭūbeh and Their Interpretation."[122] The identity of the authority named here is uncertain. Most probably, it is the same as one Andronikos (Andurīqūs) recorded elsewhere as an author of an Arabic treatise on meteorological prognostications arranged by the days of the week.[123] It is also possible that the reference is to a Late Antique, possibly Coptic, personage as yet unidentified.

In his discourse on winds that occur on New Year's Day (the 6th of Ṭūbeh) our author takes the opportunity to display his knowledge of languages, in this case the weekday names in Persian, Byzantine Greek, Coptic, "Indian," and Hebrew. An example:[124]

> When the sixth of *Ṭūbeh* falls on a Saturday—called *šanivar* in the Indian language,[125] *shambe* in Persian, *sabbaton* in Greek,[126] *pišašf* in Coptic, *she-viʿī* in Hebrew—the winds that year will be stormy, while the summer will be nice with a pleasant wind, although injurious to sheep. The fruits of palm trees will be plentiful, honey and flax will be abundant, the price of food will go up, wars between kings will be frequent, and the Nile will be low. It will be a difficult year for the people of the land, while the sea merchants will make profits, and many young boys will die. But God knows best.

The word used by our author for the Coptic name given Saturday is written in the manuscript as *b-a-sh-y-a*,[127] which is a reasonable attempt to write *pišašf*, the Coptic name for Saturday in the Bohairic dialect. This dialect of Coptic is clearly reflected in the other weekday names given in the treatise, and our author's writing the definite article that in Bohairic would be *pi-* as *b-* suggests that he may have been a speaker of Bohairic, since modern Copts (who are of course native Arabic speakers) pronounce the *pi-* as *bi-* and medieval Arabic speakers may have done the same.[128] Bohairic is the dialect of Coptic spoken in the Nile Delta, which would fit well with our author's extended discourse on the Delta city of Tinnīs in the second book of the *Book of Curiosities*.

Forecasting events based on meteorological occurrences could take other forms as well, though these are not mentioned directly in the *Book of Curiosities*. There was a popular Arabic genre of writings referred to as *malḥamah* ("forecast" or "inspiration") and attributed to the prophet Daniel.[129] In them, predictions about crops and such are based on the occurrence of lightning, thunder, winds, earthquakes, rainbows, the appearance of halos around the Sun or Moon, and similar phenomena. In some cases a *malḥamah* is restricted to occurrences on the first day of the new year, just like the wind-based prognostications given earlier, but others in the *malhamah* genre were broader in their calendrical prognostications. An Arabic text of this sort, probably dating from the beginning of the eleventh century, was still in circulation recently in Iraq, suggesting that this approach to meteorological forecasting may be a part of current folklore.[130]

Astrological Geography

While star lore and astrology fill the first book of the *Book of Curiosities*, they also play a role—though a minor one—in the second book, on the Earth. These excursions into geographical astrology provide tangible applications of the astrological theories expounded on in book 1. In the second book our author incorporates astrology into some of his geographical descriptions by assigning an ascendant (a zodiacal sign visible at the eastern horizon) to a city. For example, on the large map of the city of Tinnīs in the Nile Delta he writes:[131]

> This city was founded when Pisces was in the ascendant. The ruler of Pisces is Jupiter, the sign of ultimate felicity, while Venus was in exaltation.[132] For this reason the people of the city are full of joy and happiness. They listen to music, are always delightful, seek comfort and shun anything that causes toil and hardship.

Similarly, the building of the city of Mahdia in modern Tunisia is said to have begun on Dhū al-Qaʿdah 5, 303 (= May 11, 916), when Leo was in the ascendant.[133]

Of the island of Sicily our author says, "Its ascendant is Leo, and the Lord of the Hour is the Moon," then adding:[134]

> The astrologers claim that [when] the sign of Leo rises obliquely,[135] it exercises, despite its eminence and brightness,[136] malign influence so that in every land in which it is influential, it is difficult for the ruler to govern. And Leo rules over Samarqand, Ardabīl, Mecca, and Damascus. These cities do not suit their rulers and their rulers do not suit them.

While in this second book Leo is said to rule over not only Sicily but also Samarqand, Ardabīl, Mecca, and Damascus, in book 1 the author stated that the sign of Leo ruled (in addition to Sicily) the localities of Antioch, the Yemen, and Chalcedon, as well as the lands of the Turks, including Soghd and Nishapur. Presumably our author was reading different astrological treatises when writing these two separate entries.

In the second book our author even goes so far as to designate the ascendant for a river, in this case the River Nile:[137]

> Its ascendant is Cancer and [the Lord of] its Hour is Mars. Knowledge of its inundation comes about from observing Mars at the start of the year: if Mars is at its maximum velocity,[138] the inundation will be plentiful; if it is at its mean velocity, the inundation will reach a normal level; and if it is in its slow motion, its flow will be deficient. Take note of that.

In the same way, the Euphrates has Virgo in the ascendant, the Tigris has Leo, and Sagittarius is the ascendant of the Oxus River.[139] Curiously, such details are not given for the Indus.

While our author is an advocate of this type of astrological application, others criticized such practice. His contemporary in the eastern provinces of Persia and what is now Afghanistan, the polymath al-Bīrūnī, was one such critic.[140] In his treatise on astrology, written in 1029, al-Bīrūnī says this about such astrological associations:[141]

> As for the ascendant and the Lord of the Hour,[142] that cannot be ascertained without knowing the time of construction, and what city has such information preserved? Even if there had been a ceremonial decree for every establishment of a city, the passage of time would have obliterated knowledge of it. Even assuming that was not the case and that the time of foundation for a city might be firmly established, on what basis could one confirm for the great rivers of the world the time of their cutting a channel or the moment at which the water flowed? The wrongness of such endeavors is very obvious.

Sources and Influences

Evidence provided in the *Book of Curiosities* suggests that in Egypt during the early eleventh century a simplified form of reading the stars and skies existed alongside the more technical and mathematically demanding forms of the art employing horoscopes and astronomical calculations. Here we have seen prognostication based on the visibility of important stars, the occurrence of a planet on the eastern horizon (see the appendix for more on this topic), the rising of one of the lunar mansions, the appearance of comets, and the occurrence of earthquakes or winds. This nontechnical form of astronomy is sometimes termed "folk astronomy." This may have been the level of astronomical and astrological lore that circulated among the generally educated and ambitious classes of Egyptian society. The *Book of Curosities* was certainly not aimed at a readership comprised either of professional astronomers or astrologers.

The *Book of Curiosities* was intended to demonstrate, in an easily accessible and graphic manner, the ever-present influence of the macrocosm upon the microcosm. To achieve this end, our author drew only upon nonmathematical and nontechnical literature—a selection that probably was dictated by the purpose of the book, although it may also have reflected the limited material available to the author or the particiular interests and abilities of the author himself.

It is possible that in the first half of the eleventh century the majority of astronomical treatises circulating among the Egyptian populace who could read and write concerned folk astronomy rather than mathematical astronomy. Indeed, the term *muwaqqit,* designating what we might call a "professional" astronomer and timekeeper, well versed in mathematics, is not attested before the late thirteenth century.[143] Few treatises on mathematical astronomy or even computationally sophisticated treatises on astrology can be attributed to Egypt before the late eleventh century. So our anonymous author might have had only nonmathematical and unsophisticated material at his disposal.

Admittedly, there would also have been circulating at this time among some segments of Egyptian society a number of technical astronomical manuals and tables. Not only was the famous *Almagest* of the Alexandrian Greek astronomer Ptolemy available in Arabic translation, but Ptolemy's defense of astrology known as the *Tetrabiblos,* written not long after the *Almagest,* was also available.[144] As for Arabic technical astronomical tables, Ibn Yūnus, working in Fusṭāṭ (Old Cairo) roughly fifty years before our author, prepared important astronomical tables for the Fatimid ruler al-Ḥākim, but Ibn Yūnus was exceptional, for essentially all other Egyptian sources relating to observational mathematical astronomy were written after the eleventh century.[145]

An exact contemporary of our author also concerned himself with mathematical astronomy, although not with observational astronomy and the compilation of tables. This was Ibn al-Haytham, who died in or shortly after 1040.[146] According to later biographers, Ibn al-Haytham made a living in Egypt on what he earned from copying Euclid's *Elements,* Ptolemy's

Almagest and the *Intermediate Books* (a collection of mathematical and astronomical Greek writings available in Arabic), but he also composed a large number of treatises on mathematical astronomy, including criticisms of Ptolemy.[147] Known today primarily for his work on optics, especially theories of vision, he is also credited with *On the Configuraton of the World* (*Maqālah fī hayʾat al-ʿālām*), a nontechnical summary of basic astronomical principles in Ptolemy's *Almagest*, written for people whose main interests were in philosophy rather than for astronomers.[148] This treatise as preserved today, though nonmathematical, is considerably more technical than the first book of the *Book of Curiosities* and does not seem to have served as a source for our author.

There were a number of astrologers known to have been active in Fatimid Egypt, and their writings and prognostications may have been known to our author, though he himself does not appear to have been a practicing astrologer. One such astrologer was an exact contemporary—the prominent Egyptian astrologer, physician, and argumentative autodidact Ibn Riḍwān. The caliph al-Ḥākim appointed him chief physician about the year 1019, when he was thirty-two years of age.[149] Ibn Riḍwān wrote two books on the topic: *On the Difference between Charlatans and True Astrologers* (lost today) and a commentary on Ptolemy's *Tetrabiblos*.[150] These astrological treatises by Ibn Riḍwān may have been available to our author, but Ibn Riḍwān's commentary on the *Tetrabiblos* contains complicated techniques not to be found in the *Book of Curiosities*.

Also at the court of al-Ḥākim was a Jewish astrologer, known only as al-Isrāʾīlī. He composed a treatise on astrology titled *Chapters on the Science of the Stars* (*Fuṣūl fī ʿilm al-nujūm*), consisting of 133 short aphorisms that later circulated in Europe in a Latin translation by Plato of Tivoli.[151] In addition, the versatile scholar al-Musabbiḥī (d. 1029) wrote a history of Egypt (of which only fragments survive) and at least two astrological manuals, neither preserved today.[152]

A third astrologer exactly contemporary with our author was ʿAlī ibn Sulaymān, a physician-astrologer who lived during the reigns of three Fatimid rulers.[153] He also showed interest in the Hermetic literature on comets, for among other writings, he composed an essay titled "Enumeration of the Doubts Concerning the Star of the Tail (*Taʿdīd shukūk fī kawākib al-dhanab*)." This is likely to be a reference to the comet named The Star of the Tail (*kawkab al-dhanab*) in the *Book of Curiosities*, where it is the twenty-fourth comet attributed to the legendary Hermes.[154] It is the second comet illustrated in fig. 2.9 and was said to have three tails and to return every 107 years.

Yet another contemporary of our author is said by some sources[155] to have been an astrologer, but this is not mentioned in the early biographical sources.[156] Abū al-Wafāʾ al-Mubashshir ibn Fātik was a dedicated scholar of considerable wealth, and, judging by his titles *amir* and *maḥmūd al-dawlah,* he occupied a high position in court circles. He was personally acquainted with both Ibn al-Haytham and Ibn Riḍwān, and it has been suggested that his role vis-à-vis these scholars was as a wealthy patron and friend of scholarship.[157] In 1048 he composed his only surviving treatise, *Choice Wise Sayings and Fine Statements* (*Mukhtār al-ḥikam wa-maḥāsin al-kalim*) devoted to biographical sketches of ancient sages and collections of sayings.[158] This treatise cannot be described as astrological, but it does contain 125 sayings, more or less, attributed to the legendary Egyptian-Greek sage Hermes, and it was largely responsible for establishing Hermes Trimegistus as a source of wisdom, not only in Fatimid Egypt but in later Ismaʿili literature as well.

A striking feature of the *Book of Curiosities* is its incorporation of so much material attributed to the legendary Egyptian-Greek sage Hermes, some unique to this treatise. At the very beginning of the first chapter our author evokes Hermes, in the guise of the prophet Idrīs, commonly identified with the legendary Late Antique figure of Hermes the Wise or Hermes Trimegistus:[159]

> It is said—but only God knows His mysteries—that God revealed to Idrīs [= Hermes], may the Peace of God be upon him, the secret knowledge of the celestial bodies and the course of the shining stars [planets] in the raised-up roof [the sky] above the laid-down bed [the Earth], together with their competing movements in the orbits of their spheres, according to the plan of God, the Wise and the Omniscient. He has done that so that anyone, whether a scholar or a rascal, may observe and ponder the power of *He who made constellations in the skies, and placed therein a Lamp and a Moon giving light* [Qur. 25:61], and *contemplate the* [*wonders of*] *creation in the heavens and the earth,* [*with the thought*]: *Our Lord! Not for nought hast Thou created (all) this! Glory to Thee! Give us salvation from the Penalty of the Fire* [Qur. 3:191].[160]

A similar deep interest in the legendary sage Hermes, as well as astrology and analogies between the macrocosm and the microcosm, can be seen in the "epistles" of the Brethren of Purity (Ikhwān al-Ṣafāʾ). Some of these epistles were composed in Iraq as early as the mid-ninth century, according to recent scholarship.[161] They were influenced by Ismaʿili doctrines, though the Brethren of Purity themselves—anonymous writers in Basra and Baghdad—may not have been Ismaʿilis. Astrological determinism figures prominently in the three epistles concerned with astronomy,[162] while astral magic and talismans are the focus of the last of the fifty-two epistles.[163] The third epistle bears a striking resemblance to the material that opens the *Book of Curiosities,* while topics such as the legendary figure of Hermes the Wise, stories of India, star lore, lunar mansions, and the properties assigned the zodiacal signs and the planets are found throughout the epistles.

These epistles must have circulated widely and quickly across the Islamic world, for at least some of them were employed by a scholar working in Spain, who compiled one of the most influential treatises in the history of astrological magic—*The Goal of the Sage* (*Ghāyat al-ḥakīm*), known to Europeans as the *Picatrix.*[164] The treatise was often incorrectly attributed to the astronomer and mathematician Maslamah al-Majrīṭī, who died in 1007. Modern scholarship has demonstrated convincingly that the author of *The Goal of the Sage* was an Andalusian traditionalist and occultist by the name of Abū al-Qāsim Maslamah ibn Qāsim al-Qurṭubī, who died in 964.[165] Just as in the epistles of the Brethren of Purity, the influences of the planets, zodiacal signs, and lunar mansions figure prominently in the astral and talismanic magic advocated by this Andalusian author of *The Goal of the Sage.* Once again the influence of the Hermetic tradition is to be seen.

It is evident that in the tenth century there was a swirling mix of astrological, cosmological, and magical ideas circulating in both Iraq and Spain. This interest in Hermetic and esoteric material continued well into the next century, when our author composed the *Book of Curiosities* in Egypt. There is no concrete evidence that either the *Goal of the Sage* or the epistles of the Brethren of Purity were available to a writer working in Fatimid Egypt between 1020 and 1050.[166] Nevertheless, many of the cosmological and astrological concepts found in the *Book of Curiosities,* as well as the extensive use of materials attributed to Hermes, are common to both the epistles of the Brethren of Purity and to *The Goal of the Sage.* While no direct citations have been identified, there are striking parallels suggesting that common materials were shared by all these writers on esoteric and astrological matters.

The inclusion in the *Book of Curiosities* of so much material attributed to Hermes suggests a strong Ismaʿili association on the part of the author, possibly placing him within the context of the secretive Ismaʿili missionary *daʿwah* network. Ismaʿili missionaries were expected to be intellectuals with sufficient understanding of the exoteric and esoteric aspect of the faith, and, in particular, knowledge of the lands and peoples in their area of operation. More, however, will be said in subsequent chapters regarding the missionary *daʿwah* network and the possible, if not probable, association with the *Book of Curiosities.*

The most outstanding feature of the chapters on the Heavens, however, is the logic and thoroughness of the presentation. It is a methodical presentation of the different strata of the macrocosm that influence the microcosm. After suggesting to the reader through the opening diagram what topics would be covered, our author moved methodically from the outermost sphere down to the Earth, extracting and summarizing from an impressive range of nontechnical sources. So, for the stars, we have the classical Greek system described in terms of the zodiacal signs and their relationships to the planets (see appendix for details), but this is followed by a chapter on quite different stars drawn from a Late Antique tradition associated with the legendary Hermes. This, in turn, is followed by yet different stars recognized by pre-Islamic Bedouins, and that is then contrasted with a chapter devoted to a system derived from one originating in Asia, the "lunar mansions." In this way our author presented four different systems for stellar mapping of the skies. Similarly, with comets there are four different approaches to the topic, no doubt representing four different source texts: a generic one applicable to any comet, comets that appear in certain zodiacal signs, comets described by Ptolemy, and comets attributed to the legendary Hermes. For winds, the author does not simply present the standard Arabic wind names and meteorological theories, but provides Greek ideas and also employs two Coptic texts giving prognostications based on the occurrence of winds on certain days. He does not appear to favor one approach over another. Rather, the assemblage of all possible approaches to a topic appears to be his primary aim.

The comprehensiveness with which our author approached his subject matter has had the unintended consequence that the *Book of Curiosities* preserves for us today, roughly a millennium after its composition, several texts that were otherwise known only from fragments or lost altogether. The incorporation of so many diagrams and small maps suggests the author might have been a mapmaker. In any case, he was intent on using as many diagrams and maps as possible when asked by his unnamed patron to "write a volume encompassing the principles of the raised-up roof (the skies) and the laid-down bed (the Earth)."[167] In doing so, he provided us with a unique glimpse into the astrological and astronomical lore circulating among the educated classes of Fatimid Egypt, with its distinctive mixture of Hermetic, Coptic, Greek, and Bedouin elements.

CHAPTER THREE

The Rectangular World Map

The rectangular world map of the *Book of Curiosities* (plate 1 and fig. 3.1), the first map in the section on the Earth and the one chosen for the cover of this present volume, is the most enigmatic of all the images in the *Book of Curiosities.* This strange and curious world map contains some designs and labels familiar to us from other medieval maps, but these are brought together to construct a world that appears distorted and alien, even in comparison to other Islamic world maps of the period. The resulting image is, in fact, unlike any other world map known to us from any place or any time.

The first oddity of this map is its rectangular shape. The majority of world maps from the medieval period that have been preserved are circular in outline—such as the famous Hereford World Map[1] or the circular map associated with the geographical work of al-Idrīsī, composed in 1154 for Roger II, the Norman king of Sicily (see plate 4). Of course medieval scholars, as those before them in classical Greece, knew the Earth to be round. The fact that the Sun did not set at the same time at every location, combined with the fact that the silhouette of the Earth visible during a lunar eclipse was curved and that the tops of towers could be seen to rise from the horizon as a ship approached a Mediterranean harbor, were more than enough to convince a person that the Earth was round. The notion of a medieval flat world is a modern myth, attributed by some to the early nineteenth-century American novelist Washington Irving.[2] A world map drawn as a flat circular disk was merely a convention. Before the discovery of the Americas, medieval mapmakers limited themselves to the representation of one hemisphere only. The other hemisphere, so it was thought, was completely covered by sea. The Abbasid scholars in ninth-century Baghdad calculated the circumference of the world, refining the estimates given in the *Geography* of Claudius Ptolemy, the greatest theoretical geographer of antiquity, who lived in Alexandria in the second century AD. They concluded that the inhabited parts of the world, stretching from the Atlantic in

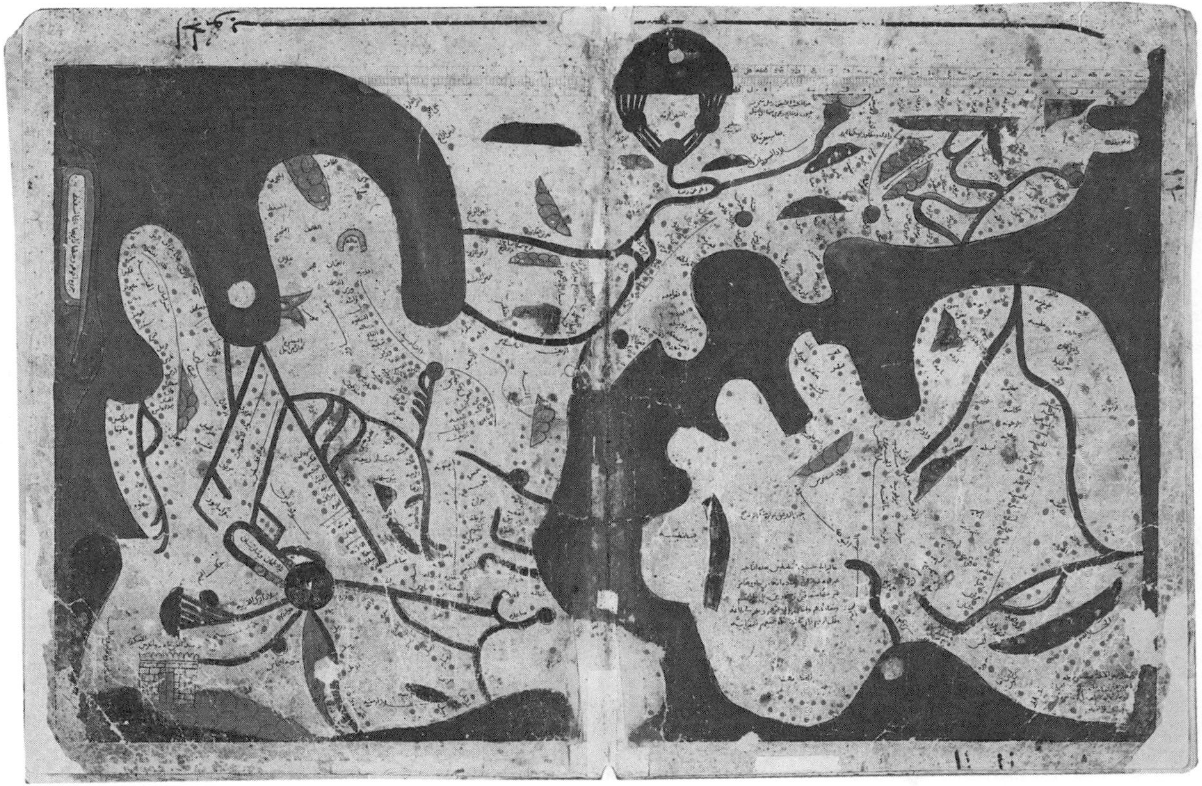

FIG. 3.1. The rectangular world map from the *Book of Curiosities*. Oxford, Bodleian Library, MS Arab. c. 90, fols. 23b–24a.

the west to China in the east, covered 180 degrees, exactly half the Earth's surface.

Not only is this map rectangular, not circular, but the layout of the landmasses is not immediately recognizable to a modern viewer. Contrary to what you might expect, south is at the top of the map, not north, but this orientation is actually typical of the majority of maps produced in the Islamic world. Yet even allowing for the shift in orientation, the relative location and size of the continents are unfamiliar. A distinctive parachute-like "Mountain of the Moon," considered the source of the River Nile, is at the top of the map and is the southernmost point shown. The rest of Africa, including those areas considered uninhabited by medieval scholars, is off the map. Europe, at the bottom right, is an enormously distorted island, mostly taken up by an oversized Iberian Peninsula. Asia, on the other hand, is tightly squeezed beyond recognition at the left side of the map, while the Arabian Peninsula is disproportionally prominent, with Mecca indicated by a distinctive yellow horseshoe symbol, one of only a few localities not indicated by a simple red dot.

The map in fact is overflowing with a dense network of red dots, each indicating a city rather than a region. Many of these dots are organized into itineraries, such as the long north-south string of red dots that cuts through the bulbous Iberian Peninsula. There are nearly four hundred place-names on this map—a nightmare to us, the modern editors. But the map is rich in other details as well. Constantinople is marked at the left extremity of the European continent behind a brown masonry wall. The design in the northeast (lower left corner of the map) represents the Wall erected by Alexander to enclose the people of Gog and Magog. There are a couple of dozen mountain ranges, complicated and detailed river systems, and coastlines that twist and turn as if this is crude attempt to represent reality, not simply a diagrammatic sketch of how the world looks.

For historians of science, however, the most exciting feature of this world map—the one single aspect that makes this manuscript as a whole so valuable—is a calibrated scale visible near the top of the map, running across the African continent and disappearing behind the "Mountain of the Moon." This is the earliest surviving example—in Islamic cartography, in European cartography, and possibly in any known mapmaking tradition—of a world map displaying such a scale bar, technically known as a graticule. It tells us something about the way the map was made: it suggests that place-names were plotted according to coordinates of longitude and latitude, or at least that this map was somehow derived from an earlier plotted map prototype.

This chapter will attempt to decipher this rectangular world map and explain its many oddities, as well as reflect on the worldview it projects. We will first argue here that the dominant feature of this map is its hybridity, for this is a map that attempts to bring together two separate cartographic traditions that were prevalent in the early centuries of Islam. In its basic constitution, and through several designs on the edges of this world map—not least the scale bar—we witness a tradition of mathematical geography that had its roots in antiquity. This tradition reached us through the work of the ninth-century mathematician al-Khwārazmī, whose geographical treatise is an adaptation of a work attributed to the Greek geographer Ptolemy. In the interior of the map, however, the map is dominated by a later, more consciously Islamic, abstract and schematic cartography focused on itineraries, which made no attempt to represent real distances or to employ any tool of mathematical geography.[3] Such abstract maps accompanied the works of three tenth-century geographers—al-Iṣṭakhrī, Ibn Ḥawqal, and al-Muqaddasī—collectively known as the "Balkhī School." The combination of these two traditions in the world map of the *Book of Curiosities* reflects the hybridity of the intellectual culture in which it was formed, indebted to antiquity and to Islam in equal measures.

Our primary focus in this chapter will be on the former layer, the layer of the mathematical geography attributed to Ptolemy. This emphasis on the mathematical tradition is not a matter of value judgment on our part. Like the author of the *Book of Curiosities* himself, we do not privilege accurate representation over legibility and accessibility, and we do not privilege Classical authors over those working under Islam. Rather, we focus on the traces of mathematical geography on this world map because they reveal a tradition that has been otherwise lost to us. While Ptolemy provided detailed instructions on how to construct a world map, no map of his survived. The influential Ptolemaic world maps of the later Middle Ages are all reconstructions. Any maps from Antiquity and Late Antiquity are rare, and none of those that do survive are derived from the Ptolemaic tradition of mathematical geography.[4]

Our second argument in this chapter is that the rectangular world map of the *Book of Curiosities* contains vestiges of a Late Antique prototype map, attributed to Ptolemy. The Abbasid era oversaw a mass movement of translation of Greek scientific and philosophical texts, usually through the medium of Syriac. The world map here demonstrates that the process of transmission of scientific knowledge included not only texts but also maps—or at least a couple of maps of the world and of the Nile—that were produced in Late Antiquity and accompanied Ptolemaic texts. Such Late Antique maps were available to several scholars working in the early Islamic centuries, including al-Khwārazmī. But it is only in the *Book of Curi-*

osities that we can detect vestiges of this Late Antique cartographic tradition displayed on a world map. It makes this rectangular world map a unique, invaluable image that visualizes the transmission and adaptation of knowledge from Antiquity to Islam.

Mathematical Geography

The rectangular world map in the *Book of Curiosities* has a clear, visible connection to the earliest set of Islamic maps known to us, which were drawn by Muḥammad ibn Musā al-Khwārazmī, who lived in Baghdad in the first half of the ninth century. Al-Khwārazmī is widely remembered today for his book *al-Jabr wa'l-muqābalah,* "reducing the terms of an equation by addition and subtraction." In this book al-Khwārazmī explained, for the first time in the history of science, how to solve quadratic equations by using geometrical constructions. The title of his book gave its name to the science we know today as algebra. The title of the translation of the book into Latin, *Liber Alghoarismi,* "The Book of al-Khwārazmī," is the origin of the term algorithm.

Not only was al-Khwārazmī seriously interested in mathematics, but he also was concerned with mathematical geography. His *Book of the Depiction of the Earth* (*Kitāb ṣurat al-arḍ*) consists for the most part of tables with lists of hundreds of place-names and their locations, expressed in degrees of longitude and latitude.[5] The places are arranged by categories, such as towns, mountains, seas, rivers, and islands, and then by climes, reflecting a division of the inhabited parts of the northern hemisphere into seven latitudinal bands, defined by the length of the longest day of the year. The first and southernmost clime was close to the equator, and its center was in Nubia, while the center of the seventh and northernmost clime was the Dnieper River.

Al-Khwārazmī tells us several times that he extracts the coordinates in his tables "from a map" (*fī al-ṣūrah*), suggesting that he had a map in front of him with calibrated scales by which longitude and latitude could be approximated; or that the map he was employing had lists of coordinates in the margins or in other available spaces.[6] The front page of the treatise declares that al-Khwārazmī "extracted" (*istakhraja*) the information from Ptolemy's *Geography.* Since most of the place-names reflect Syriac spellings of Greek names, that *Geography* was likely to be a treatise, including a map or maps, which came to him through a Syriac adaptation of Ptolemy.

Al-Khwārazmī's *Book of the Depiction of the Earth* survives in one manuscript, copied in 1037, two centuries after his death. This copy is illustrated by four maps showing the River Nile, the Island of the Jewel, the

Sea of Darkness (*al-baḥr al-muẓlim*, i.e, the Encompassing Ocean), and the Sea of Azov.[7] These four maps are primarily illustrative sketches of how maps should be drawn. Although one map, that of the Nile, has lines for the latitudinal climes, even that map is not plotted by coordinates. While the knowledge of longitude and latitude was transmitted in the form of tables and texts, al-Khwārazmī did not translate this knowledge onto a map, at least not one that has survived.

Surprisingly and quite uniquely, the rectangular world map of the *Book of Curiosities* reproduces two of al-Khwārazmī's early maps through distinctive designs in the extreme south and the extreme east of the inhabited world. One is the elongated brown landmass at the leftmost edge, shown here in detail (see fig. 3.2). It carries an inscription, encircled in red, which reads: "The Island of the Jewel. Its mountain surrounds it like a basket." A narrow portion of this island extends upward and touches what would be the top of the scale, possibly past the equator. A northern (lower) extension seems to run into an unidentified brown area filling the now damaged lower left corner of the map. This depiction of the Island of the Jewel on our world map corresponds precisely with the map of the "Island of the Jewel" found in al-Khwārazmī's treatise (fig. 3.3). Al-Khwārazmī's map shows this island close to the equator and surrounded by the Encompassing Ocean (here called the Sea of Darkness) and a nearly encircling mountain range.[8] Al-Khwārazmī also indicates the Island of the Jewel in the text of his treatise, and the label on the world map in the *Book of Curiosities* reproduces verbatim the corresponding text in al-Khwārazmī's treatise.[9]

Another distinctive visual element taken directly from al-Khwārazmī's iconography is the Mountain of the Moon along the equator. The identification of a "Mountain of the Moon" as the source of the Nile goes back to Ptolemy and is echoed by many Islamic geographers. But the iconography here can be traced to al-Khwārazmī, who not only lists the Mountain of the Moon in his tables but also depicts it in his map of the Nile—one of his four surviving maps (see plate 12 and fig. 4.1 in the next chapter). The resemblance between the depiction of the Mountain of the Moon by al-Khwārazmī and its rendering in the rectangular world map leaves no doubt that the two maps are related.[10] More will be said on the depiction of the Nile in the *Book of Curiosities* in the next chapter. But, for our purpose here, the Mountain of the Moon and the Island of the Jewel on the edges of this rectangular map show unmistakable links to al-Khwārazmī's maps. They are part of the framework, or the shell, for the depiction of the world, which both the author of the *Book of Curiosities* and al-Khwārazmī took from a Late Antique source.

The links with al-Khwārazmī's work of and with a Late Antique mathe-

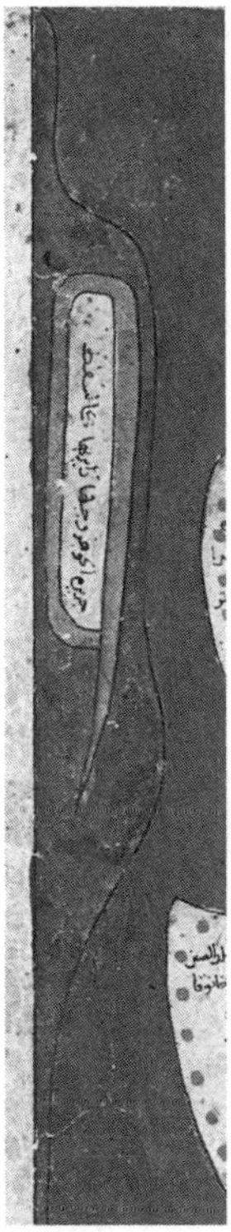

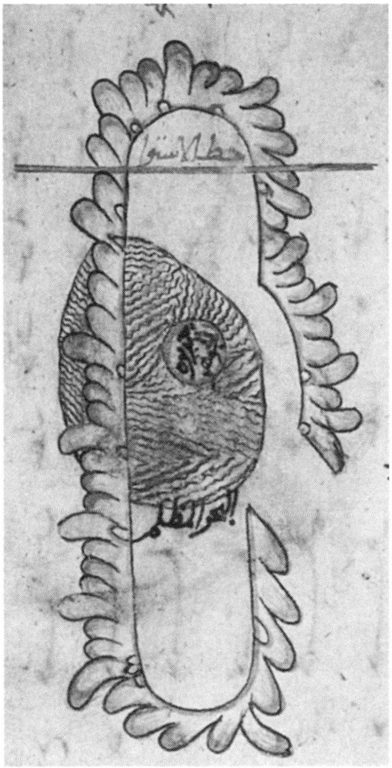

FIG. 3.2. Detail of the "Island of the Jewel" from the rectangular world map in the *Book of Curiosities*. Oxford, Bodleian Library, MS Arab. c. 90, fol. 24a, copied ca. 1200.

FIG. 3.3. Diagram of the "Island of the Jewel" (*Jazīrat al-Jawhar*) by al-Khwārazmī. Bibliothèque Nationale et Universitaire de Strasbourg, MS 4247, fol. 111b, copied 1037/429H. Reproduced courtesy of Coll. et photo BNU Strasbourg.

matical geography are reinforced by the map's very context within the treatise. The map, taking up the entire chapter 2 of the book on the Earth, is part of a section that draws heavily on Greek sources, including Ptolemy and Hippocrates. The map follows a chapter on the measurement of the Earth, where the author cites Ptolemy's *Geography* on the distance in miles of each degree of latitude, and then relates the attempts of ninth-century Muslim scholars, working under Abbasid caliph al-Maʾmūn (r. 813–33), to verify these measurements. The map is then followed by a chapter on climes, which reproduces al-Khwārazmī's method of listing categories of rivers, mountains, and springs for each clime. Portions of this chapter closely correspond with the lists produced by al-Khwārazmī, and, in some instances, the text is taken verbatim from al-Khwārazmī, or from a common source.[11]

Both al-Khwārazmī and the author of the *Book of Curiosities* assumed that source to be Ptolemy. Like al-Khwārazmī, who "extracted" data from Ptolemy's *Geography*, the author of the *Book of Curiosities* explicitly tells us that his world map is derived from the work of the Alexandrian sage. In an explanatory note, found a few folios earlier in the treatise, he explains that this world map opening the second book (on the Earth) depicts the inhabited parts of the world, from the equator to 66° N:

> This is the end of the Book One, with the blessings of God and His support. It is followed by Book Two, consisting of twenty-five chapters. The first chapter [of Book Two] is on the mensuration (*misāḥah*) of the Earth and its map (*ṣūrah*), including the seas (*ʿalā al-abḥār* ?), from the Equator to the farthest limit of the inhabited world, which is at 66 degrees, as is related by Ptolemy *al-qalūdhī* (Claudius) in his book known as *Geography*.[12]

What the author tells us here is that this world map is intended to show only the inhabited regions of the world, from the equator to the northernmost limit of the inhabited world, as related in the *Geography* of Claudius Ptolemy. That explains why Africa south of the equator, thought to be uninhabited, was left out. The map aims to show only the inhabited world, effectively only one quarter of the surface of the Earth. As we go back and look at our map, we see that the equator is likely to be right at the top, along the scale stretching across Africa. At the bottom, the Wall of Gog and Magog and northern Europe are shown, but the map does not go as far as the North Pole. This is more or less in line with the limits of the inhabited world known to al-Khwārazmī, for according to his tables the lands of Gog and Magog extended as far as 63° N.[13]

The author states that the limits of the inhabited world as defined in Ptolemy's *Geography* are from the equator to 66° N. In fact, Ptolemy gave different accounts of the northern and southern limits of the inhabited world. In his *Geography*, the *oikoumenē*—the part of the world inhabited by humans—begins south of the equator, at 16° S, and extends to the island of Thulé, at 63° N.[14] These are similar to the values given by al-Khwārazmī, who states that the inhabited world starts at 10° south of the equator, and extends to 63° N. But the values cited in the *Book of Curiosities* are closer to those Ptolemy gives in his earlier astronomical treatise, the *Almagest*. There, Ptolemy states that the inhabited lands extend from the equator to 66$^1/_6$° N. Very similar values are also repeated by the leading ninth-century astronomer al-Farghānī, who cites the limits of the inhabited world as being the equator in the south and 66° N in the north.[15] According to other statements in the *Almagest*, the *oikoumenē* extends from Taprobanē (Sri Lanka) at 4° 30′ N, to the lands of the remote Scythians at 64 ° 30′ N.[16] Therefore, it would appear that of Ptolemy's works, the *Almagest* rather than the *Geography* is a more likely source for the limits of the inhabited world as they appear in our world map. The *Almagest* had been translated into Arabic several times and would have been more accessible to Arabic scholars than the *Geography*.[17]

Most importantly, the latitudinal boundaries cited by the author provide an explanation for the rectangular shape of this world map. The *Book of Curiosities* map is unusual in that it explicitly seeks to represent only a section of the world defined by degrees of latitude extending from the equator to 66° N. Once the North Pole and any area south of the equator are excluded, a rectangular shape was simpler to execute and fitted better with the mathematical conception of the map. As anyone who has tried to plot coordinates on a map knows, the easiest and simplest way is to draw a rectangle, with scales of longitude and latitude on the margins.

This simple method was known to medieval Islamic scholars, and was described in the early tenth century by a follower of al-Khwārazmī, an otherwise unknown scholar by the name of Suhrāb. Suhrāb prepared a revision of al-Khwārazmī's tables, but prefaced the tables with a discussion of methods of projecting the coordinates onto a world map. Suhrāb advocated a simple orthogonal method of projection, placing a lateral scale of 180 degrees at the top and at the bottom of a map, and a vertical scale down each side, divided into 110 degrees, extending from 90° N to 20° S. Once the seven climes were marked on the map, the towns were to be plotted by their coordinates with the aid of a pair of weighted strings. Suhrāb did not leave us an actual world map plotted in this way, but a diagram explaining his method is preserved in the unique manuscript copy of his treatise (fig. 3.4). This crude mathematical projection discussed by Suhrāb would have resulted in a map maintaining distances along the equator and the meridians, but having a greater east-west stretch in the northern latitudes.[18] Al-Bīrūnī, a contemporary of the author of the *Book of Curiosities* working in Central Asia, was also familiar with this rectangular projection, whose description he found in Ptolemy's *Geography*. Al-Bīrūnī dismisses this projection as too distortive and offers his own innovative methods, which would have resulted in circular sky and world maps.[19]

The presence of the scale bar, the most telling indication of plotting by coordinates, at the top of the world map of the *Book of Curiosities* is obviously suggestive of a map that belonged to a tradition of mathematical geography. The scale, or graticule, has been carefully executed and drawn as part of the outline of the map. Infrared and ultraviolet lamps revealed

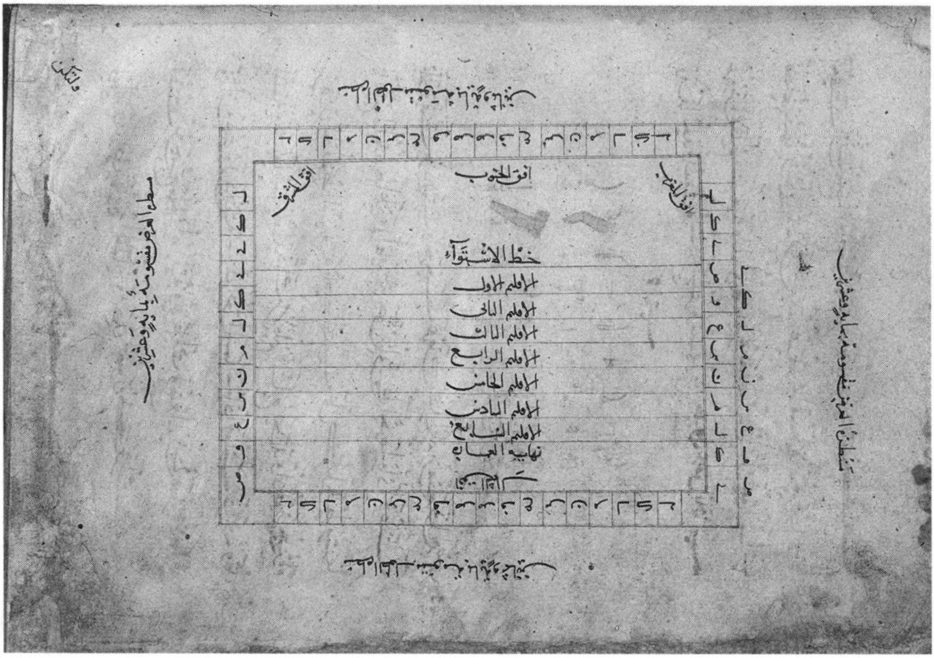

FIG. 3.4. Suhrāb's diagram for a world map. British Library, Add. MS 23379, fol. 4b. Reproduced with permission of the British Library.

that the full graticule lies under the green paint of the ocean and the brown "Mountain of the Moon."[20] The cells on the right-hand folio are numbered with *abjad* letter-numerals. These letters of the Arabic alphabet give numerical values, increasing by five-degree steps from 5° at the top right to 135°, the last visible number before the scale is overpainted with the "Mountain of the Moon" (see fig. 3.5).

But, it must be emphasized, this world map in the present copy of the *Book of Curiosities* is not plotted; it is likewise unlikely that the world map of the original treatise was constructed in the manner described by Suhrāb. For one, the scale is significantly corrupted. The most obvious error is that the scale should show 180° across the entire open bifolio, in keeping with the Ptolemaic assumption that the inhabited landmasses of the Earth, from the Atlantic to China, occupy a half of the Earth's surface. But, had the numbering continued as presented here, the last number written in the gutter of the manuscript would have been 180°, and it would have reached 360° at the end of the scale on the left-hand folio, as if it represented the entire Earth. This mistake in itself renders the scale as it appears here meaningless. There is an additional anomaly in that the scale on the left-hand page

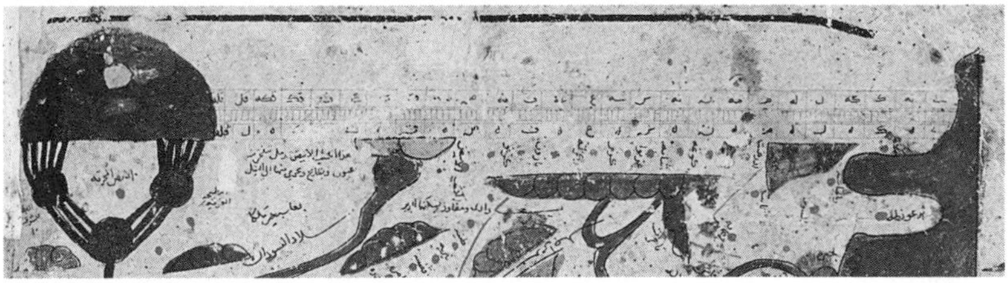

FIG. 3.5. Detail of the scale on the rectangular world map of the *Book of Curiosities*. Oxford, Bodleian Library, MS Arab. c. 90, fols. 23b–24a.

was laid out using a slightly larger proportional measure, so that it has only twenty-eight larger divisions while the right-hand side has thirty-five. The errors in the execution of this scale mean that the mapmaker, or perhaps the copyist who produced the Bodleian copy some 150 years later, did not fully understand how this longitude scale should be used.

The scale is not only corrupt, but also has no bearing on the position of any of the localities. Even after adjusting the numbering of the scale to the correct 180°, the position of prominent Islamic cities, such as Córdoba, al-Qayrawan, and Mecca has no correlation with the longitude values of the same cities in medieval coordinate tables. We have also attempted to align the same cities with latitudinal values, even though the map only has a scale for longitude, and no clime lines. We were similarly frustrated. Suhrāb may have explained how to plot a world map, but our mapmaker chose not to do so.

The author of the *Book of Curiosities* chose not to plot his maps, but it is important that he claims to have understood the basic principles of mathematical geography, principles he attributed to Ptolemy. In fact, in a key passage describing his mapmaking method, he specifically refers to the nomenclature of mapmaking associated with Ptolemy, only to consciously reject it. In a chapter that introduces our author's own approach to cartography, he explains that his maps of the seas are not accurate representations of geographical space because he opted for legibility over mathematical precision. But in doing so he also makes clear that the Ptolemaic tradition of mathematical geography was familiar to him:

> If the shape of the sea is reproduced accurately, on the basis of longitude and latitude coordinates, and any given sea is drawn in the manner described by Ptolemy in his book known as *Geography,* the [contour of the] sea would form curves in the coast (*ʿaṭfāt*) and pointed gulfs (*shābūrāt*), square (*murabbaʿāt*) and concave headlands (*taqwīrāt*). This shape of the coast exists in reality, but, even if drawn by the most sensitive instru-

> ment, the cartographer (*muhandis*) would not be able to position [literally, "to build"] a city in its correct location amidst the curves in the coast or pointed gulfs because of the limits of the space that would correspond to a vast area in the real world. That is why we have drawn this map in this way, so that everyone will be able to figure out [the name of] any city.[21]

Remarkably, two of the terms mentioned here, *shābūrah* and *quwārah*, are found on another of al-Khwārazmī's four maps—his illustrative map of the Encompassing Ocean.[22] This is a sketch map in which al-Khwārazmī is not depicting any particular part of the ocean thought to encircle all landmasses, but only illustrates the manner in which a cartographer is supposed to represent coastlines. In this sketch map, pointed triangular-like gulfs are called *shābūrah* and circular-shaped headlands are called *quwārah*. The same technical terms also occur repeatedly in the prose sections of al-Khwārazmī's treatise, referring to the images he saw on the map he was using for compiling his tables. For both al-Khwārazmī and the author of the *Book of Curiosities*, these pointed gulfs and concave headlands were the building blocks of mapmaking associated with the Ptolemaic tradition.

This means that, although the world map was not plotted, the author was aware of a Ptolemaic syntax of constructing maps. It is in this context that we should understand the rectangular world map's plethora of cities, indicated by hundreds of red dots. The indication of cities on a world map is a distinct feature that sets this map apart from nearly all other Islamic world maps, where normally only regions and provinces were indicated. While, as we shall see, most of the labels in the interior of our map are taken from the regional maps the Balkhī School, they are not found on the world maps produced by Balkhī School geographers. Let us take for example this thirteenth-century copy of the circular world map by the earliest member of the Balkhī School, al-Iṣṭakhrī, where south is also at the top (see fig. 3.6). Al-Iṣṭakhrī's world is divided into neat squares and circles, each representing a cultural or ethnic group. Europe, for example, includes boxes for Islamic Spain (al-Andalus), the lands of the Franks (Western Europeans), and the Byzantine Empire (*Rūmiyah*). No cities are indicated at all.

The use of labels for cities, rather than regions, on the world map in the *Book of Curiosities* reflects a different conception of what a world map should represent. The Balkhī world maps divided the world into cultural and ethnic units, and were, as the noted historian of cartography Gerald Tibbetts put it, "an arm-chair attempt to see all the provinces set down relative to each other."[23] The dense world map in the *Book of Curiosities* aims to do much more than that—even though it is not a plotted map, it tries

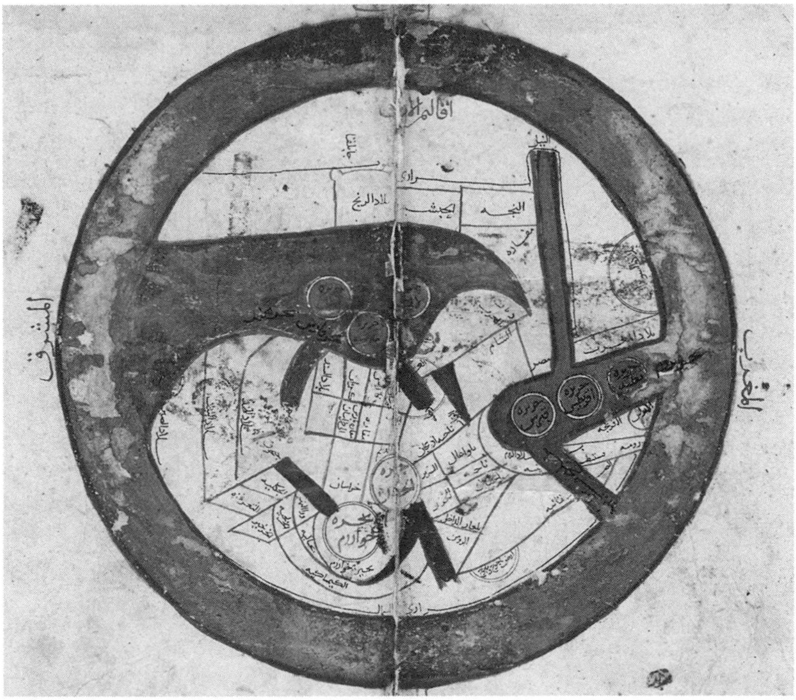

FIG. 3.6. A world map from a Persian translation of *Kitāb al-Masālik wa-al-mamālik* (Book of Routes and Provinces) by al-Iṣṭakhrī (fl. fourth/tenth century). Bodleian Library, MS Ouseley 373, copied in 1297/696H, fols. 3b–4a. South is at the top.

to give the impression of capturing our inhabited world in the manner of mathematical geography.

All this suggests that, like al-Khwārazmī before him, the author of the *Book of Curiosities* drew his world map with some version of a Late Antique world map at hand. There is strong textual evidence that such Late Antique maps, invariably attributed to Ptolemy, circulated in the early centuries of Islam and were available to other scholars as well. The Syriac Christian author Jacob of Edessa, who lived under the first Islamic caliphs (640–708), included in his theological commentary on the six days of Creation, the *Hexaméron*, a geographical chapter on the seas of the Earth that draws heavily on Ptolemy's *Geography*. At least some of the information in Jacob's chapter was taken from a map.[24] Like al-Khwārazmī, Jacob explicitly mentions two islands "visible on the map" in front of him. The Late Antique map used by Jacob of Edessa in the late seventh century had some intriguing parallels with our rectangular world map. Jacob gives the limits of the inhabited world as extending from the equator in the south to Thulé at 63° N, corresponding quite closely to the parallel boundaries in the *Book of Curiosities.* Moreover, based on the map before him, Jacob reports that a very large Sri Lanka is the only island named in the Indian Ocean, where one can also find "a peninsula situated close to China, called Chersonèse of gold, meaning the 'gilt Island.'"[25] On the basis of both location and etymology, this must be the Island of the Jewel facing China in our world map. Jacob was apparently looking at a world map of Late Antiquity that had much in common with the world map in the *Book of Curiosities.*

Pre-Islamic maps attributed to Ptolemy and Marinus were also seen by the tenth-century polymath al-Masʿūdī. In a key passage linked to his discussion of the Hellenistic division of the world into latitudinal climes, al-Masʿūdi describes three world maps he had seen, one Islamic and two pre-Islamic:

> I have seen these climes represented without labels[26] and in different colors. The best that I have seen is in the *Geography* of Marinus and in the commentary on the *Geography* of the divisions of the Earth, and in the map of al-Maʾmūn. That is the map made for al-Maʾmūn, which was constructed by a group of contemporary scholars, and in which the world is represented with its spheres, stars, lands, and seas, the inhabited and uninhabited regions, settlements of peoples, cities, and so forth. This [map of al-Maʾmūn] was better than anything that preceded it, either the *Geography* of Ptolemy, the *Geography* of Marinus, or any other.[27]

From this passage it is evident that al-Masʿūdī saw at least three world maps that indicated clime boundaries. Two came from works attributed to Ptolemy and his contemporary and scholarly rival Marinus of Tyre (ca. AD 70–130). A third was drawn for the Abbasid caliph al-Maʾmūn in the early ninth century. It seems the first two maps, explicitly attributed to antiquity, were not as rich in detail as the Abbasid one, and perhaps not even labeled, although in color.[28] Particularly intriguing is al-Masʿūdī's reference to the *Geography* of Marinus, whom Ptolemy accused of promoting orthogonal projection—the same projection suggested by Suhrāb in the tenth century, and which is likely to be associated with a rectangular world map.

In another passage al-Masʿūdī describes with more detail a treatise called *Geography*, written by a Greek sage. The author of that treatise is sometimes identified as Ptolemy, but in some manuscript copies is left unnamed. Al-Masʿūdī says that the work describes "the Earth, its cities, mountains, seas, islands, rivers, and springs" in the inhabited part of the world. The work seen by al-Masʿūdī listed 4,530 cities, clime by clime, although al-Masʿūdī comments that the names are Greek and difficult to decipher. There is no doubt the work al-Masʿūdī is describing was a version of the same *Geog-*

raphy that al-Khwārazmī used for his tables.[29] Al-Masʿūdī also states that the coastlines reflected the same cartographic nomenclature which in the *Book of Curiosities* is associated with Ptolemy: "Some of them are in the form of a *ṭaylasān* (concave), some of them in the form of a *shābūrah*, some like entrails (*muṣrānī al-shakl*), some round and some triangular."[30]

Al-Masʿūdī's writings establish that a map attributed to Ptolemy's *Geography* was circulating among Muslim scholars in the tenth century. Less than a century later, another version of that pseudo-Ptolemaic Late Antique map was available to the author of the *Book of Curiosities*, writing in Cairo in the first half of the eleventh century. The attribution to Ptolemy's *Geography* is probably imprecise, as most Ptolemaic material in the *Book of Curiosities* appears to come from Ptolemy's *Almagest*. The map is more likely to have had the orthogonal projection associated with Marinus, Ptolemy's rival. But, regardless of the precise source, what we see in this world map is the Ptolemaic tradition that permeates other parts of the *Book of Curiosities*: it is the foundation for its account of the macrocosm and the microcosm. The author's depiction of the sky that opens the first book of the treatise shows his reliance on a Ptolemaic cosmological scheme. His understanding of the influence of the stars on events on Earth is grounded, at least indirectly, in Ptolemy's popular defense of astrology, the *Tetrabiblos*. Ptolemy is also claimed as a source for names of comets. The same Ptolemaic tradition of reading the skies informs and encases our author's map of the Earth. The author viewed this world map as reproducing the basic conceptions of Ptolemy's *Geography*, and populated the edges of the inhabited world with designs inherited from Late Antiquity—including a calibrated scale bar at the top of the map.

Islamic Adaptations

All in all, the world map of the *Book of Curiosities* has unmistakable links to a cartographic tradition attributed to Ptolemy. The scale, the latitudinal boundaries, the rectangular shape and the iconographic and textual references to al-Khwārazmī's work show that the map is embedded in a shell of mathematical geography. The underlying conception of the map was mathematical, and it drew, however loosely, on the same type of Late Antique map employed by al-Khwārazmī.

But that base layer was then adapted to the Islamic era. Landmasses were stretched in order to accommodate long itineraries, while a defective grasp of the use of the scale may have led to the western quarter of the inhabited world, including Europe and West Africa, being depicted as extending over nearly half the hemisphere. The scale, as we have seen, is corrupt and useless. The author made a conscious decision not to attempt a plotted map, but rather transposed onto the base map contemporary geographical information relevant to the world of the eleventh century.

While the edges of the map are archaic and Ptolemaic, the interior of the world map demonstrates a very substantial influence from Balkhī School cartography, in particular the regional maps of Ibn Ḥawqal, who traveled around the Mediterranean in the 970s. The mapmaker took the maps of Ibn Ḥawqal—produced only fifty years or so earlier and no doubt the best and most recent maps available to him—and transposed these maps with their labels onto a rectangular world map, without any regard to the implications for longitude and latitude. Long itineraries in the regions of North Africa, Spain, and Iraq are cut and pasted from Ibn Ḥawqal's regional maps of the Maghreb, the Mediterranean, and Iraq. They are nearly a perfect match to the itineraries found on maps in an Ibn Ḥawqal copy produced in 1086 (479 H), now in the Topkapı Sarayı Müzesi, Ahmet III MS 3346. Most obvious is the long itinerary cutting through the oversized Iberian Peninsula in our world map that had been lifted in toto from Ibn Ḥawqal's map of the Maghreb (see fig. 3.7 and plate 2), which despite its name actually covers the entire Mediterranean, and occupies a tri-folio in the Topkapı copy. Of the four hundred labels on the map in the *Book of Curiosities*, the vast majority are taken from Ibn Ḥawqal's treatise.[31]

Beyond the place-names, select coastal and inland features were also lifted from Balkhī School maps and introduced onto this world map. The enlarged Iberian Peninsula precisely corresponds to the way it is depicted in Ibn Ḥawqal's map of the Maghreb that is preserved in the Istanbul Topkapı manuscript, along with the Iberian river system. Across the Mediterranean, a distinctive peninsula on the Atlantic coast of North Africa is reproduced on both maps, with a label identifying it with the Barghwāṭa tribes on the Atlantic coasts of Morocco. To the north of this peninsula, a lake depicted on the world map is most likely to be a corruption of one of the two inlets shown in this location on Ibn Ḥawqal's map of the Maghreb. There is also a very close correspondence between the North African place-names on both maps, including their location in the itineraries.

The Arabian Peninsula and Iraq on our rectangular world map (see fig. 3.8) are also very closely aligned with the regional maps of the Balkhī School. The shape of Arabia practically reproduces the shape of the peninsula in Balkhī regional maps, as, for example, in the one from al-Iṣṭakhrī below (fig. 3.9). The red, two-peaked mountain at the head of the Persian Gulf on the rectangular world map is obviously a variant of the iconography on al-Iṣṭakhrī's map, where a red band of sand stretching from the coast terminates at the foot of a prominent twin-peaked mountain range. Both

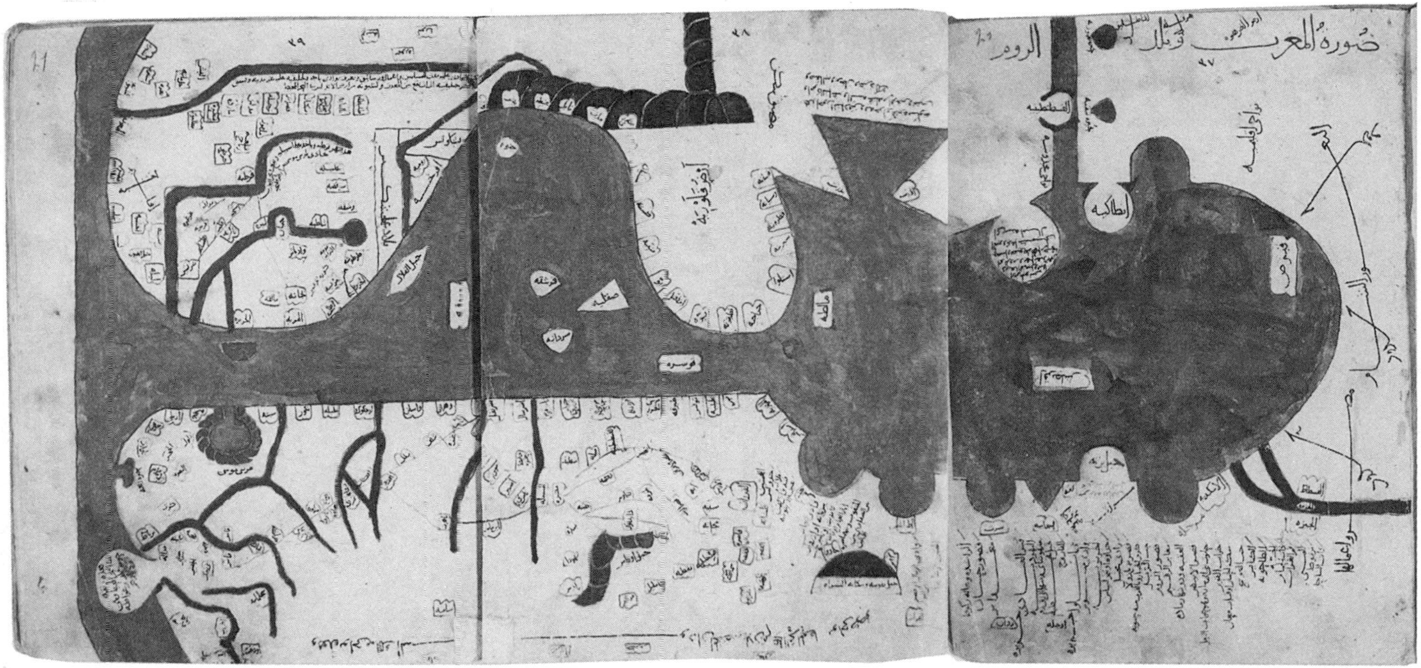

FIG. 3.7. Map of the Maghreb from Ibn Ḥawqal, *Kitāb Ṣūrat al-Arḍ*. The map is spread over three folios. North is at the top. Istanbul, Topkapı Sarayı Müzesi Kütüphanesi, Ahmet III MS 3346, fols. 19a, 19b, 20a, copied in 1086/479H. Courtesy of the Topkapı Sarayı Müzesi Kütüphanesi.

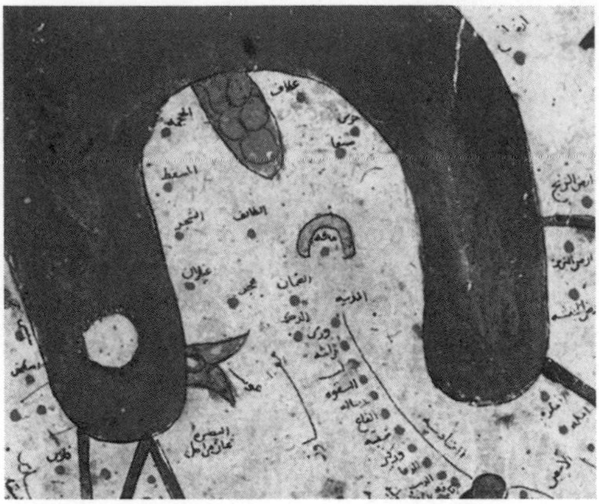

FIG. 3.8. Detail of Arabia from the rectangular world map of the *Book of Curiosities*. Oxford, Bodleian Library, MS Arab. c. 90, fol. 24a, copied ca. 1200.

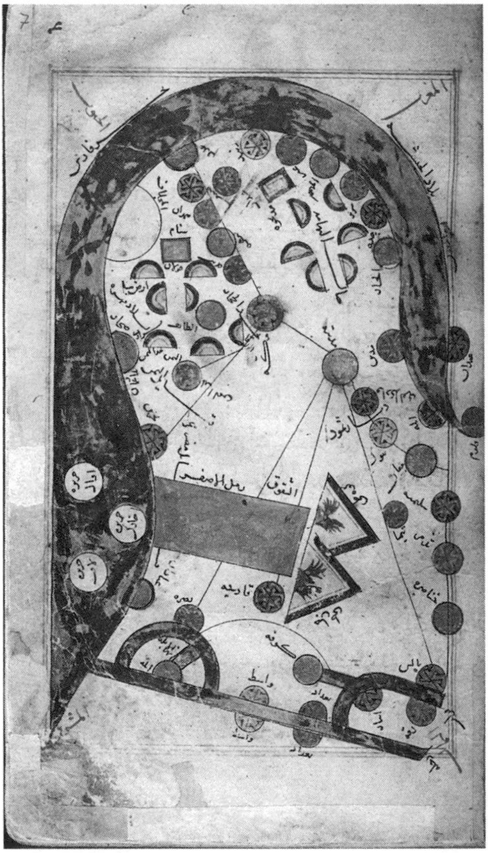

FIG. 3.9. Map of Arabia from al-Iṣṭakhrī, *Kitāb al-Masālik wa-al-mamālik*. The Nasser D. Khalili Collection of Islamic Art, MSS 972, fol. 7a. Copied 1306/706H. Copyright Khalili Family Trust.

maps show itineraries toward Mecca and Medina, although the horseshoe shape of Mecca is distinctive to our world map, and the place-names on this itinerary appear to come not from the Balkhī maps, but rather from the text of the ninth-century geographer Ibn Khurradādhbih.[32] Farther north, the depiction of the Tigris and the Euphrates reproduces the Balkhī maps of Iraq and Upper Mesopotamia.

The mapmaker made significant adaptations of the Late Antique prototype he had before him, and these adaptations come into sharper relief when we compare our rectangular world map with a modern reconstruction of the world map used by al-Khwārazmī. While the map al-Khwārazmī used has not survived, the detailed coordinate tables allow a reconstruction of what that map may have looked like. This has been tried by several modern scholars, but the one executed by Fuat Sezgin is the most complete and ambitious (see fig. 3.10).[33]

It should first be noted that the reconstruction shows the same motifs decorating the edges of the inhabited world. In the reconstructed map, the Mountain of the Moon and the Island of the Jewel are placed in a very similar position to that on the rectangular world map. The lands of Gog and Magog are similarly located in the northeastern corner of the map. In the east, both maps show a large peninsula facing the Island of the Jewel and protruding southward. In the *Book of Curiosities* this peninsula carries labels indicating localities in India and China.[34] The prominence of Sri Lanka in the reconstruction of al-Khwārazmī's map is also distinctive, and undoubtedly corresponds with the unnamed circular island, highlighted by bits of reflective gold glitter, on the *Book of Curiosities'* world map. [35]

The two maps do diverge most obviously in the relative sizes of Europe and Asia. In the map of the *Book of Curiosities*, Europe and North Africa are disproportionately wide and occupy nearly half the map, or what should be about 80 to 85 degrees of longitude. Al-Khwārazmī's coordinates, on the other hand, have the length of the Mediterranean at a much more accurate 43° 20.'[36]

One explanation for this anomaly may be a mistake on the part of the author of the *Book of Curiosities*, or perhaps the copyist of the Bodleian manuscript. The error began by depicting an oversized Europe and North Africa, perhaps when copying from a larger map, either a map for display or a map that was spread over multiple folios. Alternatively, the error may be linked to the corruption of the scale in the *Book of Curiosities* map, leading the author or copyist to misjudge the width of Europe and North Africa compared to the rest of the inhabited world. Whatever the reason, once the western quarter of the inhabited world has taken up about a half of the map, the Asian landmass had to be dramatically squeezed. The entire Asian

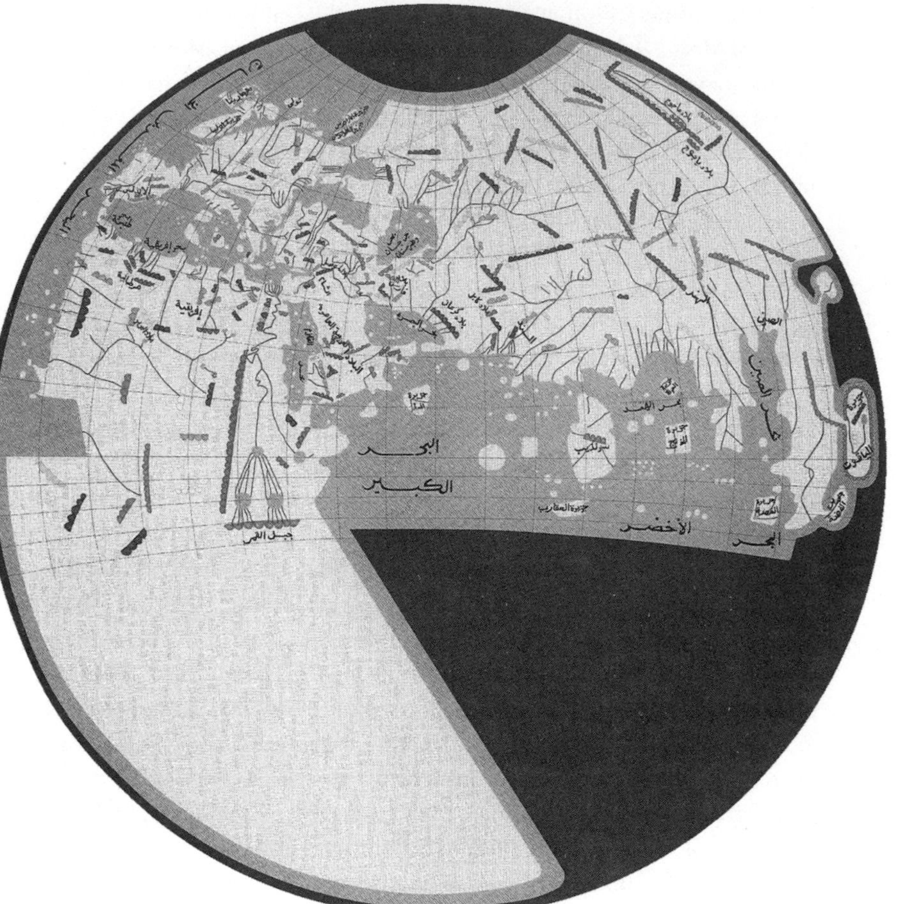

FIG. 3.10. A reconstruction of the world map used by al-Khwārazmī, using a modern projection and with north at the top. The lakes at the sources of the Nile and the "mountain of the moon" are south of the equator. The lands of Gog and Magog are in the northeastern corner of Asia. Reproduced with permission from F. Sezgin, *Mathematische Geographie*, Kartenband 4, no. 1b.

coast of the Indian Ocean between the Arabian Peninsula and China, an area which takes up much of al-Khwārazmī's map, is radically condensed and effectively omitted.

The shape of Europe on both maps offers another instructive comparison. In the *Book of Curiosities*, Europe is not only wider than in the reconstruction of al-Khwārazmī's map, but also significantly longer, stretching over two-thirds of the latitude of the northern hemisphere. Here, it seems, the mapmaker made a conscious choice to extend the European continent in order to accommodate the long itinerary he wanted to place in the Iberian Peninsula. In the rectangular world map, as in the world maps of the Balkhī School, Europe is clearly depicted as an island. Note, however, that even in the reconstruction of the map available to al-Khwārazmī, the Black Sea extends northward so as to almost cut the narrow land bridge between Asia and Europe.

We should also keep in mind that we are comparing our eleventh-century map with a modern reconstruction, one that employs modern techniques and appeals to modern expectations. Sezgin opted for a circular projection, and represents the entire hemisphere pole to pole and across 180°. That is not necessarily how the map used by al-Khwārazmī would have looked. In fact, as is evident even from the reconstruction, al-Khwārazmī's map probably showed only the inhabited world, and may have been produced through an orthogonal projection, like the one proposed by Suhrāb, with the concomitant distortion of the northern parts of Europe and Asia. If you imagine Sezgin's reconstruction to be rectangular and not circular, bounded by the Mountain of the Moon in the south and the seventh clime in the north, it would in fact appear much more like the world map in the *Book of Curiosities.*

There are other ways in which a Late Antique world map may have differed from modern reconstructions. Recent research into Ptolemy's *Geography* has revealed the important role of chorography, or nonmathematical regional maps, in Ptolemy's own work. Ptolemy had latitude and longitude data for only a limited number of localities, and he interpolated internal points within each region on the basis of nonmathematical regional maps.[37] Perhaps we are taking Ptolemy too literally in assuming that if the *Geography* was accompanied by maps, these were plotted without regard to legibility or intelligibility. Admittedly, Ptolemy admonished cartographers who tended to fiddle with the size of the landmasses in order to fit in the desired place-names, and who drew Europe larger than Asia or Africa simply because there was more information to be accommodated. But while Ptolemy evidently rejected this nonscientific representation, his account suggests that the maps known to him were often hybrid ones, marrying his mathematical geography with nonmathematical chorography and the requirements of legibility.[38] The problem of fitting in labels on a mathematically plotted map is one that our author was keenly aware of, and in the Iberian Peninsula, in particular, he adapted the contours of the landmass in order to fit in the large number of labels.

Finally, it also quite likely that the author of the *Book of Curiosities* did not work directly from the same Late Antique map seen by al-Khwārazmī, but rather from a later Arabic adaptation. A likely candidate would be a world map made for the Fatimid caliph al-Muʿizz (reg. 953–75). This map has not survived, but is described by the fifteenth-century Egyptian historian al-Maqrīzī, whose work preserves unique material from now lost Fatimid chronicles. He relates that when rebellious troops looted the Fatimid palaces in 1068, they discovered a world map on a piece of cloth of extraordinary quality, woven with gold and dyed silks. Al-Maqrīzī states that

> it had the image of the climes of the Earth, with its mountains, seas, cities, rivers and routes, similar to the *Geography* [of Ptolemy]. It also had Mecca and Medina indicated clearly. Each city, mountain, land, river, sea and route had its name written in gold, silver or silk thread. At the bottom it was written: "This was made by order of al-Muʿizz li-Dīn Allāh, out of longing for the sanctuary of God [that is, Mecca] and wishing to spread the knowledge of the milestones of the Prophet,[39] in the year 353 (AD 964)." Its value was 22,000 dinars.[40]

Given the Fatimid context of the *Book of Curiosities* as a whole, it is possible that this royally commissioned world map was a source of inspiration for the world map of the *Book of Curiosities*. Here too there is a reference to Ptolemy's *Geography* as a model, and the Fatimid world map was possibly an adaptation of the Late Antique maps associated with Ptolemy's *Geography* circulating at the time. Note that al-Muʿizz's map singled out Mecca and Medina, reminiscent of the prominent indication of Mecca in the *Book of Curiosities*. The map made for the caliph al-Muʿizz was evidently for display, and likely to be quite large. Copying it into a manuscript format would have led to substantial deformations, such as the ones found in the *Book of Curiosities*. Whether the world map of al-Muʿizz was indeed a prototype for our map is impossible to confirm, but remains a tantalizing possibility.

A Hybrid Map

This rectangular world map is a hybrid map, whose interior draws on the abstract, schematic itinerary-focused cartography of the tenth century Balkhī School of geographers. The incorporation of so much material from Ibn Ḥawqal in particular has led some historians of Islamic cartography to argue that this world map is nothing but a grotesque collage of the details found on the regional maps of the Balkhī School.[41] Andreas Kaplony viewed this map as a strange variation on the maps of the Balkhī tradition, and downplays the importance of the scale at the top of the map. According to this interpretation, the map is an attempt by a mapmaker to combine and transpose the regional maps of the Balkhī tradition into one unified world map.

Tarek Kahlaoui also points out that the color coding, with green and blue for salty water and freshwater, respectively, also follows the Balkhī tradition, and the lining up of red dots in itineraries is a characteristic feature of Ibn Ḥawqal's maps.[42]

And yet, while the interior of this world map is indeed a collage of Ibn Ḥawqal's regional maps, the framework is not. This world map is couched in a framework that does not originate with the Balkhī School, but in an older layer that retained many distinctive features of a Late Antique prototype. The divergence from the Balkhī School is conceptual as well as substantial. This is a rectangular, not circular, world map, possibly associated with a crude orthogonal projection, and one that carries a longitude scale bar. It shows only the inhabited world bounded by latitudes, from the equator in the south to 66° N in the north. Two of al-Khwārazmī's surviving maps, those of the Nile and of the Island of the Jewel, are reproduced in the southern and eastern edges. In its framework and its context within the *Book of Curiosities*, it lies squarely within the tradition of mathematical geography associated with Ptolemy, and the map as a whole is explicitly attributed to Ptolemy's *Geography*.

The mapmaker then adapted the Late Antique prototype to the Islamic context of the eleventh century. The Greek or Syriac labels, which would have been either incomprehensible or archaic, were omitted. Instead, our mapmaker chose to insert onto the map the most up-to-date information about the inhabited world, coming from the work of the tenth-century geographer Ibn Ḥawqal. The mapmaker adapted the map so he could incorporate wholesale elements of the Balkhī regional maps, such as the shape of the Arabian Peninsula or Iberian itineraries. The mapmaker or the copyist of the Bodleian copy misunderstood the purpose of the scale bar, and this may have caused the right-hand side of the map—the one containing Europe and Africa—to be so oversized, while the left-hand side is condensed. It could also be that the map was copied from a map intended for public display, like the world map made for the Fatimid caliph al-Muʿizz, or in segments from a larger, multi-folio original, such as the trifolio carrying Ibn Ḥawqal's map of the Maghreb in the Topkapı manuscript.

It is worth pointing out here the existence of some other examples of adaptations of a Late Antique world map, produced centuries after the composition of the *Book of Curiosities*. A closely related example is a pair of maps of the world and of the Nile added by a twelfth-century copyist to some of the manuscripts of Ibn Ḥawqal's treatise. Although the pair has been called "Ibn Ḥawqal III" by modern scholars, these maps do not actually originate with Ibn Ḥawqal.[43] This pair of maps is found in three copies of Ibn Ḥawqal's treatise. The world map shown in fig. 3.11 comes from a

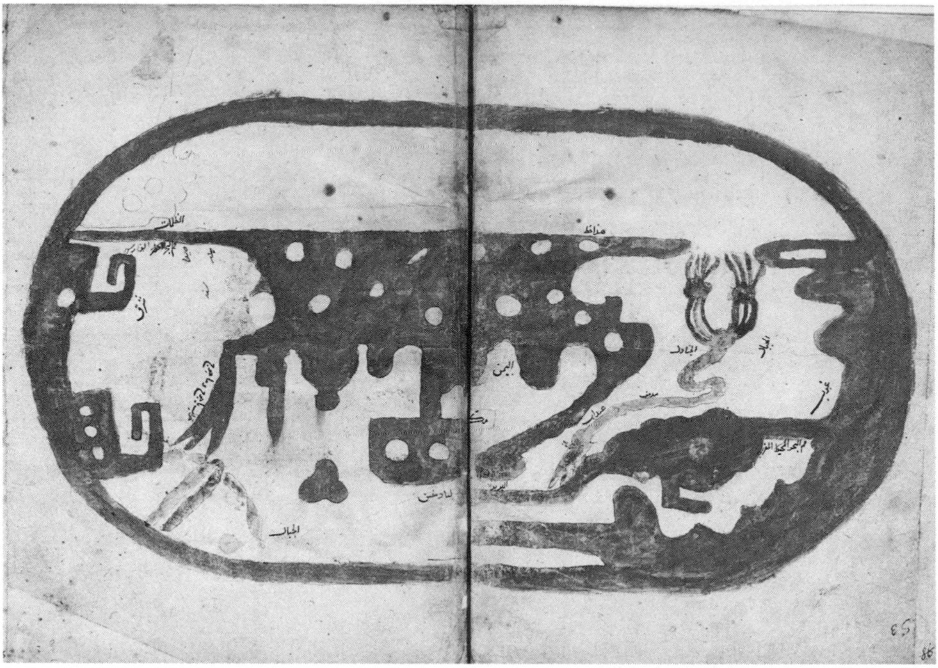

FIG. 3.11. An oval world map from the Ibn Ḥawqal III set. Paris, BnF, MS arabe 2214, fols. 52b–53a, copied 1445/847 H. Reproduced with permission of the Bibliothèque nationale de France.

Paris copy.[44] This world map has a unique oval shape, and appears to cover only the inhabited quadrant of the world, bounded by the equator in the south. Both the sources of the Nile in the south and the Island of the Jewel in the east are present, as well as the land of Gog and Magog in the northeast, all features that parallel the world map of the *Book of Curiosities.* Most importantly, the appearance of the oval "Ptolemaic" world map in a manuscript produced around 1200 is evidence for the continued circulation, as late as a century *after* the composition of the *Book of Curiosities,* of a prototype derived from Late Antiquity.[45]

An even later world map maintaining elements of mathematical geography and displaying some parallels with the hybrid world map in the *Book of Curiosities* comes from the pen of the fourteenth-century Egyptian author Ibn Faḍlallāh al-ʿUmarī. This remarkably complex circular world map has a scale running along the equator (fig. 3.12).[46] In the text of his work, Ibn Faḍlallāh repeatedly makes use of a map (*lawḥ al-rasm*) of the "author of the *Geography.*" The map as preserved today in Ibn Faḍlallāh's treatise appears to be a copy or an adaptation of the map he had before him.[47] A few

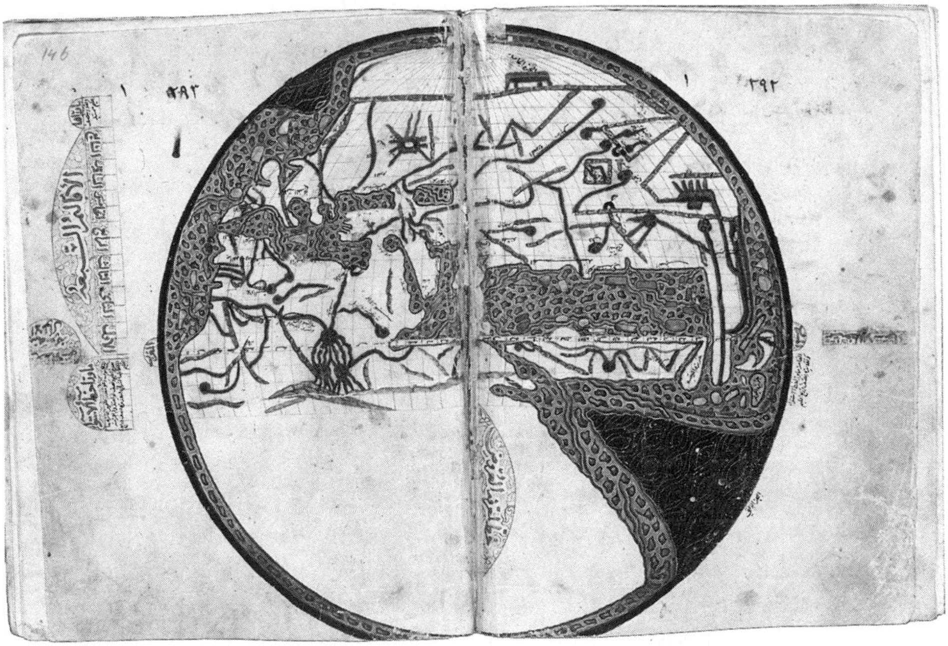

FIG. 3.12. World map in Ibn Faḍlallah al-ʿUmarī, *Masālik al-abṣār*. Shown with the north at the top. Istanbul, Topkapı Sarayı Müzesi, MS A 2797, fols. 292v–293r. Autograph copy, ca. 1340. Reproduced with permission of the Topkapı Sarayı Müzesi.

parallels between the world map of Ibn Faḍlallāh and the rectangular world map in the *Book of Curiosities* are suggestive of some shared sources. The most obvious and significant, of course, is the scale, which runs along the equator in both maps. The Island of the Jewel makes its appearance on the eastern edge of the Indian Ocean, as it does in the world map of the *Book of Curiosities*. Even though Ibn Faḍlallah's map is circular in shape, the geographical material is actually bounded by latitudinal lines, with most of Africa south of the equator and the areas close to the North Pole shown as empty spaces. The few labels that are found on Ibn Faḍlallāh's map belong to a late medieval layer, but, as with the *Book of Curiosities*, the updating of place-names is in itself unexceptional.

The oval or semicircular world maps known as the Ibn Ḥawqal III set, as well as the fourteenth-century world map of Ibn Faḍlallāh, suggest that maps based on Late Antique models, sometimes with distinctive designs or unusual non-circular shapes, continued to circulate in the late medieval period. There was a cartographic tradition in which the inhabited world was a rectangle or an oval, bounded by the equator and the seventh clime, its topography constantly adapted by superimposition of itineraries and new geographical knowledge. The rectangular world map of the *Book of Curiosities* thus joins a robust body of evidence for the continued circulation of maps derived from Late Antique sources, known collectively as the *Geography*. It is, however, the earliest surviving example of this cartographic tradition, and the strangest.

The hybrid world map of the *Book of Curiosities* is thus a visual bridge between Late Antiquity and Islam. It changes the way we understand the history of the transmission of geographical knowledge from the pre-Islamic Greek and Syriac-speaking Middle East to the Arabic scientific culture of early Islam. While modern scholarship generally assumes that the Ptolemaic tradition of mathematical geography passed to Islam via translations of texts, this world map shows that the author of the *Book of Curiosities* in the eleventh-century, like al-Khwārazmī before him, had at his disposal a version of a Late Antique map, on which labels were originally written in Syriac or Greek. The world map of the *Book of Curiosities* is our earliest preserved visual evidence for what such a map might have looked like. Islamic knowledge was then superimposed on and encased within a rectangular framework of mathematical geography inherited from Late Antiquity: a hybrid map for the hybrid early Islamic culture, at home with the Greek legacy of the ancients as it was with its imperial present.

CHAPTER FOUR

The Nile, the Mountain of the Moon, and the White Sand Dunes

In the previous chapter, the emphasis had been on highlighting traces of a Ptolemaic tradition of mathematical geography on the unique world map of the *Book of Curiosities*. The present chapter moves from the global to the particular. Here we take the evolving representation of the Nile, a dominant feature of all Islamic world maps, as a case study of the way in which new information was incorporated into an overall Ptolemaic framework. The two chapters are complementary. The depictions of the Nile in the *Book of Curiosities* furnish further proof that the author was using a prototype associated with a Late Antique map maintaining elements of mathematical geography. Moreover, a Ptolemaic world map was often coupled with a separate Ptolemaic map of the Nile. The pair formed a curious relic of the transmission of knowledge from Late Antiquity to Islam.

The *Book of Curiosities* has four unique depictions of the Nile and its sources. The most detailed representation of the river is a dedicated Nile map, one of five river maps found in the book. Another diagram of the sources of the Nile forms part of a chapter on the world's lakes. A third depiction is found in the world map discussed in the previous chapter, and a fourth is found on the circular world map of the al-Idrīsī type.[1] Despite some variance, these depictions form a coherent set that draws on a Late Antique map of the Nile, attributed to Ptolemy's *Geography* and made in the tradition of mathematical geography. At the same time, the evolving images of the Nile in the *Book of Curiosities* combine elements of the Balkhī tradition, and, in particular, introduce new material through Egyptian Coptic and North African informants.

The development of the Islamic visual representation of the Nile encapsulates the tension between the Ptolemaic (or, more generally, Late Antique) models and the firsthand knowledge acquired by Muslim travelers and geographers. The first Islamic map that has survived is a map of the Nile, found in the ninth-century work of al-Khwārazmī. Largely following

Ptolemy, it shows the Mountain of the Moon as the ultimate source of the Nile, feeding into a symmetrical system of three lakes. While the tenth-century Balkhī School maps mostly ignore the Ptolemaic model, the *Book of Curiosities* returns to it, reinstating the Mountain of the Moon and the three lakes at the southern edge of civilization. The *Book of Curiosities* also represents an eastern tributary of the Nile, which also has its origin in Ptolemy, although here it becomes associated with the coasts of East Africa.

At the same time, the *Book of Curiosities* introduces two revolutionary ideas about the origins of the Nile. The maps of the *Book of Curiosities* represent, for the first time, a western tributary of the Nile that originates in sand dunes in West Africa. The maps and their labels also depict a second mountain, strangely located in the middle of a lake, whose melting snow is said by Coptic informants to be the source of the Nile floods. That explanation is as close as medieval scholars ever got to the true reason for the Nile's annual cycle. Through extensive reworking by al-Idrīsī a century later, these two novel elements would become a permanent feature of later cartographic representations of the Nile.

Despite the long-standing fascination with the Nile and its sources, the depiction of the Nile in early Islamic cartography has only recently become a subject of systematic study.[2] Here, our aim is to trace the development of the Islamic mapping of the sources of the Nile, from the earliest map made by al-Khwārazmī in the mid-ninth century to the influential and systematic maps by al-Idrīsī in the twelfth century, with special emphasis on the *Book of Curiosities* as a key intermediate link. The importance of such a narrative of visual images is twofold. First, it contributes to our understanding of eastern and western medieval conceptualizations of the Nile. The evolving depiction of the Nile in Islamic cartography enables us to identify with more precision the links between Classical geographical ideas and the Abbasid Islamic geographical tradition, and then between the later, Fatimid and Mediterranean Islamic tradition and the Latin European tradition. Second, the evolving representation of the Nile is the story of the evolution of the entire Islamic cartographic tradition in a nutshell. This is because the Nile is such a dominant feature of all Islamic cartographic representations of the world, and also because a map of the Nile is included in the set of the oldest extant Islamic maps—the 1037 [429 H] Strasbourg al-Khwārazmī manuscript—which means that we can follow the tradition right from its earliest-known beginnings.

Al-Khwārazmī's map of the Nile

The image of al-Khwārazmī's map of the Nile is one of the best known in the history of Islamic cartography (plate 12 and fig. 4.1). It draws on the Ptolemaic model and on mathematical geography, but employs contemporary place-names. As in Ptolemy's model, the Nile flows from the now familiar parachute-shaped Mountain of the Moon into a system of three lakes arranged in a triangle. The Mountain of the Moon, at the top of the map (south) is the origin of two sets of outlets, one with four branches and another with five. These feed into twin western and eastern lakes, called marshes (*baṭāʾiḥ*) by al-Khwārazmī. The twin lakes in turn feed into another two sets of four outlets that flow northward into a third lake, labeled here "The Small Lake." This smaller lake is the source of the main branch of the Nile. To the east, the map indicates another lake, labeled as "a lake which feeds the Nile." This eastern lake is the source of an eastern tributary of the Nile, which joins the Nile north of Aswan. A large island is formed at the confluence of the main branch of the Nile and the eastern branch. Then the Nile continues northward in a curved course, passing under a massive, triangular Muqattam Mountain at the bottom left, and terminating in a schematic six-forked delta.

The link to Ptolemaic mathematical geography is apparent from the indications of climes as straight red lines that cut through the diagram. The equator is shown as running through the eastern lake, to the north of the Mountain of the Moon. Straight lines indicate the limits of the first, second, and third climes, with Alexandria and Damietta located in the fourth clime. Importantly, the clime bands are not of equal width, reflecting the latitudinal difference between climes, defined by the length of the longest day. The unequal width of the clime bands suggests that this is not a mere diagram, and that it has a direct link to a map based on mathematical plotting.

By and large, the conception is Ptolemaic: the Mountain of the Moon and the twin lakes are easily identifiable with the same features in Ptolemy. The third lake, however, is not in Ptolemy and is an indication that al-Khwārazmī was using another source, probably a Late Antique adaptation.[3] The eastern lake lying on the equator, although unnamed in the map or in al-Khwārazmī's text, is undoubtedly Lake Koloe (Κολόη), which Ptolemy claimed to be the source of the Astapos eastern tributary of the Nile.[4] The Ptolemaic Lake Koloe is also reported by Jacob of Edessa in the seventh century.[5] Another distinctive Ptolemaic feature is a canal that flows from the eastern tributary (the Ptolemaic Astapos) to the main branch of the Nile, creating a large keyhole-shaped island. This is the Ptolemaic island of Meroe, which al-Khwārazmī calls the "city of the Nubians."[6]

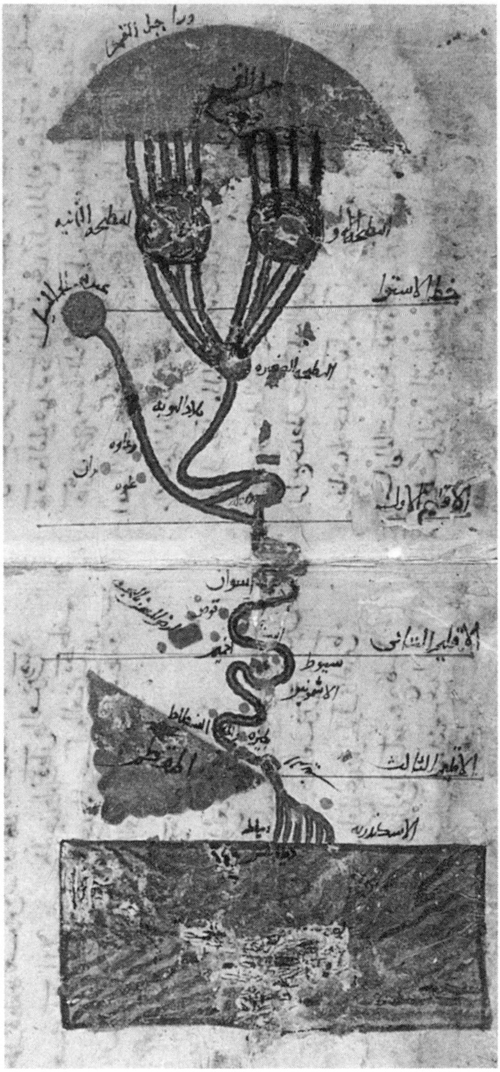

FIG. 4.1. Map of the Nile by al-Khwārazmī. Bibliothèque Nationale et Universitaire de Strasbourg, MS 4247, fols. 30b–31a, copied 1037/429H. Reproduced courtesy of Coll. et photo BNU Strasbourg.

While al-Khwārazmī's map generally retains the Ptolemaic configuration of the Nile, it lacks the Ptolemaic place-names and replaces them with place-names contemporary with the Islamic period.[7] In our discussion of the rectangular world map of the *Book of Curiosities* we have noted the superimposition of new information onto the Ptolemaic model. Al-Khwārazmī did the same in his map of the Nile. But in this case, the superimposition caused a confusion which had lasting impact on the development of the later cartographic tradition. The text of al-Khwārazmī's treatise says that the Nile of Egypt passes in the land of the Sūdān (Black Africans), through ʿAlwa, Zaghāwa, Fezzan, and Nubia, and then passes through Dongola, the city of the Nubians. The problem with this account is that while ʿAlwa and Nubia do indeed lie on the Nile, the regions of Zaghāwa and Fezzan lie far to the west, in what is today Chad and Libya. The Nile map of al-Khwārazmī then introduces another error. On the map, the labels for Zaghāwa, Fezzan, and ʿAlwa are all located to the east of the Nile. The initial error in the treatise, which was to place Zaghāwa and Fezzan on the banks of the Nile, is now compounded by an error of the mapmaker, who placed Zaghāwa and Fezzan far to the east (rather than the west) of the main branch.

Otherwise, however, al-Khwārazmī's Nile map corresponds almost perfectly with the latitude and longitude data given in the tables in his main text. Modern reconstructions of a world map based on the treatise's coordinates, by von Mžik, Dzhafrii, and Sezgin, show the Nile almost exactly as it appears on the extant separate Nile map. For Sezgin's reconstruction of al-Khwārazmī's world map, see fig. 3.8.[8] The differences between the reconstruction based on the coordinate values and al-Khwārazmī's Nile map are slight: the map condenses the upper regions of the Nile compared to the northern Egyptian areas, and introduces a massive Muqattam mountain range that is not so prominent in the textual data. Still, the resemblance between the Nile map and the reconstructions of the world map used by al-Khwārazmī is remarkable, and even the clime lines are placed at precisely the same position.

Al-Khwārazmī's Nile map reproduced a Late Antique map of the Nile, attributed to Ptolemy even if it differed from the Ptolemaic account known to us. Such a Late Antique map of the Nile was also available to al-Masʿūdī in the tenth century. Al-Masʿūdī, whose account of a Greek world map from the *Geography* was translated above, reports elsewhere that he has also seen the *Geography*'s map of the Nile.[9] His account of this map suggests that it was a variation of the Ptolemaic model, not precisely identical with either Ptolemy or al-Khwārazmī. He was viewing a map where the Nile emerges from twelve springs in the Mountain of the Moon, feeding two large lakes or marshes. The two marshes then flow into the main Nile—with no third lake (we have noted the third lake, found in al-Khwārazmī, is not mentioned by Ptolemy). Then, curiously, the map showed an eastern arm of the Nile that splits from the main branch and flows to the Indian Ocean. This was probably a corrupt version of the Ptolemaic Astapos tributary, here transformed into an eastern arm flowing into the sea.

The Balkhī School: A Straight Nile

While al-Khwārazmī clearly draws on a Late Antique source, the depiction of the Nile in the Balkhī School maps shows an entirely different cartographic tradition. The Nile map by al-Iṣṭakhrī (fig. 4.2), in its various medieval manuscripts, is an abstract and schematic diagram, with no attempt to represent the actual course of the Nile, and none of the Ptolemaic features that characterize al-Khwārazmī's map. There is no Mountain of the Moon and no trace of the symmetrical, tripartite lake system at the sources of the Nile, nor are there any clime lines. The Nile flows in a straight line from south to north, with two parallel mountain ranges on both sides. Rather than ending with a forked delta, the river ends with a semicircular bay, in which lie the parallel, oversized islands of Tinnīs and Damietta. The representation of the Nile in al-Iṣṭakhrī's world map (fig. 3.6) replicates that of his separate map of the Egypt.[10] The maps are based on the text of al-Iṣṭakhrī's treatise, which ignores Ptolemaic assumptions on the sources of the Nile. Instead, al-Iṣṭakhrī states that the source of the Nile is not known, save that it emerges in an impassable desert beyond the land of the Zanj (East Africa).[11]

The depiction of the Nile in the maps of the other Balkhī geographer, Ibn Ḥawqal, bears a close resemblance to al-Iṣṭakhrī's maps. The Nile is again schematic and nonmathematical, with no mountains or lakes at the sources of the Nile, no curves or tributaries. There is, however, an evident attempt to follow the branches of the Egyptian Nile, as it was known in the late tenth century. Much additional detail in the northern, Egyptian, section of the Nile was added by Ibn Ḥawqal through his extensive travels in Egypt.[12] Ibn Ḥawqal also introduces into his map the Fayyum arm of the Nile, known as the Baḥr Yūsuf, which returns to the Nile after irrigating the lands of the Fayyum depression. As shown by Jean-Charles Ducène and John P. Cooper, the detailed and asymmetrical delta reproduces rather accurately the contemporary distribution of canals and branches in the area.[13] These new details were added by Ibn Ḥawqal to his map of the Nile and also found in the text of his treatise.[14]

Ibn Ḥawqal's world maps offer an interesting variation on al-Iṣṭakhrī's, perhaps incorporating some Ptolemaic influences.[15] The Nile is again represented largely as a straight line, with no tributaries and no delta. Yet at its southern tip, which curves eastward, there is a rudimentary system of mountains and twin parallel lakes, all unlabeled, but reminiscent of the Ptolemaic system. This depiction seems to be derived from the simplified account of the Ptolemaic Nile system found in the geographical treatises of Ibn Khurradādhbih (d. ca. 912) and Ibn al-Faqīh (fl. ca. 903), who report

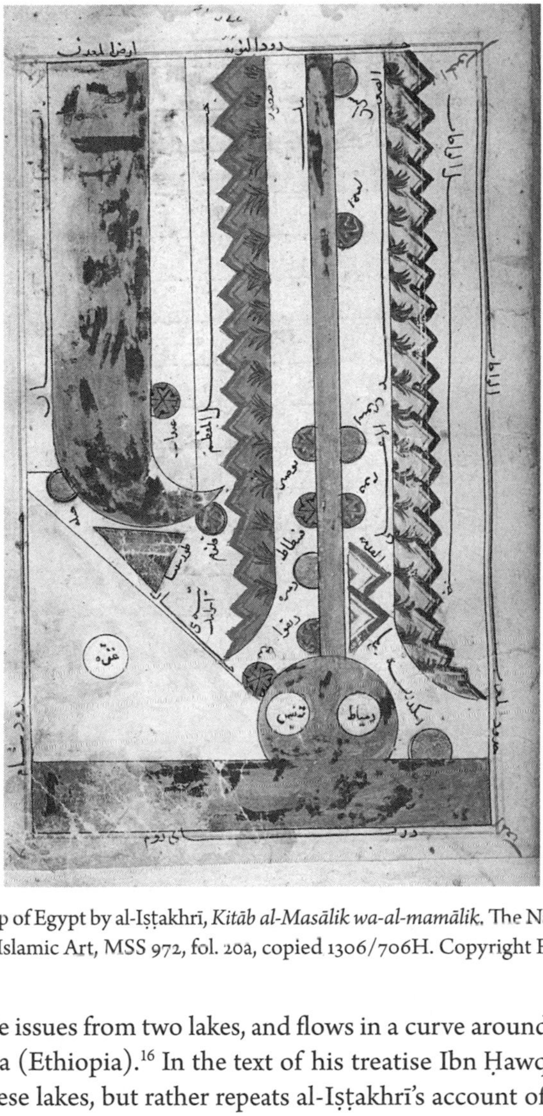

FIG. 4.2. Map of Egypt by al-Iṣṭakhrī, *Kitāb al-Masālik wa-al-mamālik*. The Nasser D. Khalili Collection of Islamic Art, MSS 972, fol. 20a, copied 1306/706H. Copyright Family Trust.

that the Nile issues from two lakes, and flows in a curve around the land of the Ḥabasha (Ethiopia).[16] In the text of his treatise Ibn Ḥawqal does not mention these lakes, but rather repeats al-Iṣṭakhrī's account of an impassable desert, adding that Ptolemy in his *Geography* discussed the origins of the Nile, but has not ascribed it to a specific location.[17]

Map of the Nile in the Book of Curiosities

The map of the Nile in the *Book of Curiosities* (plate 11 and fig. 4.3) is, beyond doubt, a variant of the map of the Nile by al-Khwārazmī (fig. 4.1 and plate 12). The map is included in chapter 18, on the major rivers of the world. This chapter includes also river maps of the Euphrates, Tigris, Indus,

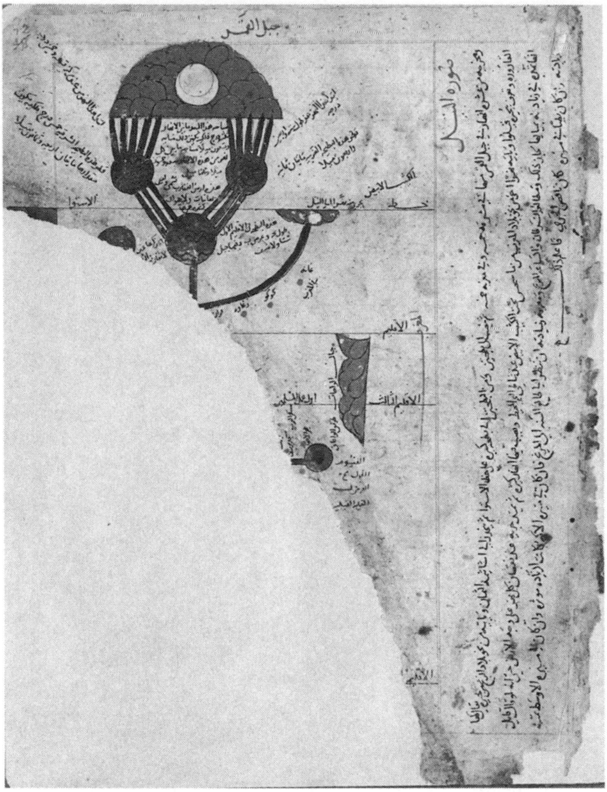

FIG. 4.3. The River Nile in the Bodleian copy of the *Book of Curiosities*. Oxford, Bodleian Library, MS Arab. c. 90, fol. 42a, copied ca. 1200.

and Oxus, which borrow significant elements from the Balkhī School regional maps. But the Nile map is markedly different in style and offers a clear return to the Ptolemaic notion of the origins of the Nile. The lower-left part of this Nile map has been lost in the Bodleian copy due to damage to the folio. Nonetheless it can be partially reconstructed with the help of the text on the side panel, which describes the course of the Nile as it appears on the map. In addition, an abstracted and sparsely labeled diagram in the Damascus copy of the *Book of Curiosities* contains a few labels in the otherwise blank rectangular space, and these labels also prove useful in reconstituting the missing part of the map as preserved in the Bodleian copy (fig. 4. 4).

The Nile maps in the *Book of Curiosities* and in al-Khwārazmī's treatise are obviously closely related: the dominant feature of both maps is the parachute-shaped Mountain of the Moon and the system of three lakes at the origins of the Nile, connected by parallel sets of outlets.[18] Another very

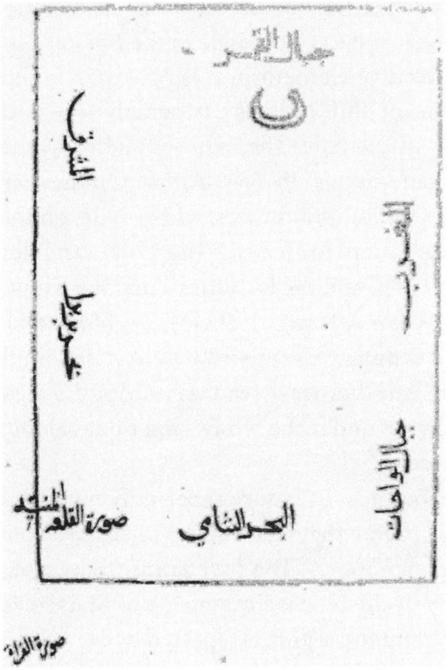

FIG. 4.4. The River Nile in the Damascus copy of the *Book of Curiosities*. Damascus, Maktabat al-Assad al-Waṭanīyah, MS 16501, fol. 119b, copied 1564/972H.

visible link to al-Khwārazmī is the clime lines. In both maps the equator passes between the twin lakes and the third, northern lake, and first clime is noticeably wider than the second clime. The upper half of a blue lake (blue indicating sweet water) on the left is undoubtedly the same as the eastern lake of al-Khwārazmī's map, and the Ptolemaic Lake Koloe. At the bottom left of the Damascus copy, in the area lost in the Bodleian map, the text refers to an image of a citadel: this is where we would expect, following al-Khwārazmī, to find the prominent Muqattam Mountain.

The similarities between the *Book of Curiosities*' Nile and that of al-Khwārazmī are so evident that it takes time to realize the differences. One significant departure from al-Khwārazmī's model is that the relative size of the three lakes has changed. While al-Khwārazmī has the eastern and western marshes flow into a "Smaller Marsh," the Nile map in the *Book of Curiosities* depicts this northern lake, the bottom point of the triangle of lakes, slightly larger than the two lakes to its south. The label next to this northern lake indicates that within the lake "there is a mountain covered with snow in winter and summer."[19] The significance of this reference to a mountain found within the northern lake, not visually indicated here, will become clear when we will discuss in the next section the separate set of diagrams of the marshes of the Nile in the *Book of Curiosities.*

The most innovative element in this Nile map—and the one that anticipated a significant shift in later cartography—was the depiction of a western tributary of the Nile. The map shows this western tributary flowing from "white sand dunes" in West Africa. On the map the Sand Dunes are indicated by a red mountain west of the Nile, on the right-hand side. The label near the mountain reads: "The white sand dunes from which a river flows to the Nile," and the localities along the tributary are all in West Africa, including Ghāna, Kawkaw (Gao), Zaghāwa, and Fezzan. The side panel describes a tributary issuing from sand dunes on the Atlantic coasts of the Maghreb: "Another river reaches it from the area of the Maghreb, from a spring flowing under the white sand dunes along the shores of the Encompassing Ocean."

Classical geographical literature repeatedly mentions a western tributary of the Nile, a theme that has recently received a comprehensive treatment by Robin Seignobos.[20] The first author to be cited as referring to a Western tributary of the Nile is Euthymène of Marseilles, who lived in the sixth century BC, and the report is repeated by Herodotus in the following century. The most detailed account is found in the *Natural History* of Pliny the Elder who relates that the Nile rises in Lower Mauritania, not far from the Western Ocean. After terminating in a lake formed by it behind the Great Atlas, Pliny continues, it reemerges from the sands of the desert as the Niger. This view, however, was not incorporated into Ptolemy's *Geography,* nor included in al-Khwārazmī's map of the Nile.

However, the immediate source for the representation of a western Nile on the map of the Nile in the *Book of Curiosities* is an Islamic tradition regarding the conquest of North Africa. In Arabic geographical literature, the first mention of a West African Nile is a report cited by Ibn al-Faqīh in the late ninth century. Following his simplified account of the Ptolemaic Nile, Ibn al-Faqīh cites an Arab military commander who claims to have reached the source of the Nile in West Africa. The report reads: "Abū al-Khaṭṭāb related that al-Mushtarī ibn al-Aswad said: I have raided the land of Anbiya twenty times from the furthest Sūs, and I have seen the Nile. Between the river and the brackish sea (*al-ajāj*) is a dune of sand and the Nile issues from beneath it."[21] Much of this report about the western origins of the Nile is obscure, and it may well be a re-dressing of Classical material in an Islamized garb. The "furthest Sūs" is Morocco, but the identification of Anbiya, a locality or people of West Africa also mentioned by al-Masʿūdī and al-Yaʿqūbī, has been subject to speculation.[22] Abū al-Khaṭṭāb has been speculatively identified as a Shiʿa leader in mid-eighth century North

Africa. The source cited for this information, a general called al-Mushtarī ibn al-Aswad, is not otherwise known.[23]

As with the account of the mission to the find the Wall of Gog and Magog, which Travis Zadeh has linked to Abbasid imperial ideology,[24] this is an imperial discovery, with the Muslim conquerors pushing and testing the boundaries of civilization at the edges of the known world. The account of the Nile emerging from beneath a dune of sand in West Africa did have a mythical flavor, forcing Ibn al-Faqīh to explain, on the authority on unnamed sages, that water can spring from the desert.[25] The report has similarities with the semilegendary accounts of the Muslim conquest of West Africa by ʿUqba ibn Nāfiʿ, who is reported to have reached the shores of the Atlantic in 61/681.[26] The account by Ibn ʿAbd al-Ḥakam (d. 871) of ʿUqba's conquests blends motifs from the Alexander Romance, specifically the scientific inquiry regarding the nature of the people who live in the lands beyond the edge of civilization.[27]

The notion of a tributary of the Nile emerging from a sand dune in North Africa is taken up in the *History of Nubia* by Abū Sulaym al-Uswānī, one of the few Fatimid treatises of geography known to us.[28] Al-Uswānī was sent by the Fatimid general Jawhar to the king of the Nubians sometime between 969 and 973, and later wrote an account of travels there.[29] His account of the origins of the Nile is largely based on firsthand experience, and not on the literary geographical tradition. He notes an eastern source of the Nile, which can be identified as the Atbara, and mentions the two arms of the Nile as the White Nile and the Green Nile. With regard to the sources of the White Nile, al-Uswānī reports: "I have asked the Maghribis who travel the land of the Sudan about the Nile which flows in their lands and its colour. They report that its waters come from mountains of sand, or a mountain of sand, and that they collect in the lands of the Sudan in huge pools, before flowing in an unknown direction."[30] There is no evidence of a direct reliance of the *Book of Curiosities* on the *History of Nubia*, and there is no further correspondence between both works in their accounts of the Nile and its wonders. But al-Uswānī's discussion of mountains of sand as a North African source of the White Nile shows that the notion circulated in Fatimid or Ismaʿili circles in the century preceding the composition of the *Book of Curiosities*.

The mapmaker of the *Book of Curiosities* incorporated these reports—whether coming from Ibn al-Faqīh, from Ibn Sulaym, or another source—by visualizing a western tributary that emerges from sand dunes in West Africa and joins the main branch of the Nile. Positing a western tributary also solved another problem, which was al-Khwārazmī's confusion regarding the location of Zaghāwa and Fezzan. As we have seen, al-

Khwārazmī located Zaghāwa and Fezzan on the banks of the Nile (in his text), and on the eastern tributary (in his map). But by the Fatimid period, at the time the *Book of Curiosities* was composed, these localities were known to lie on the route to West Africa. The presumption of a western tributary of the Nile in the *Book of Curiosities* enables the mapmaker to locate Zaghāwa and Fezzan on the banks of this Western Nile, which also flows through Ghāna and Gao.

Alongside the western tributary of the Nile, two other tributaries are indicated. The lost part of the map included as eastern tributary, the equivalent of the Ptolemaic Astapos. This is known to us not only from comparison with al-Khwārazmī, but also from the the side panel, which states: "[The Nile] is joined by a river coming from the land of Zanj (East Africa), from a lake which is called the flask (*al-qārūrah*) and is also known as Lake Qanbalū." While Ptolemy was familiar with an eastern tributary of the Nile emerging from a lake in East Africa, the association of this lake with the island of Qanbalū (Pemba) in the Indian Ocean is novel. The final tributary of the Nile in the *Book of Curiosities* is the Fayyum arm to the west of the Nile, which is not found in al-Khwārazmī's map. Though some of the Fayyum arm is lost in our copy, the remnant is remarkably similar to the representation of the Fayyum in the Nile maps of Ibn Ḥawqal.[31]

Beyond its new conception of the Nile, this map is of paramount importance to the history of mathematical geography. Very unusually, some of the labels on the Nile map in the *Book of Curiosities* include data on longitude and latitude. Thus, the rivers that flow from the Mountain of the Moon are said to be located between 46° and 59° longitude. Longitude and latitude data is also given for the Fayyum. Both the diameters of the lakes at the source of the Nile and the distances between the tributaries of the Nile are indicated by celestial (*falakīyah*) degrees, which are then converted into miles. For example, the diameter of the eastern marsh is given as "five celestial degrees, equivalent to 284 miles."

Indications of geographical coordinates on medieval Islamic maps are very rare. A later example comes from the regional maps in one of the manuscripts of al-Idrīsī's work, where longitude and latitude coordinates are provided for 34 localities, almost all in the first clime, just north of the equator. Some, although not all, are similarly associated with the sources of the Nile.[32] The values given on the map of al-Idrīsī are broadly similar to those in the *Book of Curiosities*. In the Nile map of the *Book of Curiosities* the center of the largest marsh is given as 58° (possibly a mistake for 53°) longitude and 2° latitude, whereas al-Idrīsī gives the longitude of the Nile at the equator at 53°, and the longitude of the Smaller Lake is given at 56°.[33]

The indications of longitude and latitude on the map of the Nile in the *Book of Curiosities* offer an undeniable link with mathematical geography. They do not, however, directly derive from the work of al-Khwārazmī, who does not include coordinates on his map of the Nile, and the values in his tables differ from those on the *Book of Curiosities*' map. For example, the coordinates for the Fayyūm given on the Nile map are longitude 48° 5′ and latitude 30°, but al-Khwārazmī in his tables gives longitude 61° 55′ and latitude 28°. Neither do the coordinate values for the Fayyum in the *Book of Curiosities* correspond to the values in Ptolemy's *Geography*, although they are not too far from the values for the Fayyum (or Arsinoe) given in the commentary by Théon of Alexandria (4th century) on the Handy Tables of Ptolemy (longitude 48° 20′, latitude 31° 20′).[34]

Like the rectangular world map, the Nile map is itself not plotted, but it is likely to be copied from at least a partially plotted Late Antique prototype. That prototype reproduced the Ptolemaic conception of the sources of the Nile, his division into climes, and some plotting of localities according to known coordinates. The author of the *Book of Curiosities* had access to that prototype not through al-Khwārazmī, but through the continued circulation of variants of this prototype well into the eleventh century. Such a map of the Nile was also known to al-Masʿūdī. In his own account of the Nile, he also cites a set of coordinates not derived from al-Khwārazmī, and converts them into distances in a manner that parallels the *Book of Curiosities*. Thus, for example, al-Masʿūdī locates the Mountain of the Moon at 7.5 degrees beyond the equator, which he converts to 141 2/3 farsakhs, and then to 425 miles.[35]

All this suggests that the map of the Nile in the *Book of Curiosities* was based on a Late Antique Nile map that maintained elements of mathematical geography, and that it was often linked to a world map of a similar provenance. This is unexpectedly confirmed by the presence of another version of this Nile map in the twelfth-century so-called "Ibn Ḥawqal III" set. As said above, this set includes an oval or semicircular world map introduced by the copyist of some Ibn Ḥawqal manuscripts no earlier than the sixth/twelfth century.[36] The only other map introduced by this copyist is a map of the Nile, shown below (fig. 4.5).[37] As Gerard Tibbetts observed, this diagram is a variation on al-Khwārazmī's map of the Nile and owes little to the Balkhī School. We find here the Mountain of the Moon, from which twin sets of five parallel branches feed into two parallel lakes and then a third, northern lake. This third lake is not smaller, as in al-Khwārazmī, but equal in size to the other two lakes, as in the *Book of Curiosities*. We also find an eastern, unlabeled lake as the source of an eastern tributary that meets the main branch north of Aswan, creating the now familiar trian-

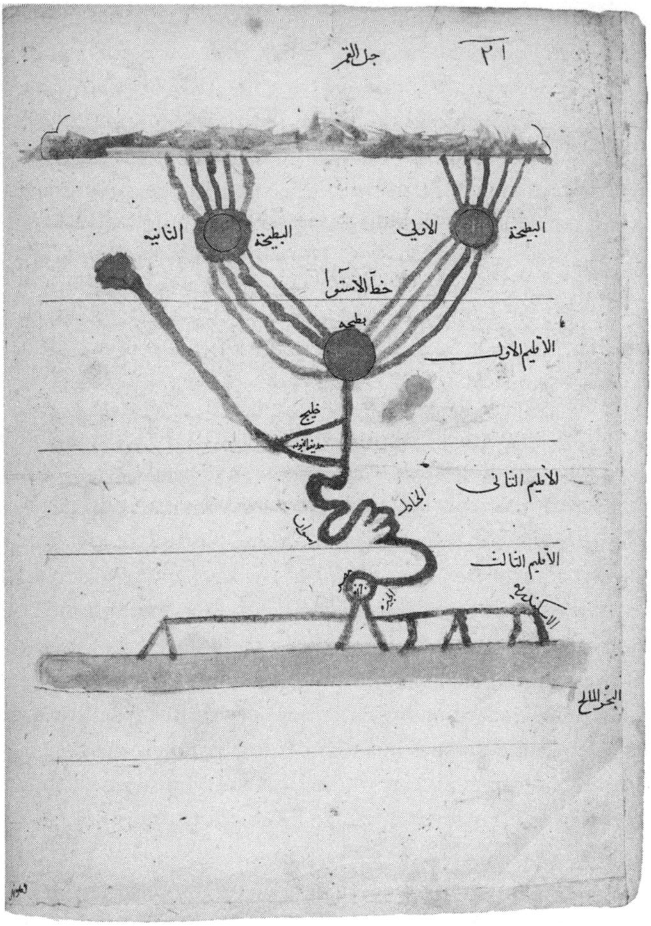

FIG. 4.5. Map of the Nile from Ibn Ḥawqal III set. Paris, BnF, MS arabe 2214, fol. 13b, copied 1445/847H. Reproduced with permission of the Bibliothèque nationale de France.

gular island, here labeled "City of the Nubians." The clime lines are also indicated in more or less the same position as on the Nile map of the *Book of Curiosities.*

The connection to the Nile map in the *Book of Curiosities* is confirmed by the addition of coordinate data that matches the values on the *Book of Curiosities* Nile map.[38] The copyist says he drew this map of the Nile on the basis of information found in a work called the *Geography*, from which he also drew longitude and latitude data. According to this, the center (*markaz*) of the third lake is at 53 (نج) degrees longitude and 2 degrees latitude. This is surely the same as the 58 (نح) degrees longitude and 2 degrees latitude reported as the location of the third lake in the *Book of Curiosities*, the difference being yet again the mere placement of a dot. The location of the eastern spring (the Koloe Lake) is given as 59 degrees longitude, on the equator—giving the same value as a label found in the Map of the Nile in the *Book of Curiosities.*[39] The copyist of the Ibn Ḥawqal III map continues with more coordinates for the course of the Nile further downstream, including for the delta, then refers the reader to the map of the Nile he has drawn on the opposite page. These similarities strongly suggest that the copyist of Ibn Ḥawqal III was using the same source used by the author of the *Book of Curiosities* some hundred years earlier. Since the value of the coordinates given in both sources is not the same as that found in al-Khwārazmī's work, they are clearly not derived from it, but rather come from an independent source.

We have seen in the previous chapter that the Ibn Ḥawqal III set, like the *Book of Curiosities,* preserves a world map that has elements that date back to Late Antiquity.[40] The same is true of the Nile maps preserved in these works. The maps of the Nile of Ibn Ḥawqal III are not directly derived from the *Book of Curiosities,* since they do not have a western tributary of the Nile. Yet there are striking similarities that suggest a common source independent of al-Khwārazmī. The Ibn Ḥawqal III distinctive pairing of an oval world map and a map of the Nile parallels the distinctive pair of maps—the rectangular world and Nile map—in the *Book of Curiosities.* This pair, a remnant of a Ptolemaic tradition of mathematical cartography, must have been circulating together well into the eleventh and twelfth centuries.

A Mountain in a Lake

The mapmaker of the *Book of Curiosities* is not content to merely reproduce Classical ideas. His most exciting maps are those in which he adapts inherited models to new information that arrived in the Islamic world. One of the folios of this manuscript includes a set of diagrams that illustrate the lakes at the sources of the Nile, part of a chapter on the lakes of the world, which visualizes a revolutionary conception of the annual cycle of the Nile flood (fig. 4.6 and plate 13). The melting snows on a large mountain placed in the middle of the large lake at the top are accorded a key role in the summer rise of the Nile waters, a concept that is explicitly credited to Coptic informants. The open-mouthed crocodile in the bottom lake represents the association between the eastern tributary of the Nile and the island of Pemba in the Indian Ocean. The significance and meaning of these diagrams, which are unique to this treatise, can only be understood by careful reading of the text and labels that surround them.

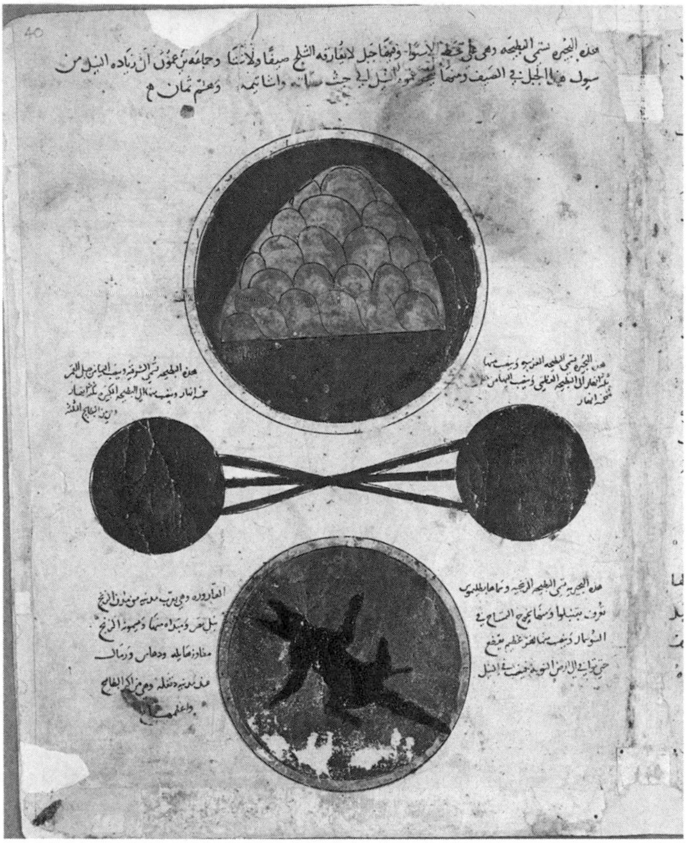

FIG. 4.6. Diagram of the Sources of the Nile from the *Book of Curiosities.* Oxford, Bodleian Library, MS Arab. c. 90, fol. 40a, copied ca. 1200.

The two central, smaller lakes on this folio are immediately recognizable as the twin, parallel lakes of the Ptolemaic system. The labels near them name them as the Eastern and Western marshes. The labels also state that each of the lakes is fed by five rivers that flow into it from the Mountain of the Moon. The labels then add that each of the two lakes issues three rivers that flow into the "Great Marsh." These twin sets of rivers are depicted in the diagram; the intention, in spite of the visual impression, is not to connect the two lakes to each other, but to show that they issue parallel sets of rivers that flow into the same locality. This is in general agreement with the al-Khwārazmī-Ptolemaic model, although al-Khwārazmī (map and text) has the two eastern and western marshes flow into what is described as the "Smaller Marsh," while here they flow into the "Great Marsh."

The "Great Marsh," which is the top, large lake, is described here as "the largest lake on the face of the Earth." According to the label, it is a marsh (*baṭīḥah*) on the equator, and the source of the Nile and its floods. Within it—as depicted in the diagram—there is "a mountain that is covered with snow during winter and summer. Most Copts maintain that the Sun, when in the summer it is at its zenith over this mountain, melts the snow away and causes the rise of the Nile and its continuous flow."[41] As noted above, a label on the Nile map of the treatise also describes a lofty mountain in the middle of this lake.

This "Great Marsh" is certainly the third, northern lake of the Ptolemaic system, but its function in the system of the Nile sources has been transformed. While in the Ptolemaic model it was the smallest lake of the tripartite system, it is here "the largest lake on the face of the Earth." This is connected to a shift in the function of this body of water. Rather than a collecting point for the waters that come from the Mountain of the Moon farther south, the northern lake now contains what appears to be the major source of the Nile, a mountain perennially covered with snow. This is clearly not the Ptolemaic Mountain of the Moon, but rather a mountain reported by Coptic informants. Remarkably, the Coptic informants claim that the Nile floods are a result of the summer melting of the snow over this mountain—an explanation that is as close to the actual Nile system as ever been achieved by medieval Islamic scholarship. This additional data is made to fit, rather uncomfortably, with the Ptolemaic system. The snow-covered mountain in the middle of the third lake effectively obviates a need for the Mountain of the Moon as a source for the Nile. Yet, the two mountains are mentioned together—although, interestingly, not shown together visually, perhaps because that would make the discrepancy between the two models too obvious.

The label describing the fourth, bottom, lake in this set of diagrams identifies it as the eastern lake of the Ptolemaic system, but links it to evolving contemporary knowledge of East Africa. The label states that this lake was known to Ptolemy as the Flask (*al-qārūrah*), but now it is known as the "Marsh of the Zanj." The lake, the label states, is found near one of the cities of the Zanj called Qanbalū, near the coasts of East Africa. It is also the source of the Nile crocodile, called by the Zanj *sūsmār* (which is in fact Persian for crocodile).

In another awkward attempt to reconcile the Ptolemaic model with more recent geographical notions, the eastern lake is associated with the island of Pemba (Qanbalū), but without showing an actual connection between the Nile and the Indian Ocean. As noted above, al-Masʿūdī claims to have seen a map of the Nile where—instead of an eastern tributary coming from a lake—an eastern arm of the Nile reached the Indian Ocean near East

Africa. He used this map to argue that an eastern arm of the Nile flows into the Indian Ocean near the island of Qanbalū, known to him as a major trading emporium.[42] Al-Masʿūdī repeats the claim of an eastern arm of the Nile flowing to the Indian Ocean later on in his work, citing a Coptic informant, and also claims that the Nile floods cause the water of the Indian Ocean to swell.[43] While al-Masʿūdī has the eastern arm of the Nile flow *from* the main arm into the Indian Ocean, the *Book of Curiosities* retains here the Ptolemaic conception of an eastern tributary that flows to the Nile from an eastern lake. Nonetheless, the association with new knowledge about the East African trading emporium of Qanbalū is visualized by the crocodile—the only depiction of a living organism in the treatise.

The Nile in the Rectangular World Map

In the rectangular world map of the *Book of Curiosities*, the new conceptualization of the Nile is set within a global framework (fig. 4.7). The "Mountain of the Moon," located on the equator, is again represented by a semicircular, parachute-like mountain from which ten streams diverge, five on either side. This is practically identical to its depiction in the separate Nile map in the treatise. Both maps share a decorative element, a circle enclosing a lunar crescent, drawn into the mountain itself, which is not found in al-Khwārazmī's map. A set of rivers flows from the Mountain of the Moon and pours into twin parallel marshes, or lakes, which in turn feed into a third lake before emerging as the River Nile. The third lake is as large as the other two, in line with the diagram of the sources of the Nile elsewhere in the treatise.

The western tributary of the Nile, flowing from sand dunes in West Africa, on the equator, is also depicted. The visual rendering of the sand dunes and the western tributary is virtually identical to that found on the separate Nile map, and the label is very similar: "This is the white sand dune from which springs and marshes gush forth, flowing into the Nile." It seems certain that the two maps—the Nile map and the rectangular world map—were made at the same time and in relation to each other.

Curiously, the world map shows two arms of the Nile that flow into the Red Sea or the Indian Ocean—a feature not found in any other extant map of the Nile. The upper—that is, the southern—branch is undoubtedly the eastern tributary of the Nile, the Ptolemaic Astapos. As in al-Khwārazmī's map, a distinctive large island is formed at the confluence of this river with the main branch of the Nile. The label on this island has "Island of Suwaydah," the latter word likely to be a mistake for *al-Nūbah*, "Island of the Nubians," as on Ibn Ḥawqal's map of the Nile.[44] As Seignobos points out,

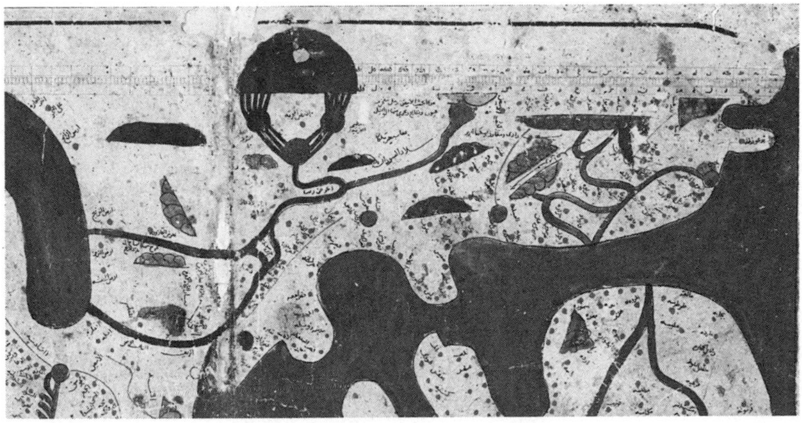

FIG. 4.7. Detail of the Nile in the world map of the *Book of Curiosities*. Oxford, Bodleian Library, MS Arab. c. 90, fols. 23b–24a, copied ca. 1200.

this label must refer to the "City of the Nubians" located on this island by al-Khwārazmī.[45]

Although the rectangular world map shows this eastern river to flow directly to the Indian Ocean, the labels suggest that an eastern lake (the Ptolemaic Koloe) is assumed, even if it is not presented visually. One label describes the river as "a river from the *Qārūrah* Lake," and another label has "the crocodile comes from it to the lands of the Zanj," referring to the "Lake of the Crocodile" in the lake diagrams discussed above. But, alternatively, it is also possible that the mapmaker was here laying out visually al-Masʿūdī's novel idea of a direct link between the Nile and the Indian Ocean. A notion of an arm of the Nile flowing into the Indian Ocean sits here uneasily with a Ptolemaic notion of an eastern tributary.

Most peculiarly, the rectangular world map has also another branch of the Nile, flowing in the northeast into—or from—the Red Sea, near Qulzum (modern Suez). This branch has no parallel in al-Khwārazmī's map or in any of the other maps in the treatise. It may well be a copyist mistake. But if it is not a mistake, then it may be an attempt to depict the Nile–Red Sea canal, originally excavated by Greek and Roman rulers, and re-excavated by the Arab conquerors in the seventh century. It was finally blocked under orders of the Abbasid caliph al-Manṣūr (AD 754–75).[46] Such a possibility is tantalizing, as no other medieval map known to us shows this man-made fit of engineering, which became obsolete in the very early Islamic period. In the absence of any textual evidence in the rest of the treatise, and given the corrupt nature of the rectangular world map as a whole, this must remain a speculation.

The Western Nile and the Circular World Map

The medieval image of the sources of the Nile crystallizes with al-Idrīsī, the most influential of all Islamic mapmakers, who worked for Roger II of Sicily in the mid-twelfth century. Like our anonymous author a century earlier, al-Idrīsī was heavily indebted to the Ptolemaic tradition of cartography, but attempted to adapt the Ptolemaic conception of the Nile to new contemporary data available to Muslim geographers. [47] Some of al-Idrīsī's cartographic solutions to the contradictions of the Ptolemaic tradition are similar to those found in the *Book of Curiosities*, even if there is no evidence to suggest that the *Book of Curiosities* was available to him. Both envisage a large mountain located within the third (northernmost) lake, and a western Nile in West Africa. But al-Idrīsī's most distinctive contribution, which has shaped much later cartography in the Middle East and in Europe, is to reimagine the western Nile as an arm that flows from the sources of the Nile toward the Atlantic Ocean.

The depiction of the Nile in al-Idrīsī's sectional maps can be seen in the collation of the sectional maps of the Istanbul Ayasofya 3502 manuscript (fig. 4.8),[48] and in the online resource offered by the Bibliothéque nationale de France.[49] The Mountain of the Moon and the tripartite lake system lie just beyond the equator. An eastern lake, also on the equator, is the source of an eastern tributary of the Nile, whose confluence with the main branch creates a large island. The depiction of the lower, Egyptian sections of the Nile, the two parallel mountain ranges on each side of the river and, especially, the detailed delta, owe much to Ibn Ḥawqal. Following Ibn Ḥawqal and the *Book of Curiosities*, the Fayyum arm (Baḥr Yūsuf) is also prominently indicated.

The most distinctive aspect in al-Idrīsī's depiction of the Nile, as in the maps of the *Book of Curiosities*, is a Western Nile. But while in the *Book of Curiosities* the Western Nile was assumed to originate in sand dunes in West Africa (a visualization of a tradition attributed to the early Arab conquerors of the seventh century), al-Idrīsī has no mention of this tradition or of the sand dunes. Instead, he makes this western Nile flow all the way to the Atlantic Ocean, as an arm of the main Nile, not a tributary. This transformation of a West African tributary into an Atlantic arm is linked to a change in the function of the third, northernmost lake of the tripartite Ptolemaic system. In the *Book of Curiosities* the third lake, which used to be the Small Lake of the Ptolemaic tradition, has become "one of the largest on Earth," and a large mountain now appeared in the middle of the lake. Al-Idrīsī's maps, like those of the *Book of Curiosities*, show the third lake as larger than the other two, and also introduce a mountain that protrudes into the lake.

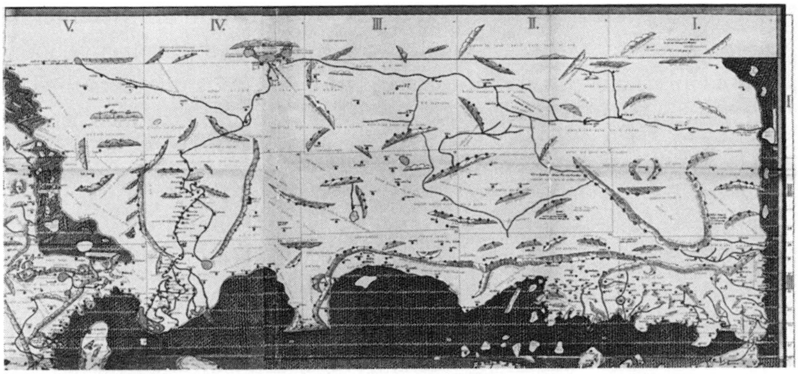

FIG. 4.8. The Nile in the regional maps of al-Idrīsī. South is at the top. This composite map draws together the different regional maps of al-Idrīsī's *Nuzhat al-mushtāq li-ikhtirāq al-āfāq*. It was prepared by Konrad Miller as *Charta Rogeriana—Weltkarte des Idrisi—vom Jahr 1154 n. Ch.* (Stuttgart, 1928).

But al-Idrīsī, who is far more systematic than the author of the *Book of Curiosities,* transforms the function of this mountain. In his text, the mountain is not a source of the Nile, but rather a barrier that splits the lake in two. As a result, the Nile bifurcates into the Western Nile, Nīl al-Sūdān, in the west and the main Nile to the north.[50] Al-Idrīsī's maps show a mountain range protruding into the third Ptolemaic lake at a northwestern angle, causing the Western Nile to flow to the west and the main branch to the north.[51]

Al-Idrīsī solves the unexplained contradictions caused by the western Nile by suggesting that it originates from the same source as the main Nile, and not from beneath sand dunes. Al-Idrīsī also alters the function of the prominent mountain range that protrudes into the third, northern lake. In the *Book of Curiosities,* this mountain is, as reported by contemporary Coptic informants, a source of the Nile, whose melting snow causes the annual floods. Its presence makes the Ptolemaic Mountain of the Moon redundant. This contradiction is solved by al-Idrīsī by retaining the Mountain of the Moon, while transforming the new mountain range from a source of the Nile to a barrier that separates the Western Nile from the main branch.

The circular world maps that accompany the manuscripts of al-Idrīsī's work have become the most familiar images of medieval Islamic cartography (see plate 4 and fig. 1.10). They appear in several copies of al-Idrīsī's treatise, although the text of the work does not mention such a world map, and they are always stand-alone artifacts placed at the front of a volume, before the start of the text.[52] The Nile on the circular world maps of the Idrīsī tradition is in keeping with its depiction on his sectional maps (fig. 4.8).[53] In all copies, the Mountain of the Moon is immediately recognizable at the top of the map, beyond the equator, together with the tripartite lake system. The third lake is as large as the other two, and the distinctive northwestern mountain range—al-Idrīsī's invention—protrudes into the lake and splits it up, with a western Nile flowing toward the Atlantic. A western Fayyum arm is also invariably depicted, terminating in Lake Qārūn, as it is shown in the sectional maps of al-Idrīsī's treatise. The eastern tributary, however, which was part of the original Ptolemaic concept, is for some reason absent from most extant copies of the al-Idrīsī world maps made in the fifteenth century or later.[54]

The developing features of the Nile in the circular world maps of the al-Idrīsī tradition offer an interesting comparison with the circular world map in the Bodleian copy of the *Book of Curiosities* (see fig. 1.11 above and plate 3). This map may not be an integral part of the original treatise, but added no later than the twelfth or thirteenth centuries, when the Bodleian copy was made. (This is a matter of a long-standing debate between the two coauthors of this book, as discussed above in chapter 1, "A Discovery.")[55] Even if a later addition to the Fatimid *Book of Curiosities*, this would still be the earliest surviving copy of this type of world map (the previously earliest copy, in the Bibliothéque nationale de France, was probably copied around 1300). The circular world map of the *Book of Curiosities* shows the three distinctive features of al-Idrīsī's Nile: the Mountain of the Moon and its three lakes, an Atlantic Nile, and a diagonal mountain range that splits the third lake into western and northern rivers. Here, however, the diagonal mountain range is named "the mountain of the dune sands,"[56] possibly in an attempt to link this circular world map to the other maps of the *Book of Curiosities*, which include "white sand dunes" as a source of the western Nile.[57]

*

Al-Idrīsī presents the most systematic attempt to resolve the tension between a Ptolemaic conception of the Nile and the knowledge of the African interior available to Muslim scholars. It is worth pointing out how much influence the Ptolemaic—or, to put it more precisely—Late Antique models continued to exert on the Islamic cartography of the Nile. The Ptolemaic parachute-shaped Mountain of the Moon, with its three lakes, is the most identifiable design on most medieval Islamic world maps on display, as well as on the earliest extant Islamic map found in a copy of al-Khwārazmī's treatise. While the Mountain of the Moon was mostly ignored by the Balkhī School mapmakers in the tenth century, this was just a short interlude. By the end of the tenth century Ibn Ḥawqal already shows interest in the Ptolemaic configuration of the Nile, and by the first half of the eleventh century the anonymous author of the Fatimid *Book of Curiosities* reintroduces the Mountain of the Moon, the three lakes, and the eastern tributary (the Ptolemaic Astapos). The adoption of the Ptolemaic model by al-Idrīsī a century later is therefore a continuation of an existing trend.

New geographical data led to adjustments of the Ptolemaic model. The ever-changing delta of the Nile was subject to rather detailed mapping, and the Fayyum arm, first shown by Ibn Ḥawqal in the tenth century, was incorporated by al-Idrīsī in the twelfth. But the new data could also be reinterpreted in order to fit into the Ptolemaic model. A most intriguing example is that of Coptic informants quoted in the *Book of Curiosities*, who report that the actual source of the Nile is a snow-covered mountain in the middle of a large lake, and that the melting of the snow is the cause of the Nile's flood. The mapmaker in the *Book of Curiosities* attempts to incorporate this information into his maps, but he does so in addition to—rather than instead of—the Ptolemaic Mountain of the Moon. Al-Idrīsī solves the apparent contradiction by accepting the existence of a mountain in a lake, but assigning it a different role: it is not a source of the Nile, but rather a barrier between the main branch of the Nile and the imagined western arm.

The Western Nile, the *Nīl al-Sūdān*, is the most significant Islamic adaptation to the Ptolemaic model. It is first imagined by Islamic cartographers within the imperial framework of Muslim conquest. A report attributed to an Arab general on expedition in West Africa establishes the emergence of the Nile from under sand dunes in the desert, in a way that parallels accounts of the expeditions to find the Wall of Gog and Magog. Both accounts also have links to pre-Islamic myths.[58] The parallels are clearly visible in the rectangular world map of the *Book of Curiosities*, where these two designs—the sand dunes at the far southwest and the Wall of Gog and Magog in the far northeast—represent the edges of civilization.

In the *Book of Curiosities*, the rectangular world map and the map of the river Nile stand out as the two maps that retain the most elements of mathematical geography—in particular, the scale in the former and the longitude and latitude data in the latter. These two maps are also very closely related to the maps produced in the ninth century by al-Khwārazmī, especially the map of the Nile and his diagram of the Island of the Jewel. Al-Masʿūdī in the tenth century gives a textual account of this pair of maps that he had seen in a work entitled the *Geography*—one world map in Greek, and another map of the Nile. The pair of maps in the *Book of Curiosities* also anticipates the maps of the world and of the Nile introduced into several copies of the Ibn Ḥawqal corpus (a set known as Ibn Ḥawqal III) no earlier than the mid-twelfth century. While the maps in these treatises may be independent of each other, they all employ tools of mathmatical geography and appear to be derived from a common source, a Ptolemaic work called the *Geography*. This pair of two Late Antique maps survived into the Islamic era—the world map came with a map of the Nile, a distinct couple that belonged in the tradition of mathematical geography and that continued to be available to Muslim scholars at least up to the twelfth century.

CHAPTER FIVE

The View from the Sea

NAVIGATION AND REPRESENTATION OF MARITIME SPACE

Chapter 16 of the *Book of Curiosities*, "On the Depiction of Inlets—that is, Bays—in particular the Bays of Byzantium," opens with an account of five bays in southwestern Anatolia. The sequence begins with two bays facing the island of Rhodes and goes up the western Anatolian coast to the Bay of Miletos, near the mouth of the river Meander (Menderes). The account of the five bays is as follows:[1]

> The Small *Ṭrakhīyah* [Tracheia] Bay. This bay is 12 miles long and its entrance is 3 miles wide. One enters it with southern winds proceeding northeast.
>
> The Large *Ṭrakhīyah* [Tracheia] Bay. This bay is 30 miles long and its entrance is 6 miles wide. One enters it from the south proceeding northwest. There is an uninhabited island at its end.
>
> Bay of *Kāramū* [Kerameios]. This bay is 70 (?)[2] miles long and is 20 miles wide. One enters it with southern winds proceeding northward.
>
> Bay of *Mūlaṣā* [Mylasa]. This bay is 50 miles long, and its entrance is 25 miles wide. One enters it with southern winds proceeding northward.
>
> Bay of *Miyāṭayū* [Miletos]. This bay is 6 miles long, and its entrance is 20 miles wide. The fortress of *Malīṭayū* [Miletos] is in the middle of the bay. To its west there is a river that flows into the sea (River Meander).

This is evidently a navigation guide. The description of the coast is from the point of view of a mariner, and any maritime historian would find here broad similarities with other premodern navigation guides, such as the Greco-Roman *periploi* and late medieval portolan texts. For each bay, the text provides essential information regarding length, given in miles, the width of the entrance, also in miles, and the direction of the bay according to an eight-part wind-rose. The information regarding size and direction is

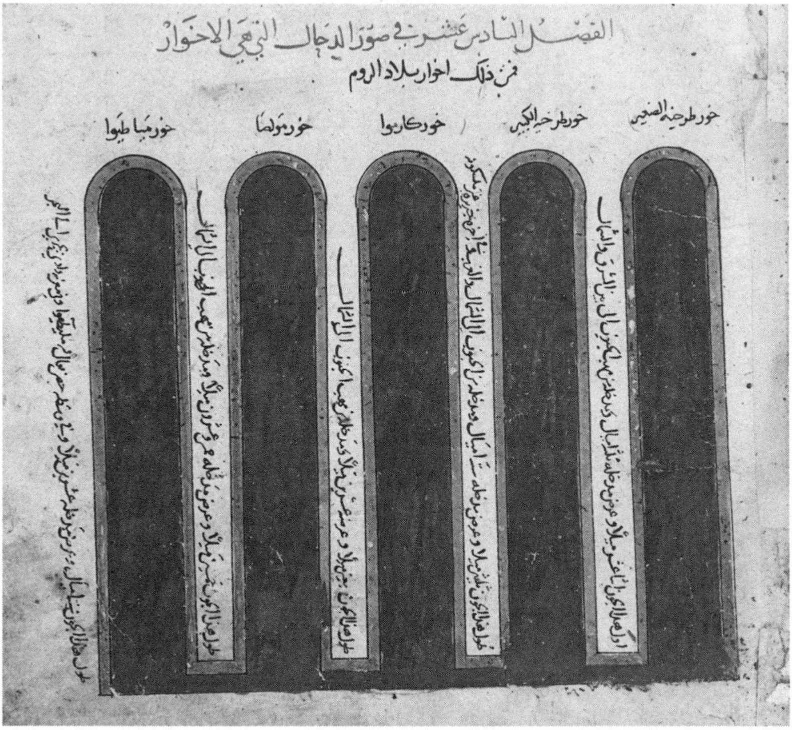

FIG. 5.1. Illustration of the Bays of Byzantium from the *Book of Curiosities*. Oxford, Bodleian Library, MS Arab. c. 90, fol. 38a, copied ca. 1200.

so systematic you could imagine it translated into a diagram of the curves of the Anatolian coast.

In fact, the author of the *Book of Curiosities* does present this data visually, but not in the manner we would expect him to (fig. 5.1). In a move that captures the difference between our modern expectations of maps and that of eleventh-century audiences, the mapmaker has drawn five perfectly identical fingerlike bays. Although the size and direction of the coasts are systematically listed in the labels of the diagram, they are not transformed into visual scale and orientation. Then, by the next folio, the author gives up on the visualization of the data. He continues with accounts of another twenty-three bays in the Aegean Sea and along the Peloponnesus, with data on size, scale, and key topographical features, all presented as text only, and not accompanied by any diagram.

This account of the bays of the Aegean provides us with the earliest and richest navigation guide that has come to us from any medieval Muslim source. While other medieval texts that discuss sea travel are written from the perspective of merchants, this account foregrounds navigation at the expense of any other information. There is nothing here about goods or markets, nothing of political or administrative divisions, very little on the cities and people who inhabit lands that are not immediately visible from aboard a ship. The *Book of Curiosities* and its lavish illustrations were not meant to be taken to sea, but the text here could only have originated with crews of eastern Mediterranean vessels.

The *Book of Curiosities* is a treatise about the sea, about maps of the sea, and in particular about navigation and travel in the Mediterranean. One of the more iconic maps of the *Book of Curiosities* is a carefully executed oval diagram of the Mediterranean, dense with hundreds of harbors and islands, but unrecognizably abstract (fig. 5.3). It is closely related with another very abstract map, one that presents the island of Cyprus as a perfect square (fig. 5.5). In what appears at first to be a paradox, this absolute abstraction comes with unprecedented wealth of material on the quality and size of anchorages and harbors, sailing distances, water sources, and wind directions. Other examples of such material on navigation in medieval Arabic geographic literature, such as the later maritime itineraries by the Andalusi al-Bakrī (d. 1094) and the North African al-Idrīsī (d. 1165), are both less detailed and purely textual.[3]

Taken together, these texts and diagrams are of major interest for the history of navigation and maritime charts before the portolan charts of the later Middle Ages, not only in Islam but in the Mediterranean at large. The *Book of Curiosities* gives us an unprecedented glimpse into records made by medieval Muslim mariners. What had they found most important to record, and for what purposes? How did they arrange the navigation material they recorded? And, most intriguingly for the history of maritime cartography, how did they represent the material visually? As is demonstrated by the example of the diagram of the Aegean Bays, but is also apparent in the maps of the Mediterranean and Cyprus, the navigational data was presented in the form of schematic diagrams. The omission of scale and orientation from the maritime diagrams—even though data on distances and directions was readily available—is our key to understanding these images.

A Mediterranean of Ovals, Circles, and Squares

The mapmaker of the *Book of Curiosities* was not the first Muslim scholar to draw a map of the Mediterranean. Many maps by Balkhī School geographers, like the one by al-Iṣṭakhrī shown here in fig. 5.2, have the Mediterranean as a circle with a wide opening to the Atlantic, with indications

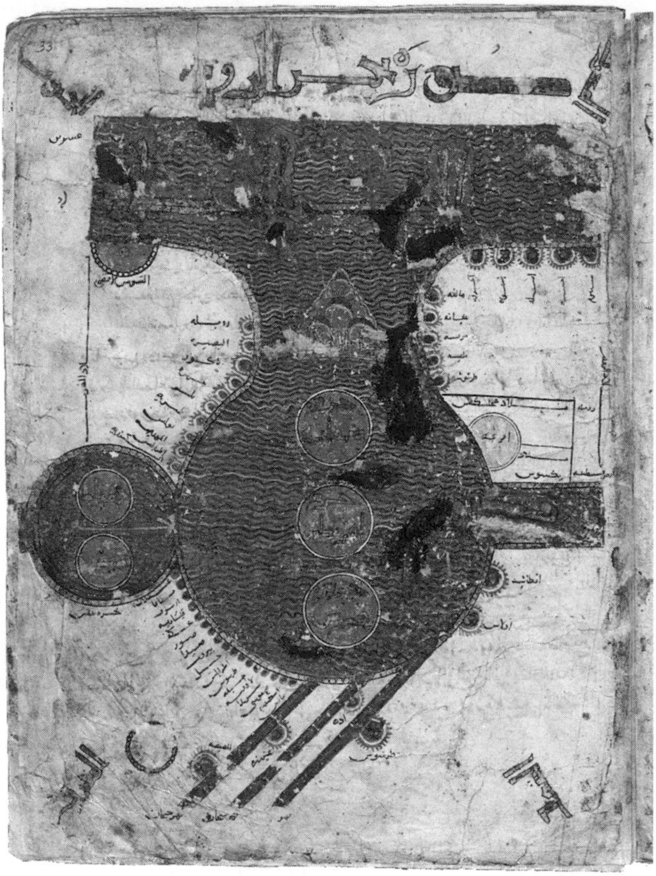

FIG. 5.2. Map of the Mediterranean from al-Iṣṭakhrī's *Kitāb al-Masālik wa-al-mamālik*. West is at the top. Leiden University Library, MS Or. 3101, fol. 33a, copied 1173/569H. Reproduced courtesy of the Leiden University Library.

of key prominent gulfs and islands. The maps of Ibn Ḥawqal, in different recensions, depict the Mediterranean in a more recognizable form. His representation of the Mediterranean is more elaborate, more detailed and less schematic than that of al-Iṣṭakhrī, with numerous peninsulas and islands indicated, and attention paid to both scale and orientation (see fig. 3.5 above).[4] We know that the author of the *Book of Curiosities* had access to Ibn Ḥawqal's map of the Mediterranean (which Ibn Ḥawqal actually calls the Map of the Maghreb), as he copied most of its labels into his rectangular world map.

The map of the Mediterranean in the *Book of Curiosities* is completely different (fig. 5.3 and plate 5). It entirely eschews Ibn Ḥawqal's attempts to

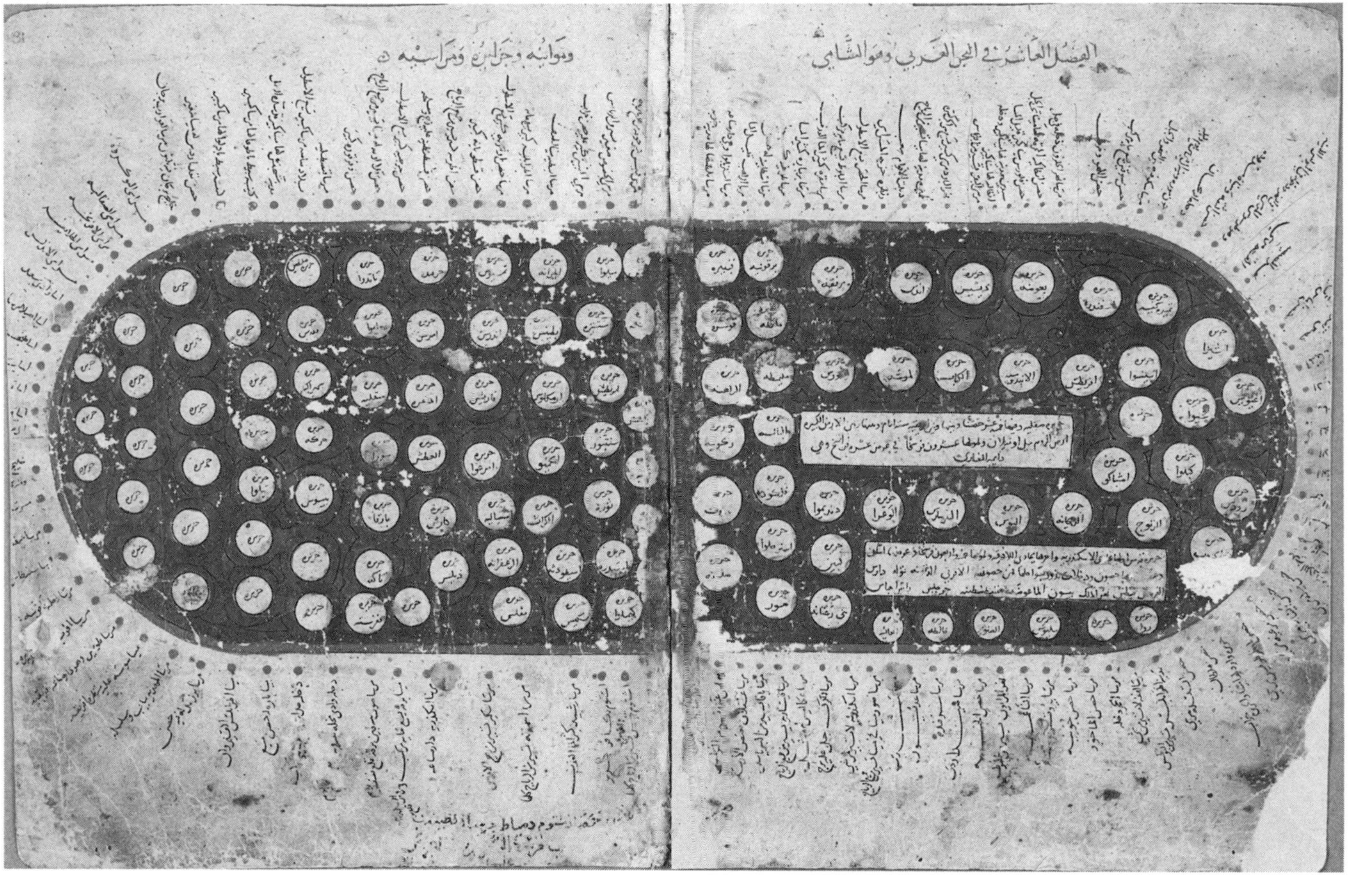

FIG. 5.3. Map of the Mediterranean from the *Book of Curiosities*. Oxford, Bodleian Library, MS Arab. c. 90, fol. 30b–31a, copied ca. 1200.

draw the Iberian Peninsula or Italy, and its Mediterranean is even more abstract than that of al-Iṣṭakhrī's. In line with almost all other maps in the treatise, there is no attempt to trace any of the actual coastlines. Not even the Iberian Peninsula, which is very prominent on earlier Islamic maps of the Mediterranean, is represented.[5] But it is not only the design which is dissimilar. The difference goes to the very purpose of these maps. Al-Iṣṭakhrī and Ibn Ḥawqal's interest is primarily the land, not the sea, and their labels carry no navigational material whatsoever. The map of the Mediterranean of the *Book of Curiosities*, on the other hand, is a depiction of maritime space. The focus is the coasts to the exclusion of any inland features, and, even more remarkably, on navigation material provided in the text of the labels. While earlier maps of the Mediterranean are armchair attempts to give general geographic orientation, this unique map is the first surviving example of a map looking at the coasts from the sea, rather than the other way around.

The red title across the top of this map reads: "The Tenth Chapter: The Western Sea—That Is, the Syrian Sea—and Its Harbors, Islands and Anchorages." As the title indicates, there is nothing else in this Mediterranean but islands and mooring points: there are 118 islands within the dark-green sea, and 121 harbors and anchorages on the rim. The oval seems to be orientated to the north. The Straits of Gibraltar, indicated by a thin red line, are at the far left of the oval. The harbors and anchorages of Western Europe and Anatolia are in the upper half, while those of Palestine, Egypt, and North Africa are at the bottom.

Proceeding clockwise from Gibraltar on the far left, the next seven harbors above the Straits are anchorages on the Atlantic coast of Morocco, such as Tangier. Each of these seven labels begins with the word "toward," indicating that the localities are not actually on the Mediterranean rim. Thereafter, the mapmaker briefly alludes to the harbors of Muslim Spain and Europe. "The Gulf of Burjān," that is the Black Sea, is indicated by label only.[6]

What follows is the most detailed list of Byzantine anchorages found in any medieval source before the appearance of the Italian portolan charts in the late thirteenth century. The harbors in Western Anatolia and the Dardanelles appear first, and are difficult to identify with certainty. The sequence probably includes Atarneus, Erythrai, and Trogilion, each of them facing one of the major Micro-Asiatic islands—Lesbos, Chios, and Samos, respectively. We are on firmer ground once we reach the Byzantine harbor of Strobilos, on the southwestern tip of Anatolia. Strobilos, opposite the island of Kos, is known to have been an important naval and military post in the middle Byzantine period.[7] After Strobilos, continuing clockwise, the map lists more than thirty anchorages and harbors on the coasts of south-

FIG. 5.4. The Mediterranean circa 1020–1050.

ern Anatolia, including the major Byzantine ports of Attaleia and Tarsos. A few thin red lines in between indicate rivers flowing into the sea. The southern coasts of Anatolia, it should be emphasized, were during this period under firm Byzantine control.

The detailed list continues with ports and harbors under Muslim control, starting with the region around Latakia. This was the frontier between the Byzantines and Fatimids during the first half of the eleventh century, but the mapmaker made no visual or textual attempt to indicate this political boundary (fig. 5.4). The dense account of anchorages and harbors continues, listing thirty-five place-names from Latakia down the Syrian and Palestinian coasts to the port of Alexandria. West of Alexandria, however, the North African coast is represented with far less detail. Few major North African ports are mentioned, such as Barka, Tripoli, and Mahdia. Taken as whole, the anchorages and harbors of Latin Europe and Byzantium occupy nearly the entire upper half of the oval and the Islamic ones the lower half.

The labels for the harbors provide unique information on capacity of harbors and anchorages, defensive installations, and protection from prevailing winds. With regard to capacity, many harbors are described as large or small. In a few cases, the information is more specific. One harbor in the western coasts of Anatolia, possibly Trogilion, is said to be able to accommodate one hundred ships. The anchorages "of the Oak" and of Syke in southern Anatolia, as well as Barka on the Libyan coast, have similar capacity. Two unidentified harbors on the North African coast are said to accommodate two hundred ships. Four Byzantine harbors are said to be able to accommodate a fleet (*usṭūl*, from the Greek *stylos*). These are Sāsah (probably Sestos in the Dardanelles), Jurjiyah (probably Agios Georgios at the head of the Gallipoli peninsula),[8] Erythrai in the Aegean coast of Anatolia, and an unidentified anchorage north of Patara. While the term *usṭūl* sometimes designated an individual heavy galley, here it is certainly used in its original meaning of a fleet or convoy of military ships. Such usage is also attested in the contemporary chronicle of Yaḥyā of Antioch.[9] Indications of size or capacity are provided for anchorages and harbors on the coasts of Anatolia and North Africa, but not for the harbors of Egypt and Syria, which were under Fatimid control.

The labels also mention defensive aspects of the anchorage points, and especially the existence of a fort (*ḥiṣn*), found in most Byzantine and Muslim harbors. In the case of Mahdia, the existence of a gate and a chain is mentioned. This system was common to a number of the major ports of the medieval Mediterranean, and is illustrated visually in the *Book of Curiosities* on the maps of Mahdia and Palermo, to be discussed in the next chapter.[10] Arsenals, used for the construction and storage of military galleys, are mentioned for Byzantine Strobilos in southwest Anatolia, as well as for Alexandria and Tunis. The listing of arsenals on this Mediterranean map is not meant to be comprehensive. The author himself was aware of the existence of arsenals in Tinnīs and Palermo, which he indicates on their separate maps.[11] The map of the Mediterranean also mentions an armory situated at Constantinople and a watchtower at Sousse, probably referring to the well-known *ribāṭ* of the city.

Protection from prevailing winds is mentioned for a large number of harbors and anchorages, using Greek nomenclature for wind navigation not otherwise attested in any Arabic medieval sources.[12] Some of the best anchorages—such as Tyre, Caesarea, and Tinnīs—are said to offer protection from all winds. Most, however, offer protection only from certain winds, whose names are drawn from a four-point Greek system of wind navigation. For example, the mapmaker notes that the anchorages of Iasos and Jaffa offer protection from the *Boreas* (north) wind, while the harbors of Tripoli (Syria) and Gaza offer protection from the *Notos* (south) wind. The direction of wind required to enter a harbor, on the other hand, is only noted for the harbor of Tarsos, which is to be entered "with a gentle *Boreas* (north) wind."[13] This is in keeping with the accounts of al-Idrīsī and al-Bakrī, who occasionally mention the protection from the wind, or lack thereof, offered by harbors on the North African coast. But their accounts are less systematic, and they employ Arabic terms, not Greek ones.[14]

The Mediterranean map in the *Book of Curiosities* also occasionally indicates quality of harbors and availability of freshwater. A couple of harbors are described as being blocked: the harbor of Sh.j.n.s, probably Sigeion at the southern entrance to the Dardanelles, has been filled with sand, while the harbor of the Fortress of al-Thiqah, perhaps Thekla, near Tarsos is said to be blocked. The quality of the harbors of Tyre and Acre is praised, while Sidon's is said to be poor. Information on the availability of freshwater is provided for a few harbors, mostly in southern Anatolia. The anchorage of "the Monk," next to Strobilos in southwest Anatolia, is described as having "little water," while Marmaris and Makre, farther to the east, have much. Kalonoros, modern Alanya, is described as "a large harbor with little water." In Fatimid lands, Ascalon has access to a stream of water, and Rosetta has freshwater in abundance.

Sailing distances are given for travel in the open seas, measured in days and nights of sailing, but distances between neighboring anchorages are never given. Mylai, on the southern Anatolian coast, is said to be one day and one night sailing from Cyprus. This is obviously a case of sailing the open seas rather than coast-hugging. Sailing distances are also given for a westward North African route, starting at Alexandria. Barka is said to one's day sailing from Alexandria, Sirte a further half a day's sailing from Barka, and the inlet of *Wādī Maḥlah* is one day's sailing west from Sirte. In this North African route too, the sailing would presumably be away from the coasts, although the sailing times cited for the routes Alexandria-Barka and Barka-Sirte are incredibly short, and cannot be correct. In all these examples, the direction of travel is conceived as clockwise, from Anatolia to Cyprus, and from Egypt toward North Africa. This is certainly because prevailing winds and currents in the Mediterranean require counterclockwise vessels to sail along the coasts, but allowed some direct travel in the opposite direction. Going from Alexandria to Constantinople required coasting, but one could travel in the open seas from the Aegean to Alexandria, or, as is the case here, from Anatolia to Cyprus, or from Alexandria toward North Africa.[15]

As for the islands in the interior of the oval, their position bears no relationship to the sequence of harbors on the rim. The Aegean islands are clustered to the left of the gutter, while the groups of islands surrounding Sicily are clustered to the right. The Aegean islands were entered by columns, as it is clear from several groupings of geographically adjacent islands which appear one on top of the other. For example, Amorgos is on top of

Astipalaia; Agathonissi on top of Lipsi (these are two tiny islands near the coasts of western Anatolia); and Lesbos, Tenedos and Samothrace are all in the same column. Malta and several islands of the Aeolian, Egadi, and Palagean groups of islands are all located above or to the left of the large rectangle which indicates Sicily, although there is no clear pattern to the sequence of the islands. The major islands of the western Mediterranean, such as Corsica and Sardinia, are not on this map at all.

Unlike the wealth of information on harbors and anchorages on the rim of the oval, almost all the islands in the middle of the map are merely mentioned by name and represented as small circles of equal size. Only Sicily and Cyprus, represented as rectangles, have longer descriptive labels, which correspond to some of the material on the individual maps of these two islands in the following folios. The abstraction is extreme on the far left: The islands in the four left columns are each labeled merely "island" and given no names.

The marked discrepancy between the rich description of mainland ports and the terse listing of island names makes little sense for the purpose of navigation, as many of the islands had serviceable harbors of their own. There is no doubt that the author was familiar with the importance of ports located on islands. Elsewhere in the treatise he provides short descriptions of Rhodes, Crete, and the small islands of Khalki and Tilos, just north of Rhodes. For Rhodes, the information pertains to navigation in the same manner as most labels on the Mediterranean Map: "The harbor is found in the west of the island, and it gives protection from every wind. There is water in the harbor."

One striking visual account of an island's harbors and anchorages is a separate diagram of Cyprus, which is the earliest detailed map of the island known from any cartographic tradition (fig. 5.5 and plate 10).[16] It contains data for navigation that is similar in format to that given on the map of the Mediterranean, an extension of the general account of Mediterranean to the particulars of this one strategically important island. This map of Cyprus belongs to a chapter on "The Islands of the Infidels," devoted to Mediterranean and Indian Ocean islands in the hands of non-Muslims. That included Cyprus, which used to be shared between Muslim and Christian forces until it fell under full Byzantine control in 965.

The island is represented by a square surrounded on all four sides by a strip of green-colored sea. Twenty-five cells along the edges, or "coasts," are meant to represent anchorages or harbors, although some have been left empty. Nine additional anchorages or harbors are represented by a strip of nine cells in the middle of the square, the strip entitled "the names of the remainder of its harbors." There seems little doubt that the mapmaker was

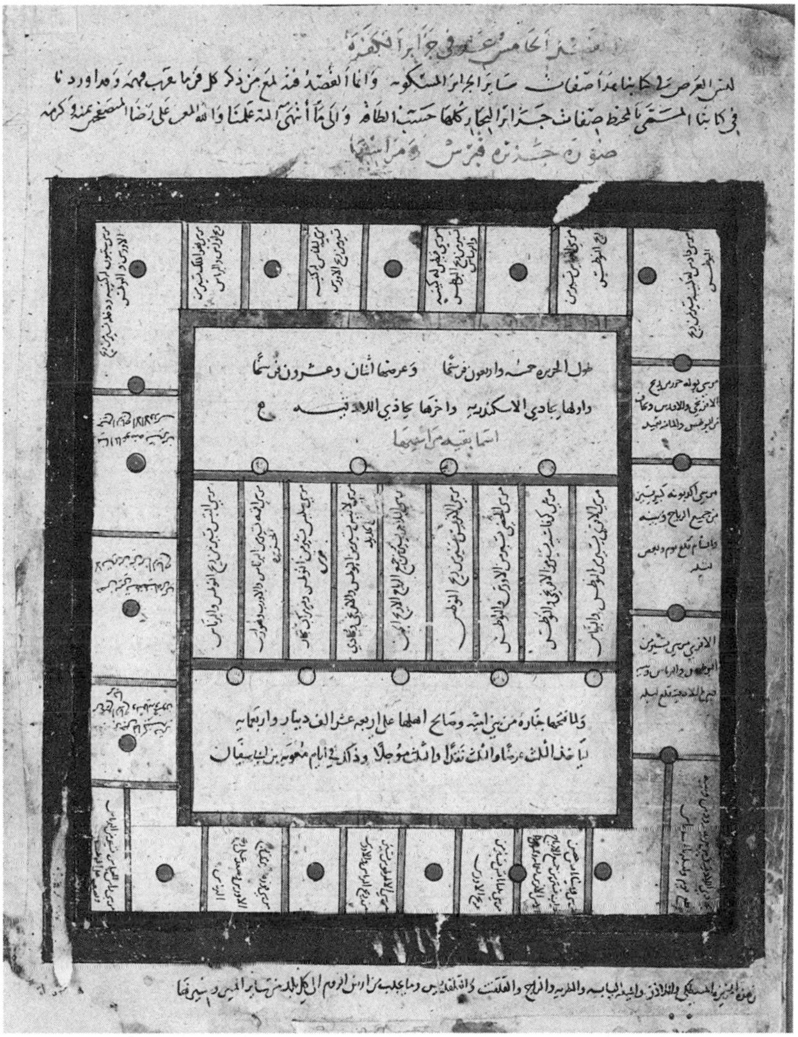

FIG. 5.5. Map of Cyprus from the *Book of Curiosities*. Oxford, Bodleian Library, MS Arab. c. 90, fol. 36b, copied ca. 1200.

forced to place these harbors in the middle of the square as a result of lack of space along the edges. This peculiar schematic representation of Cyprus is attested also in the Damascus manuscript, which has a simplified diagram with the same title but with no labels, indicating that such a schematic map of Cyprus was indeed part of the original treatise.[17]

In spite of the strange design, the harbors are listed in approximately correct sequence. The harbors on the northern coasts of Cyprus are represented on the middle strip, starting from the northwestern Acamas Peninsula on the left-hand side of the strip. Acamas is described as the "beginning of the island," and is also the starting point for two coastal itineraries in the Late Antique *Stadiasmus* navigation guide, about which more will be said shortly.[18] The sequence proceeds along the strip until it reaches the anchorage of Akraia, located at Cape Apostolos Andreas in the northeast tip of Cyprus. The list then proceeds in counterclockwise order along the four edges of the rectangle, representing harbors on the eastern, southern, and western coasts of the island. It culminates with Paphos, on the eastern coasts, on the bottom right. All in all, twenty-five separate harbors are recorded, on par with the account of Cypriot ports in the *Stadiasmus.*[19] The author also added a few lines of historical information and a brief account of the principal exports from Cyprus, related to the summary account of Cyprus by Ibn Ḥawqal.[20]

The type of information provided for each of the harbors and anchorages of Cyprus is very similar to that provided in the map of the Mediterranean, though often more expansive. In particular, protection from prevailing winds is systematically recorded. The winds named are the *Boreas* (north), *Notos* (south), and *Euros* (east or southeast), as well as a wind called "the Frankish" (*al-ifranjī*). This term clearly refers to westerly winds, but it is not attested in Greek sources.[21] Again, the information generally concerns protection from winds, not their propelling power. For a few anchorages there is a warning against specific winds. The *Notos* wind is to be feared in the anchorage of Būlah, and in the anchorages of Curias and Raʾs al-ʿAbbās there is an indication of winds which "fill the sails," apparently also a risk to the vessels. The anchorage of Makaria is said to protect from all winds save the "wind of the bay."

Capacity, water sources, and prominent landmarks are again frequently mentioned. The ports of Paphos and of Jurjis (probably Hagios Georgios, a monastery east of modern Limassol on the southern coasts) are said to accommodate 150 vessels each. The harbor of Soloi on the northern coasts is said to contain the ships of the merchants of the island, but no other harbor is associated with commercial activity. As in the Mediterranean map, there are occasional references to water sources. These include the Basileus River on the southeast coast of Cyprus, appearing here as "river of the King" in Arabic translation. On the other hand, freshwater is said to be "far" from the anchorage of Būlah. There is again frequent mention of prominent churches and fortresses, either for their military significance or because they would have been visible from aboard a ship.

Sailing distances are here too given only for sailing on the high seas, with no distances given between adjacent anchorages or harbors. The sailing distance from the unidentified anchorage of *Akrubūnah*, in eastern or southern Cyprus, to Syria is one day and part of a night. The sailing from Akraia (Cape Apostolos Andreas in the northeast) to Latakia is said to be one night, but is likely to be a copyist mistake, for reckoning a sailing distance merely by nocturnal travel would be counterintuitive. Ibn Ḥawqal mentions a sailing distance of one day and one night between Cyprus and Jabala, not far from Latakia.[22] The sailing from Akraia to Rhodes, cited at the not unreasonable one day and one night, required however propulsion by the *Boreas* wind. This rare indication of the use of winds for sailing is related to the difficulty of traveling away from the coasts in a counterclockwise direction. All other indications of sailing distances in the maps of the Mediterranean and Cyprus concern clockwise open seas travel.

All in all, the two maps of the Mediterranean and of Cyprus share the same features and were most probably derived from the same source. Both images are most obviously abstractions, diagrams more than maps, where the coastlines and even the major gulfs are not represented at all. At the same time, anchorages and harbors are listed in geographical sequence. The key data for each anchorage or harbor is the protection from four prevailing winds, indications of size and quality, as well as availability of freshwater. Distances between adjacent anchorages are not given at all, while distances of sailing in the open seas are given for select clockwise routes. The labels in the map of Cyprus are slightly more expansive, while the labels for the Mediterranean islands pretty much contain nothing but their names. This is possibly because there is simply more space on the map of Cyprus map to fit in longer labels, although other explanations may also be advanced.

But what is the purpose of this pair of maps? Why did the author so radically break away from the Mediterranean maps of Ibn Ḥawqal and al-Iṣṭahkrī, both in form and in content? We have already said that the maps of the Mediterranean and Cyprus are drawn from the perspective of the sea. This does not mean the maps as we have them were meant to be taken on board a ship. But, in all probability, the mapmaker was collecting information for these diagrams from records kept by sailors. He also made a conscious choice of depicting only information relevant for seafarers, and of opting for absolute abstraction. These radical choices of both form and content are interrelated, but we shall deal with content first and form later.

The labels on the maps of the Mediterranean and Cyprus are clearly suggestive of a military, naval context. The indications of capacity of ports and harbors could only have a military purpose, and the reference to the ability of an anchorage to accommodate an *usṭūl*, a naval fleet or squadron, makes it obvious that this list of anchorages has military implications. A merchant has no obvious interest in noting that a certain harbor can fit a hundred or two hundred ships. But for a military officer in command of a fleet consisting of a large number of galleys, on the other hand, such information would be of utmost importance. As John Pryor and Elisabeth Jeffreys remark in their study of the Byzantine *dromon* galley, fleets "could not simply be parked bumper to bumper." Each vessel needed space to allow it to swing around its anchor, a diameter of eighty meters if anchored in shallow waters, more if deeper. Limiting this diameter, for example by adding a second anchor, risks the stability of the galley. Pryor and Jeffreys calculate that a fleet of one hundred ships would require at least three kilometers of shoreline, and much time to arrange the ships properly. In fact, they point out that there are but few Aegean harbors with such long coastlines.[23]

There are other features that fit in well with the needs of naval warfare at the time. The indications of forts and other defensive structures are obviously useful for an attacking fleet, and so would be information on water sources, important for all vessels but crucial for the oarsmen rowing in the Mediterranean summer. Even protection from gales was more important for a galley than for larger, heavier ships. While commercial ships would often travel long distances without any stopover, as is attested in the records of the Geniza,[24] galleys were used only for short distances, their range limited above all by physical limitations of the oarsmen. A dense network of anchorages, ports, and bays is indicated, as one would expect in naval records. The inclusion of material on Fatimid ports, and not only Byzantine enemy bases, could also be explained by contemporary military tactics. Raiding fleets had to coast their own territory before making short passages to targets of attack. The limited range and the restrictions of visibility also meant that no navy could completely block off any of the major sea lanes. The lines of demarcation between one's own territory and the enemy could easily be crossed.[25]

In a preliminary chapter on cartography of the seas, the author of the *Book of Curiosities* claims that he collected his maritime information from sailors, merchants and ship captains. The terms ship captains (*rubbān*) and sailors (*baḥrīyīn*) were associated in this period with mercantile shipping, while the captains of military galleys are known to be called *raʾīs*.[26] However, claiming to collect maritime information from merchants and captains may be somewhat of a literary topos. There is little here that would be of relevance to merchants, such as local products or custom houses. Moreover, the formulaic nature of the information given in these maps suggests it originated in some sort of a ledger, not a collection of oral reports by seamen.

The compilation of information on the ports and inlets, especially in enemy territory, was one of the responsibilities of Mediterranean naval commanders. This is expressly stated in an appointment decree for a governor of a Mediterranean port town, dating from the Abbasid period and preserved in the administrative work of Qudāmah ibn Jaʿfar (d. 948 ?). In this appointment decree the commander is told, inter alia, to build new galleys to high standards and maintain the upkeep of existing ones, beaching them during the winter. He is also to carefully select his crews of sailors and throwers of Greek fire (*naffāṭūn*), and prevent the enemy from purchasing military equipment in the lands of Islam. The gathering of intelligence regarding the ports and bays of the enemy receives special attention:

> It is the commander's responsibility to see that the reconnaissance forces and spies which are sent to gather information about the enemy, are trustworthy, able to give good advice, of good religion, and honest. They must have experience of the sea, its harbors (*mawānīhi*), inlets (*dakhalātihi*), and hiding places, so that they bring only correct and reliable reports. In case you are overtaken by enemy ships which you cannot resist, then withdraw to the places which they know well and where they know that you will be safe.[27]

This Abbasid appointment decree highlights the importance accorded to the collection of information of harbors and inlets, using the same terms as in the *Book of Curiosities.* The emphasis here is finding places of refuge for one's fleet when it is overtaken. This corresponds very well with the type of data given on the maps of the Mediterranean and Cyprus, which concerns almost exclusively the mooring possibilities: the size, quality, wind protection, and water supply of the anchorage point.

Such information was of course collected by other Mediterranean navies. In an almost contemporary Byzantine military manual, the *Taktika* composed around 1000–1011, the Byzantine governor of Antioch Nikephoros Ouranos states that a commander of a fleet should acquire knowledge of the wind patterns, water supply, hazards to ships, such as hidden rocks or shoals, and distances between harbors.[28] The Byzantine manual's emphasis on risks to vessels and distances between adjacent harbors goes beyond the focus on mooring facilities that characterizes the comparable Abbasid appointment decree. It is interesting to note that this additional information is not generally found in the *Book of Curiosities'* maps, and may reflect different recordkeeping by Muslim and Byzantine navies.

Beyond the naval context, the labels on the maps of the Mediterranean and Cyprus also have a thematic connection to the genre of the maritime itinerary, well known since antiquity and revived in medieval Islam by the Andalusian al-Bakrī in the eleventh century. Al-Bakrī, writing a few decades after the composition of the *Book of Curiosities*, lists ninety anchorages and ports in a maritime route from Tangier to Alexandria. Ports on the shores of Muslim Spain are paired with the North African ports that lie straight across the Mediterranean, usually with the distance between them in days of sailing (*majrā*). Distances between adjacent ports, in miles, are only given for the westernmost North African ports. For most anchorages along the route, and especially for the stopping points west of Mahdia, al-Bakrī indicates safety from adverse winds—the term used is "sheltered" (*maʾmūn*), allowing mooring over the winter. Sweet water sources, if available, are also indicated, as well as prominent landscape features visible from the sea.[29] There is a mention of galleys using the port of Bone to raid the northern shores of the Mediterranean.[30] It is possible that al-Bakrī himself used official military records in Córdoba when compiling his geographical treatise. [31]

A century later, al-Idrīsī provides a maritime itinerary for circumnavigating the island of Sicily, and information relating to navigation along the coasts of North Africa.[32] The itinerary is a list of coastal localities, proceeding eastward from Oran. The distances between them are given in miles, both in direct line (*ruʾsiyah*) and along the contour of the coast (*taqwīrān*). There is little else on navigation, despite a passing reference to the anchorage of Waqūr as giving shelter from easterly winds only.[33] In other sections of his *Nuzhat al-Mushtāq*, the information on navigation and harbor facilities is incorporated in the account of the locality. Like al-Bakrī, al-Idrīsī also reckons the distances between ports in North Africa to ports in al-Andalus in day sailings (*majrā*), and occasionally adds a brief note on water supply or on protection from winds.[34]

The pre-Islamic antecedents for these maritime itineraries are the late Roman *periploi*, or accounts of coastal navigation. The *periplus* of Arrian (second century AD) is ostensibly a letter to the Emperor Hadrian following his inspection of the military installations in the Black Sea. It contains quite a lot of narrative unrelated to navigation, but also terse accounts of coastal sailing from one harbor to another, reminiscent of the *Book of Curiosities.* Arrian mentions sources of water and availability of mooring for different sizes of vessels, as well as shelter from winds.[35] More closely related is the *Stadiasmus maris magni*, "Measurement in stades of the Great Sea [Mediterranean]" by an anonymous late Roman author, which survives only in a single tenth-century manuscript. It is a comprehensive account of the distances of sailing along the coasts of the Mediterranean and its major islands, taking Alexandria as the starting point. It consists only of navigation material, with no literary narrative or information that would be of interest to merchants. The primary point of interest is the distances in *stadia* between harbors, followed by information on the size of the harbor and the availability of freshwater. The *Stadiasmus* also provides additional information addressed directly to the mariners, such as the right winds for entering a harbor and warnings of shoals.[36]

There is much in these maritime itineraries, whether from Late Antiquity or from al-Idrīsī and al-Bakrī, that reminds us of the navigational material in the *Book of Curiosities.* Information on protection from winds seems to be the most important aspect of an anchorage point, alongside notes on sources of freshwater and prominent landscape features. Both the maps and the maritime itineraries give distances on the high seas between ports that are further away from each other, usually in days of sailing. There are also some differences. The maritime itineraries sometimes provide distances between adjacent ports, data notably absent from the maps of the *Book of Curiosities.* Compared to the maritime itineraries, the maps of the *Book of Curiosities* have an overt military tone, referring to the capacity of ports to accommodate large number of ships, as well as fleets or squadrons, although al-Bakrī's maritime itinerary at least may have also been set within a semiofficial context.

The form of the maritime itinerary was evidently familiar to the author of the *Book of Curiosities,* and he included one such maritime itinerary on the top left of his map of Mahdia (plate 9 and fig. 6.3 in the following chapter). The itinerary is written in the open space within the walls. It lists stopping points from Mahdia to Palermo along the North African coasts, then on smaller Mediterranean islands, and then the final stages along the western coasts of Sicily. This itinerary consists of the names of anchorages along the way and the distance between them in miles. No further detail is provided. Thus, the format of the itinerary is: "From Mahdia to *a-l-B-r-ṭ-w-l,* 30 miles. Then to Sousse, 15 miles. Then [to . . .]*īyah,* 16 miles. Then to Hergla, 12 miles."[37] There is no visualization of the itinerary, even though it is placed within a map of the city of Mahdia.

The Mediterranean and Cyprus maps are different from these maritime itineraries because of the effects of visual presentation. Compared to maritime itineraries—such as the one from Mahdia to Palermo in the *Book of Curiosities,* or those by al-Bakrī and al-Idrīsī—the *Book of Curiosities* maps do not provide guidance on how to travel along the coasts. The anchorages, harbors, and bays are presented in sequence, but without an obvious starting point, unlike the narrative of a maritime itinerary, which develops in a linear way, from point A to point B. The map of the Mediterranean depicts the sea as a perfect oval, with no opening to the Atlantic Ocean, no gulfs or bays. The geometrical shape does not have any edges that would hint at a visual start or terminus point—only the sequence of the possible stops, and information about the quality, size, and protection from the wind.

Paradoxically, compared to the maritime itineraries, this visualization of the maritime space—that is, the transformation of textual narrative to a map—removes any sense of scale or direction. It also severs any geographical relationship between the mainland coasts and the Mediterranean islands. Like the harbors along the rim, the islands are also presented in their own sequence, with nearby islands placed next to each other, and are all indicated similarly by small circles (apart from the rectangles of Sicily and Cyprus). The mapmaker, however, is unable to relate them to the ports on the mainland coasts. From reading this map, it is impossible to know which harbor is nearest to any individual island, nor what is the distance or direction of an individual island from the mainland. This is again related to the loss of scale and orientation that resulted from the visualization of maritime space. In a narrative maritime itinerary, the presence of islands poses no particular problem, as they can easily be incorporated as stopping points along a maritime route (as the author of the *Book of Curiosities* does when he includes the islands of Qūsrah and Favignana in his textual maritime itinerary from Mahdia to Palermo). But when these same islands are presented on the Map of the Mediterranean, they are not part of an itinerary, and since the map has no scale or orientation, their location within the oval sea is arbitrary.

These shortcomings should not distract us from the revolutionary step being taken by the mapmaker here. No earlier Islamic, Roman, or European map known to us visualizes navigation material in such a way; it is the first extant attempt to draw the Mediterranean from the perspective of the seagoing mariner. The mapmaker has tried to visualize the maritime itineraries that were available to him. One could compare this attempt with the Balkhī School's visualization of land routes, where the textual narrative of the itinerary and its linear progression are transformed and abstracted to a series of straight lines, with only minimal orientation and no indication of scale. As argued by Emilie Savage-Smith in earlier publications, the Balkhī School maps' abstraction serves a purpose. The maps present the main routes and the stops on the way in an accessible way, allowing the planning of journeys and a grasp of the links between routes and cities. The Balkhī maps of the land routes, like the modern Underground maps, remove scale and precise orientation, but allow us to grasp the interchanges, and to create relationship between data in a way not possible in a narrative text.[38]

In the map of the Mediterranean of the *Book of Curiosities*, the total abstraction of the coastline—the perfect oval—is chosen because the maps are intended to simplify the presentation of maritime space. The omission of all gulfs was not done out of a landlubber's ignorance. On the contrary, these maps carry more information relevant to navigation than any other medieval Islamic document before the age of the portolan charts. But when planning a maritime journey that is by default a voyage along the coasts, the key information is the sequences of the harbors, their quality, and any recognizable markers which would have been visible from aboard a vessel. The distances between adjacent harbors are not indicated, perhaps because they were usually short. This information is sometimes omitted in the narrative maritime itineraries too. On the other hand, sailing distance is mentioned for travel in the high seas, usually between ports that are not next to each other. Here the map has the potential to surpass the textual narrative by creating possible connections between ports that are not located next to each other—whether distances are given or not. A map, unlike a text, can allow the viewer to read the navigation material in ways that are not only sequential or linear.

Circumnavigating the Aegean

It is time to return to the account of the Aegean Bays we had mentioned at the beginning of this chapter. Clearly, this Aegean section, too, looks at the coast from the perspective of a mariner. It joins the maps of the Mediterranean and Cyprus as an outstanding example of a navigational text coming to us from the medieval Islamic world. Unlike the maps, however, the interest here is in bays, not in harbors or islands, and the recourse to visual presentation is short-lived. As said above, the section opens with a schematic diagram of the first five bays, starting with two bays facing Rhodes on the southwestern shores of Anatolia, called here the Smaller and Larger Tracheia Bays. But in the next folios the author drops the visual diagram, and the remaining twenty-three Byzantine bays are described in text only, and in much more detail.

The map below shows the area of the Aegean Sea covered in this section (fig. 5.6). The author moves counterclockwise along the west coast of Anatolia up to the mouth of the Dardanelles, but does not enter the Straits. Instead, the list of bays continues westward to Thessalonica, then down the coast of modern Greece to Corinth. The bays continue along the coasts of the Peloponnesus as far as Patras, only for the author to strangely return to the Bay of Corinth and repeat in more expanded form the description of most of the Peloponnesus' bays before he cuts the text abruptly.

This account of the Aegean bays provides an astonishing amount of detail on the bays and gulfs of the Aegean, all of which lay deep inside the Byzantine domains. The information regarding each bay concerns width and length, direction of entry according to an eight-point wind-rose, major forts, small islets and prominent landmarks. As an example, this is the entry for the eighth bay in the sequence, Smyrna, modern Izmir: [39]

FIG. 5.6. The Aegean Sea in the eleventh century, with some of the place names indicated in the *Book of Curiosities*.

> Further to the north is the Bay of Izmirnah [Smyrna]. This bay is thirty miles long, and in its widest place it is ten miles wide. At the head of the bay is the fortified settlement of Izmirnah [Smyrna], located three miles from the sea. At the entrance to the bay there is a small and uninhabited island called Jurjis. The fortress of Qlazūmnī [Klazomenai] is to the south of the bay and the fortress of Fūqīyah [Phocaea] is to the north. One enters it from the west to the east. It has also an inhabited island.

This formulaic account of the length, width, direction, and prominent landmarks is repeated for all bays along the Aegean. Additional details may refer to other aspects of navigation, or to elements of naval interest. In the account of the Bay of Miletos, for example, the author mentions that the wide Byzantine *shelandia* galleys can enter the river Meander.[40] In the account of the Argolic Gulf of the Peloponnesus, it is mentioned that Cape Maleas marks the halfway point along the maritime routes between Constantinople and Sicily.[41] In contrast, inland features are only infrequently alluded to. Of special interest are several references to Slav populations in the northern Aegean and the Peloponnesus.[42]

While the Mediterranean and Cyprus maps are chiefly concerned with harbors, the account of the Aegean bays rarely makes any reference to mooring facilities.[43] None of the place-names on the map of the Mediterranean is repeated in this chapter, save when a harbor gives its name to a bay or a gulf.[44] Nor is there any mention of the islands of the Aegean, so carefully listed on the interior of the map of the Mediterranean.[45] The account of the Aegean bays neatly complements the maps of the Mediterranean and Cyprus—all three are concerned with navigation to the exclusion of any interest in inland topography, but they never overlap in the information they provide.

Again, as in the Mediterranean and Cyprus maps, there are often references to strongholds or forts along the coasts, and especially at the entrance to bays. The reference to the Byzantine *shelandia* galleys in the River Meander resonates again with military concerns, and the lack of any commercial information also suggests that this was not written in the service of trade. The unusual references to Slav populations may also have had military significance, since the Fatimid navy of the late tenth and early eleventh century was dominated by Slav commanders, all manumitted slaves who were originally bought or captured in Europe.[46]

This chapter takes up only Aegean bays, and no other bays or gulfs in the Mediterranean. This is partly due to military considerations: all of these bays were under Byzantine control at the time, so this chapter charts enemy territory, and this account would have had obvious benefits to a commander of a Fatimid fleet planning a raid on the Aegean. But it is notable that the bays of Byzantine southern Anatolia are not included in this account, nor is the maritime space surrounding the biggest military and political prize of all, Constantinople. Rather, the author is following here a division of space familiar to the men of the sea. The region he describes in this chapter is an area of deep, narrow bays, more so than in any other section of the eastern Mediterranean coasts. Strictly from the point of view of navigation, the Aegean is a coherent unit.

This is an account of the bays of the Aegean. It is not, contrary to first impressions, an itinerary of navigation. There is no mention of navigational hazards, such as shoals and reefs, and the reader is not addressed in the second person, as in the Roman *periploi* or the European portolan guides.[47] Furthermore, the geographical organization of the chapter, proceeding from the southwest tip of Anatolia to the mouth of the Dardanelles and then turning south toward the Peloponnesus, does not follow any likely maritime route. Commercial vessels traveling to Byzantium from Fatimid lands would almost certainly try and reach the markets of Constantinople, rather than turn west at the mouth of the Dardanelles. Vessels en route to the Adriatic or to Sicily would go directly from Rhodes to Modon in the Peloponnesus, sometimes making stops on islands in the southern Aegean.[48] In general, medieval ships traveling in the Aegean would almost always resort to island-hopping, using anchorages on Chios, Samos, and Lesbos. We know that vessels avoided the toothed coastline of the shallow and narrow bays, even when taking the difficult northward journey against the summer *meltemi.* The island route, through Lesbos, Chios, and Samos, was taken by Byzantine fleets traveling to Cyprus or to Crete.[49] A ship may sometimes enter one of the bays, but not all of them. In 1102, the English Crusader Saewulf passed through the mainland port of Smyrna on his way from Palestine to Constantinople. But his other Aegean stops—Rhodes, Chios, Lesbos, and Tenedos at the mouth of the Dardanelles—were all islands.[50] None of these large micro-Asiatic islands is mentioned in the account of the Aegean bays in the *Book of Curiosities.*

The primary purpose of the account is to describe the mainland coast in detail, with an almost cartographic sensibility. This is about mapping the Aegean rather than traveling in it. Like the ports and islands on the map of the Mediterranean earlier in the *Book of Curiosities,* the bays here form a sequential list. Each of them is described in a way that would be useful for a mariner, especially a commander of a galley: entrance direction, supporting winds, islets, fortresses, and visible landmark features (although there is no mention of shoals and other hazards to the vessels, as we have come to expect from European portolans). Since they are not part of an itinerary, it is possible to skip to the next gulf, or even to hop to the other side of the Aegean. The narrative of the bays in this chapter allows the reader to mentally imagine them as forming together the Aegean Sea, and it seems that this was the ambitious intention of the author.

Indeed, this chapter almost seems like a template for a crude portolan chart. The key elements are there: the width and length of each bay, the direction of entry according to an eight-point wind-rose, as well as major forts, small islets, and prominent landmarks. We could even speculate that the author was copying his information from some sort of a visual guide, and transforming it into a textual narrative. One intriguing piece of evidence does very tentatively suggest that this chapter was compiled by reference to a chart. This comes at the last section of the chapter, when the author reaches the Bay of Patras to the northeast of the Peloponnesus, thus completing his account of the Peloponnesus' bays which started with the

Bay of Corinth in the northwest corner of the peninsula. Surprisingly, after his account of the Bay of Patras, the author returns to the Bay of Corinth, describing it in a slightly more expanded form, and then continues clockwise, repeating the information for the Bays of Nauplia, Cape Maleas, and the Bays of Kalamata and Methone before stopping abruptly. This unusual repetition is also found in the Damascus manuscript and therefore not a copyist mistake. A possible, speculative explanation would be that the author was copying from a chart, in which the Peloponnesus was diagrammatically represented as a circle. Going through the bays in sequence, the author did not notice that he had already completed copying the entire circle representing the peninsula of the Peloponnesus, and thus continued to repeat the entry for the Bay of Corinth.

But in the *Book of Curiosities,* when the author attempts to present navigation material visually, his diagrams look nothing like a portolan chart, not even a very crude one. In a pattern similar to his treatment of the Mediterranean and Cyprus, the author's visualization of the navigation material involves stripping it of any scale or orientation. The five bays are equal in size, and all point in the same direction. This is so startling because the information on length, width, and direction is listed on the map itself. Yet the principle governing the visualization of maritime spaces, here and elsewhere, is that of presenting uniform units of analysis (ports, islands, bays) sequentially, not as they are in the real world.

The Book of Curiosities *and the Emergence of the Portolan Charts*

In his cartographic introduction to the maps of the seas, including the Mediterranean, the author explains that "these sea maps (*al-ṣuwar al-baḥrīyah*) are not accurate representations" of coastlines. He gives two reasons for this decision. First, he argues that land can turn to sea and sea to land, citing several examples extracted from the work of al-Masʿūdī. Therefore there is no need to bother with tracing the coasts in detail. Second, the Ptolemaic method does not satisfy the needs of those using the maps of the seas:

> This shape of the coast exists in reality, but, even if drawn by the most sensitive instrument, the cartographer (*muhandis*) would not be able to position [literally, "to build"] a city in its correct location amidst the curves in the coast or pointed gulfs because of the limits of the space that would correspond to a vast area in the real world. That is why we have drawn this map in this way, so that everyone will be able to figure out [the name of] any city.[51]

Thus, even if it were possible to produce an accurate map of a sea in the manner described by Ptolemy by employing precise drawing instruments, the irregularity of the coastline would not leave room for labels. Accuracy matters less than legibility.

This approach to sea maps stands in complete contrast to the late medieval portolan charts, such as the earliest surviving example, the Carte Pisane of the thirteenth century (fig. 5. 7). The emergence of the portolan charts in the second half of the thirteenth century has transformed navigation and had far-reaching implications for world history. Like the maritime maps of the *Book of Curiosities,* the focus is navigation, the meeting between the sea and the land; inland areas are usually blank. But unlike the maps of the *Book of Curiosities,* even the earliest portolan charts represent the coastlines of the Mediterranean far more accurately than any maps that preceded them.

Portolan charts erupt rather suddenly into our records, without any cartographic parallel in preceding European or Islamic maps. Their origins are still widely debated. The manner in which they comprehensively eclipsed all earlier navigation aide-de-memoire, whether visual or textual, means we know very little about how mariners found their way in the pre-portolan era. The *Book of Curiosities* offers a glimpse into that world. The fact that it comes from medieval Islam matters little: in the Mediterranean, a shared and contested space, the problems facing Muslim and Christian navigators were the same. Much of the maritime nomenclature was common across the boundaries of faith, and the technologies Christian and Muslim mariners used were broadly similar.

Medieval Muslim sources make it clear that long-haul navigation was not simply oral, not merely a matter of memorization by the captain of the ship.[52] This was certainly true for the Indian Ocean. The tenth-century geographer al-Muqaddasī tells us that, with regard to the Indian Ocean, the seamen have "in their possession ledgers (*dafātir*) which they study carefully together and on which they rely completely, proceeding according to what is in them."[53] Al-Muqaddasī adds that these ledgers were used by the experienced ship captains and merchants who circumnavigated the Arabian Peninsula, from al-Qulzum (Suez) to ʿAbbādān.[54] But the ledgers did not contain maps. When al-Muqaddsī did consult maps, he found them in literary works or in the treasuries of rulers. He reports seeing a map of the Indian Ocean in the geographical work of Abū Zayd al-Balkhī (d. 934), another one drawn on paper in the treasury of the amir of Khorasan, one in treasury of the Buwayhid ʿAḍud al-Dawlah (reg. 936–83), and one on a piece of cloth in Nishapur.

Finding these erudite maps of the Indian Ocean contradictory and un-

FIG. 5.7. Carte Pisane, dated 1258–91. Paris, BnF, Res. Ge. B1118 (Dep. des Cartes et Plans). Undated, ca. thirteenth century. Reproduced with permission of the Bibliothèque nationale de France.

satisfactory, al-Muqaddasī asked for clarifications from a leading merchant in Aden, whose ships traveled to every corner of that maritime space. The merchant then "rubbed the sand with his palm and drew the sea on it. It did not have the shape of a *ṭaylasān* or of a bird. He drew winding coastal features and many gulfs, then said: 'This is the shape of this sea, there is no other form (*ṣūrah*).'"[55] Evidently, a leading merchant in Aden did not have a map at hand, although he had in his mind a mental map he could draw on sand. In his turn, al-Muqaddasī, who insists that this detailed map drawn on sand was made by the most well-informed person at the time, decides not to use it in his own work. Instead, he tells us that he decided to simplify his map as much as possible. His map of the Indian Ocean left out all the bays and gulfs except the Red Sea, which was, he says, too well known and not disputed.[56]

Like the author of the *Book of Curiosities,* al-Muqaddasī struggled with the transition from written records of navigation to a map format. Al-Muqaddasī's account suggests that the seamen, who had written records, did not have charts. Though they were able to draw them when requested by a geographer, they did not feel the need to use maps for navigation. Maps of the seas were found in literary works or in the treasuries of rulers, but those bore little resemblance to the firsthand knowledge of captains and merchants. Even after al-Muqaddasī traveled the Indian Ocean himself, and collected extensive data on winds and sea hazards, the map he decided to draw was, as he says, "abstract" (*sādij*). We can see the author of the *Book of Curiosities* making a similar decision. After collecting a tremendous amount of information on ports, islands, capes and bays, his visualization of the data is intentionally abstract, leaving out scale and direction, bays and headlands—the actual contours of coast—from his maritime maps. The author of the *Book of Curiosities,* like al-Muqaddasī before him, chose abstraction over the detailed contours of the coast.

Set against the portolan charts, the intentional abstraction by al-Muqaddasī and by the author of the *Book of Curiosities* is at the opposite end of the spectrum of cartographic representation. In view of the evidence of the maritime maps of *Book of Curiosities,* it seems highly unlikely that some early version of portolan maps was known in Fatimid Egypt. This is worth emphasizing because some have argued that portolan maps first emerged in the Islamic world. According to this hypothesis, a direct line leads from al-Khwārazmī and ninth-century mathematical geography in Baghdad to the portolan chart of the late thirteenth century.[57] The proponents of this view excuse the chronological and geographical gap by the scarce survival of Islamic sources. Were we to find maps of the Mediterranean from the intervening centuries, so the argument goes, they would be crude versions of the earliest portolans. Yet now we have found such maps in the eleventh-century *Book of Curiosities,* and they do not resemble portolan charts whatsoever. The prefect oval shape of the Mediterranean and the Indian Ocean is intentionally divorced from accurate representation of the coastlines. There is no trace of longitude and latitude coordinates in these maritime maps. If maps were to be judged only by their ability to accurately represent reality, the maps of the *Book of Curiosities* are a step backward when compared to the more elaborate tenth-century Mediterranean map of Ibn Ḥawqal.

The maritime diagrams of the *Book of Curiosities* also shed doubt on a rival hypothesis, which is that portolan charts emerged from an earlier genre of Latin navigational texts. P. Gautier-Dalché has discovered several examples of such texts, which demonstrate an increasing interest in the systematization of maritime knowledge, often spurred by the participation of clergy on naval missions to the Holy Land in the wake of the Crusades.[58] One such text, *De viis maris,* composed in England at the end of the twelfth century, presents a list of ports and distances from York to Egypt and further to India.[59] Another text, the *Liber de existencia riveriarum,* composed in Pisa in the following century, gives more complex navigation data. It opens with a list of ports and coastal configuration features, which begins and ends at Gibraltar. Next it divides the coast into sectors—first describing the coastal features of the littoral, then a succession of place-names with the distances between them expressed in miles.[60] The author says that he consulted a map (*cartula mappe mundi*) in order to compile the text, and Gautier-Dalché argues that this map would have been an early version of the later portolan charts. In his view, European maritime mapmaking was developing gradually over the twelfth century, with early portolan charts, which have not survived, serving as the basis for the information collected in navigation texts.[61]

While the Arabic *Book of Curiosities* presents broadly similar categories of Mediterranean navigation material, the diagrams it offers are nothing like the portolan charts. It should remind us that a transition from text to map is not as intuitive as it may seem to us today. Admittedly, there are some important differences between the Latin texts and the *Book of Curiosities.* For one, the distances between adjacent ports, pivotal to the Latin texts, are not regularly mentioned in the maps of the *Book of Curiosities.* The Latin texts also tend to be specific about hazards to ships, something which is missing completely from the *Book of Curiosities* and from other Islamic maritime itineraries. But the information on ports, their size and quality, the protection they offer from prevailing winds and the water supply, are all comparable. The account of the Aegean Bays in particular, which offers information on the size and direction of the bays, provides quite accurate data on the Aegean coastlines. Were it to be drawn to scale, it would have provided the kind of crude portolan prototype that scholars have been looking for.

But, again, it was not. The author of the *Book of Curiosities* has provided a map of the Aegean Bays, but it is an abstract diagram, in which all five bays depicted are of identical size and shape. The argument for a direct link between the Pisan *Liber* navigational text and the portolan charts rests on the assumption that the scale and direction described in the text would have been visually drawn in the map. But the map that accompanied the *Liber* may have been similar to the diagrams of the *Book of Curiosities,* in which scale and orientation play no role. Such simplification of the contours of the coastlines had precedents in the Islamic cartographic tradition, as we have seen with al-Muqaddasī. Although rich navigational material was available, the maps of the *Book of Curiosities* were part of a cartographic tradition that valued accessibility over accuracy and precision. In the contemporary Latin context, similar priorities may have applied—we do not know, as no such map survived. But the *Book of Curiosities* visibly draws our attention to the numerous possibilities, and difficulties, of visualizing maritime itineraries.[62]

The discovery of the maps of the *Book of Curiosities* lends support to those, like Ramon J. Pujades, who argue that the portolan charts *were* a radical break with cartographical tradition, whether European or Islamic.[63] This break was, most likely, brought about by the introduction of the magnetic compass during the thirteenth century in both European and Muslim vessels. As indicated by the rhumb lines, which are the distinctive feature of all portolan charts, the compass was the one piece of technological innovation that was essential and indispensable for the production and consumption of portolan charts. It was the compass that transformed the ability of mariners to follow a course in the high seas, and allowed cartographers to collect much more accurate data about the directions of the coasts.[64]

Precisely because they are so different from portolan charts, the maps of the *Book of Curiosities* emphasize the revolutionary impact of the compass on Mediterranean seafaring. Even though these particular maps occur within a literary compilation, never meant to be used on board a ship, they still allow us to glimpse the use of cartography for navigation in the pre-compass era. This is no different than many of the extant luxurious European portolan charts, which survived because they were included in literary compilations or as pieces of display. As we assume that decorative portolan charts available to us are close relations of more practical charts, we can also assume that the maps of the *Book of Curiosities* tell us something of significance about the way Muslim mariners recorded and presented navigation material before the emergence of the portolan charts.

The maps of the *Book of Curiosities* demonstrate that eleventh-century mariners lacked neither interest in recording navigational data, nor means to collect rich information on distances and directions. But, unlike the producers of portolan charts, they did not have the instrument—the compass—that would allow them to determine the direction of the coast in a sufficiently precise way. Moreover, at sea, captains, merchants, and naval commanders were unable to make sufficiently accurate decisions about distances and directions even if they had such precise charts. Without the compass, they were restricted to coast-hugging in most circumstances, only seldom making short crossings on the high seas. The coast, which is never too far away in the Mediterranean, was thus a continuous feature, a line whose bends and twists were somewhat less important. As in the maps of the *Book of Curiosities*, there was no need to visualize the coastlines in precision. The Mediterranean could become a perfect oval, Cyprus a precise square.

The navigation material of the *Book of Curiosities* should be taken seriously. In the absence of other maritime charts dating from the centuries preceding the portolan charts, these maritime maps are the best we have. The recording of navigation data in a systematic way is unique and distinctive. So is the interest in projecting this data onto maps, as abstract and ineffectual as they may seem to us. We find here systematic navigation data and maps together, and that is unlikely to be a coincidence. While the maritime maps in the *Book of Curiosities* are very different from the later portolan charts in their execution, they herald a growing interest in thinking differently about maritime spaces. The narrative account of the Aegean bays is particularly novel. It is not a maritime itinerary, but a conscious attempt to follow the contours of the coast in a way that is not present in other Islamic sources. The mere endeavor of creating maritime maps exclusively devoted to navigational material is in itself a breakthrough. While the attempt to integrate different categories of navigational data was mostly unsuccessful, it remains a very exciting and bold effort in imagining the relationship between medieval societies and the sea.

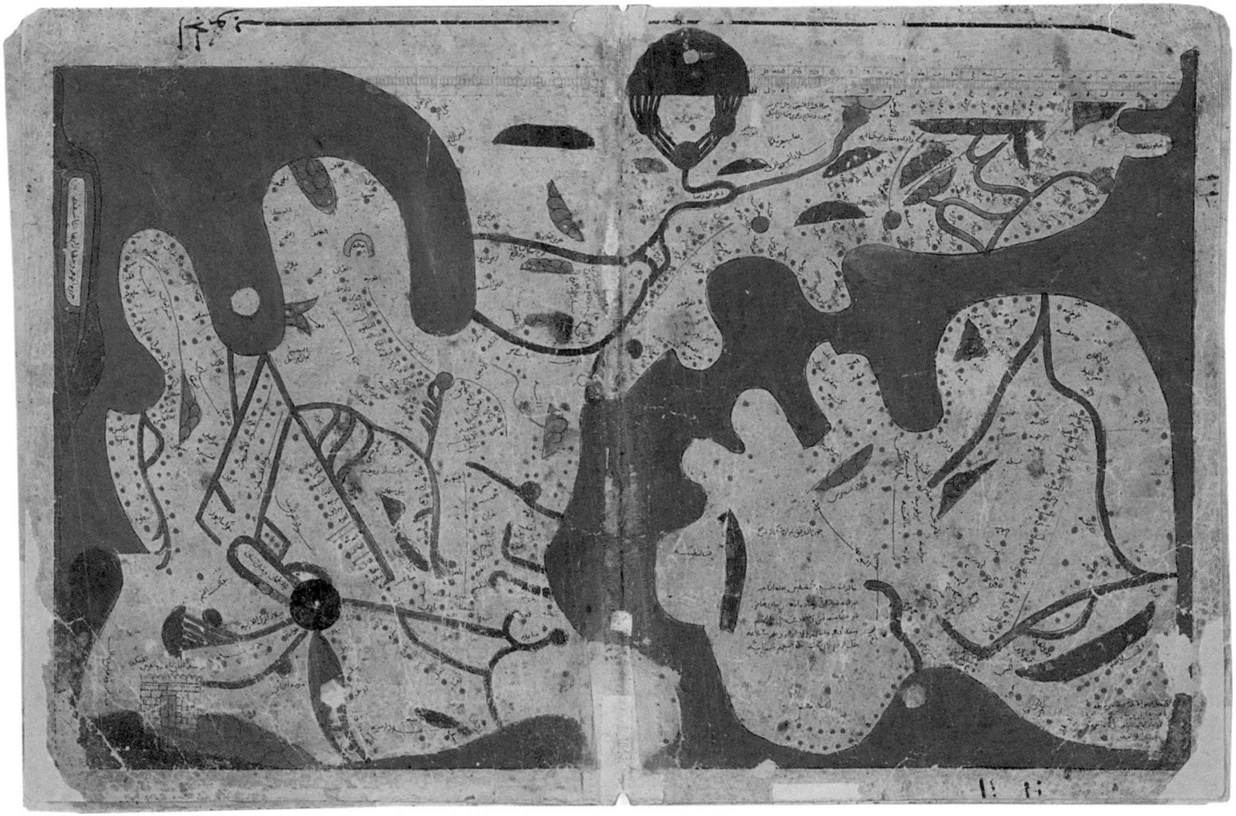

PLATE 1. The rectangular world map. Oxford, Bodleian Library, MS Arab. c. 90, fols. 23b–24a, copied ca. 1200. South is at the top.

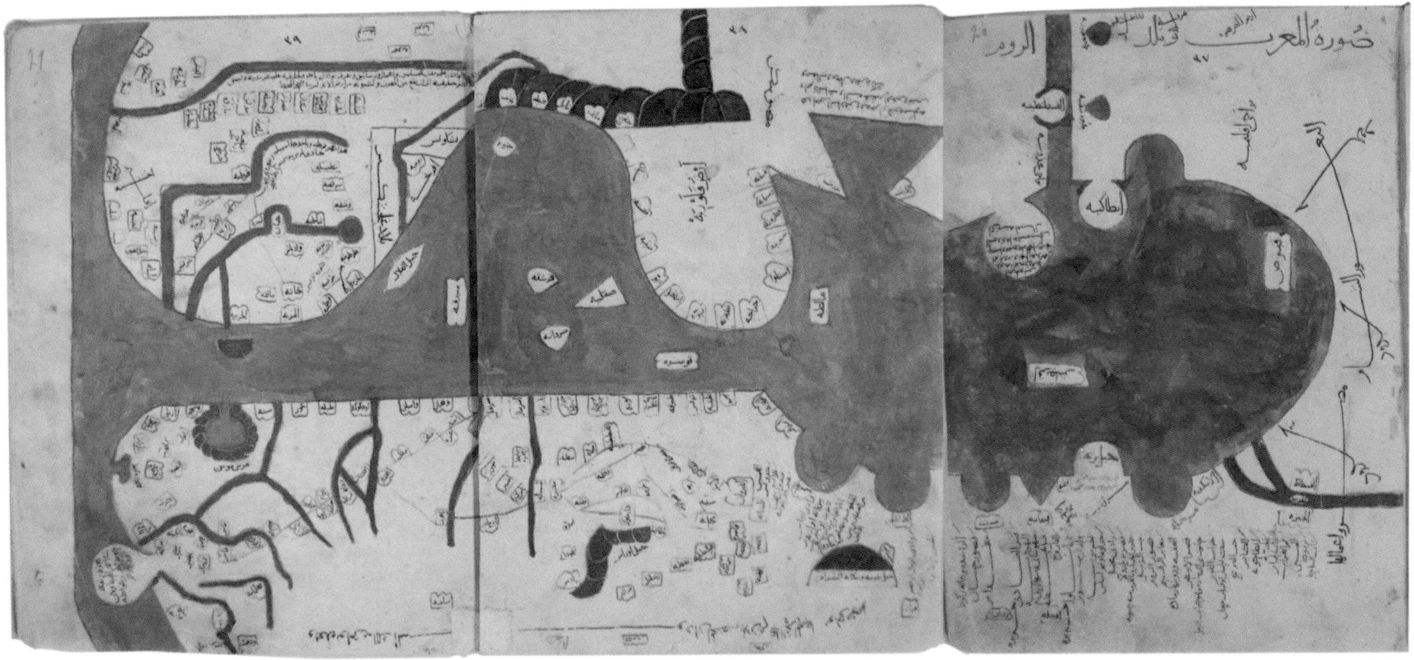

PLATE 2. Map of the Maghreb from Ibn Ḥawqal, *Kitāb Ṣūrat al-Arḍ*. The map is spread over three folios. North is at the top. Istanbul, Topkapı Sarayı Müzesi Kütüphanesi, Ahmet III MS 3346, fols. 19a, 19b, 20a. Copied in 1086/479H. Reproduced with permission of the Topkapı Sarayı Müzesi.

PLATE 3. Circular world map from the *Book of Curiosities*. Oxford, Bodleian Library, MS Arab. c. 90, fols. 27b–28a, copied ca. 1200.

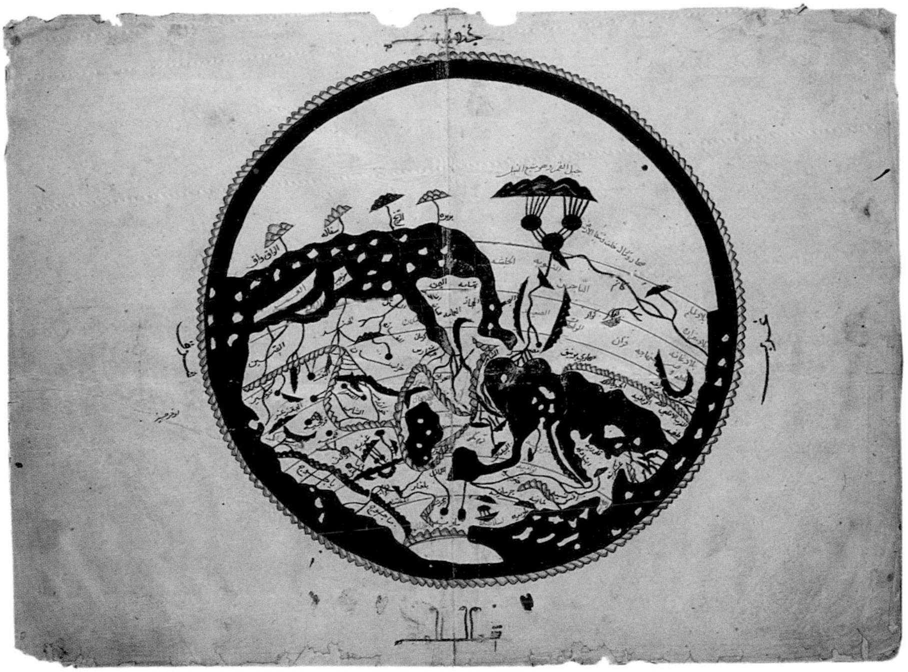

PLATE 4. Circular world map, from Idrīsī. Oxford, Bodleian Library, MS Pococke 375, fols. 3b–4a. Copy completed July 25, 1553/13, Shaʿbān 960H, by ʿAlī ibn Ḥasan al-Hūfī al-Qāsimī.

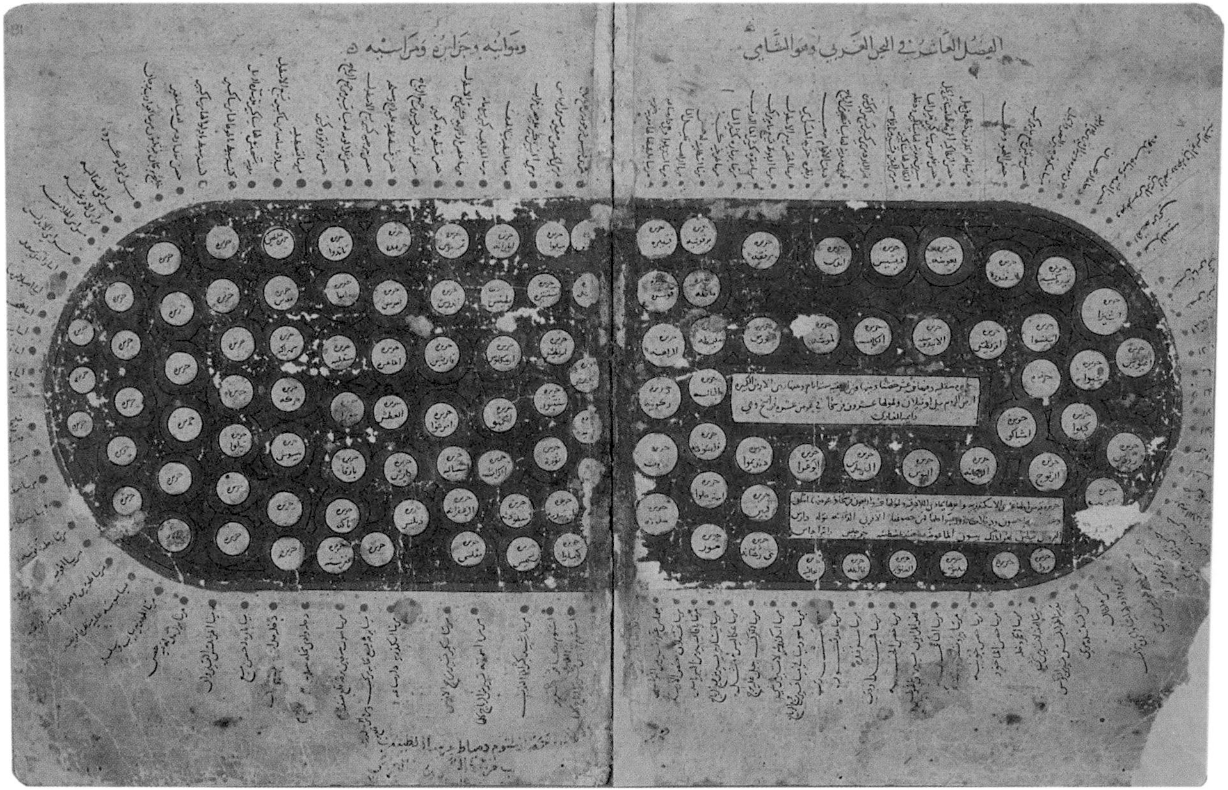

PLATE 5. Map of the Mediterranean from the *Book of Curiosities*. Oxford, Bodleian Library, MS Arab. c. 90, fols. 30b–31a, copied ca. 1200.

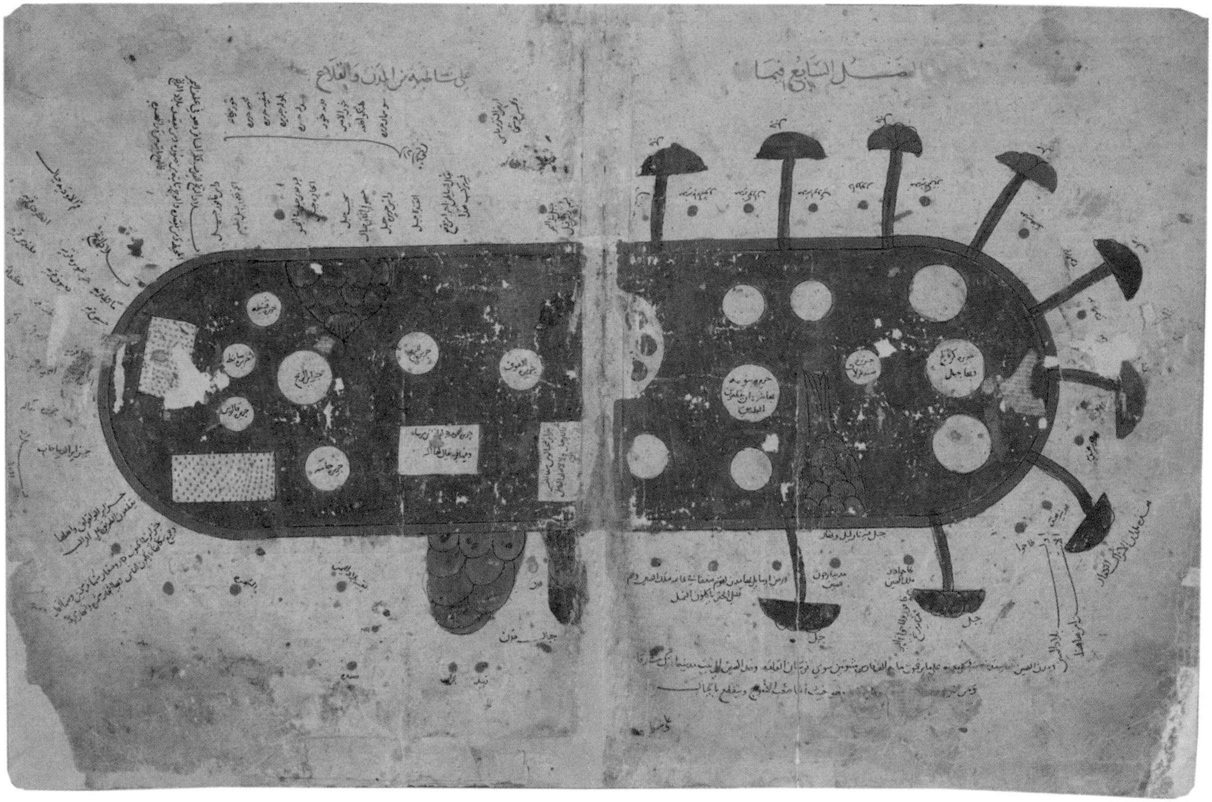

PLATE 6. Map of the Indian Ocean from the *Book of Curiosities*. Oxford, Bodleian Library, MS Arab. c. 90, fols. 29b–30a, copied ca. 1200.

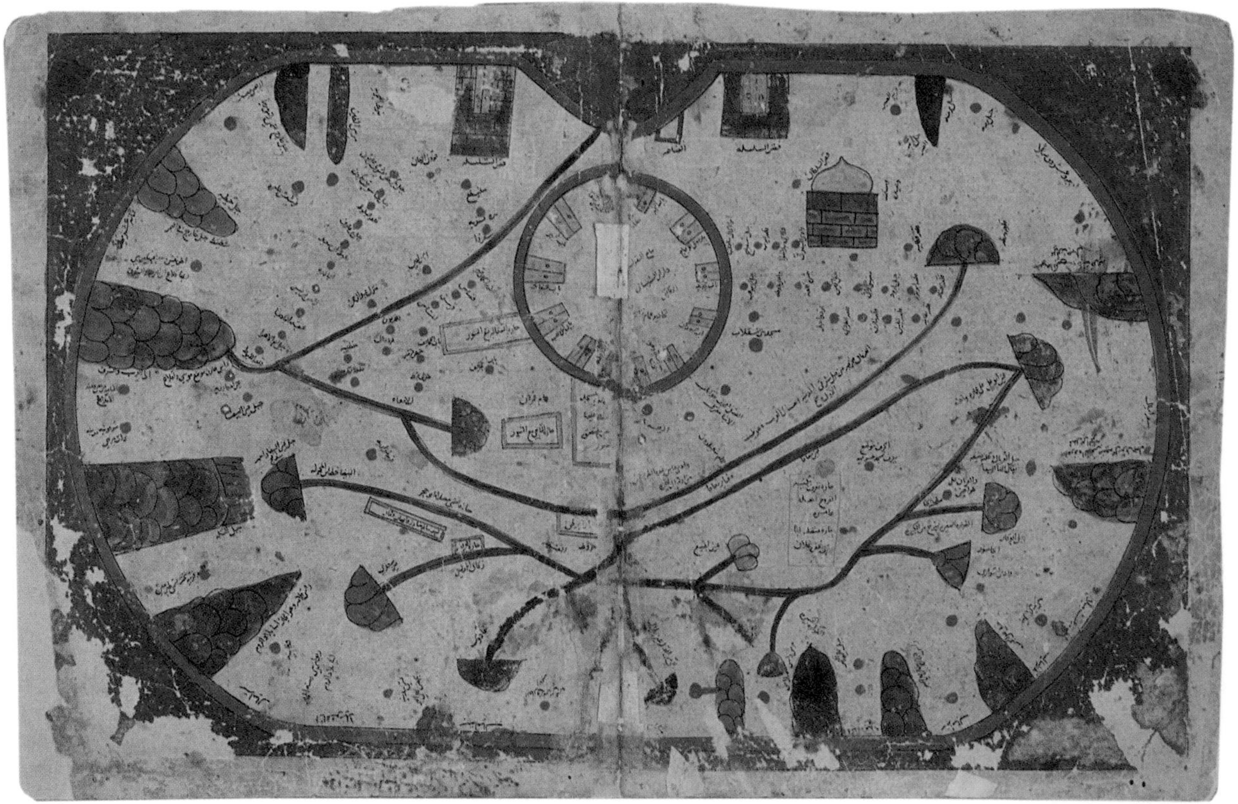

PLATE 7. Map of Sicily from the *Book of Curiosities*. Oxford, Bodleian Library, MS Arab. c. 90, fols. 32b–33a, copied ca. 1200.

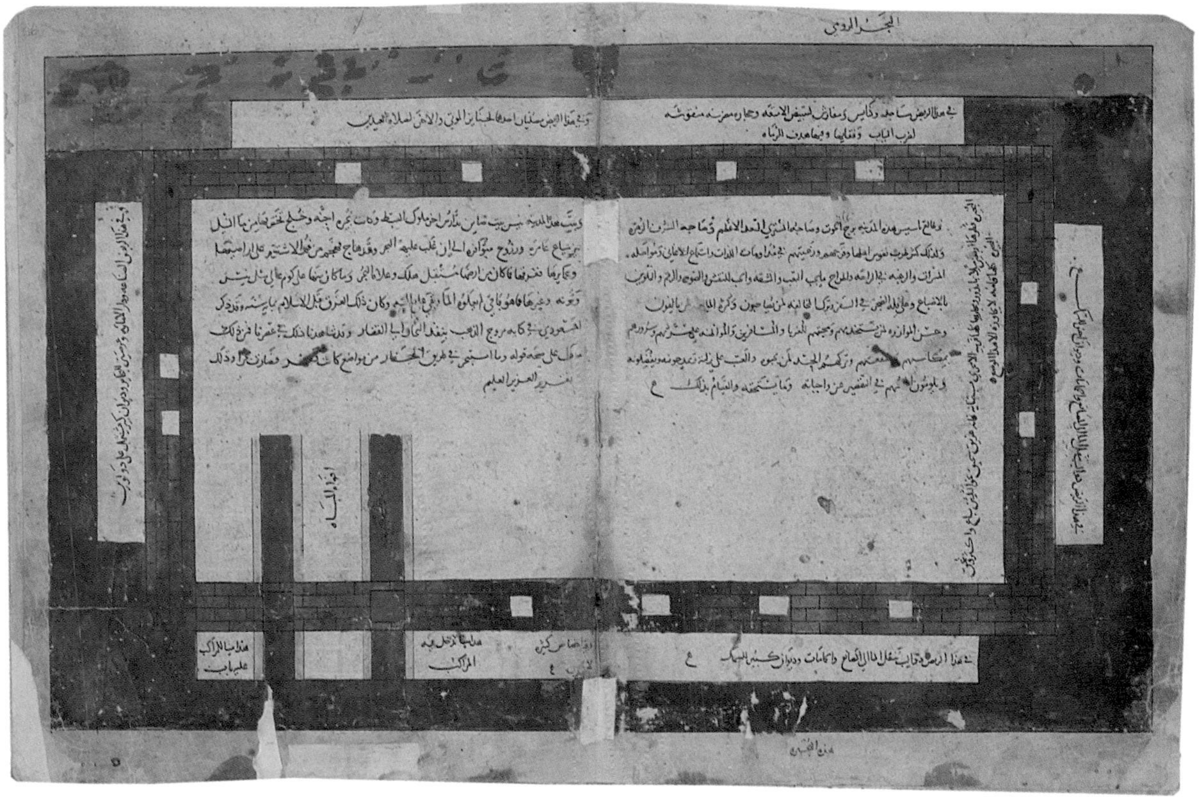

PLATE 8. Map of Tinnīs in the *Book of Curiosities*. Oxford, Bodleian Library, MS Arab. c. 90, fols. 35b–36a, copied ca. 1200.

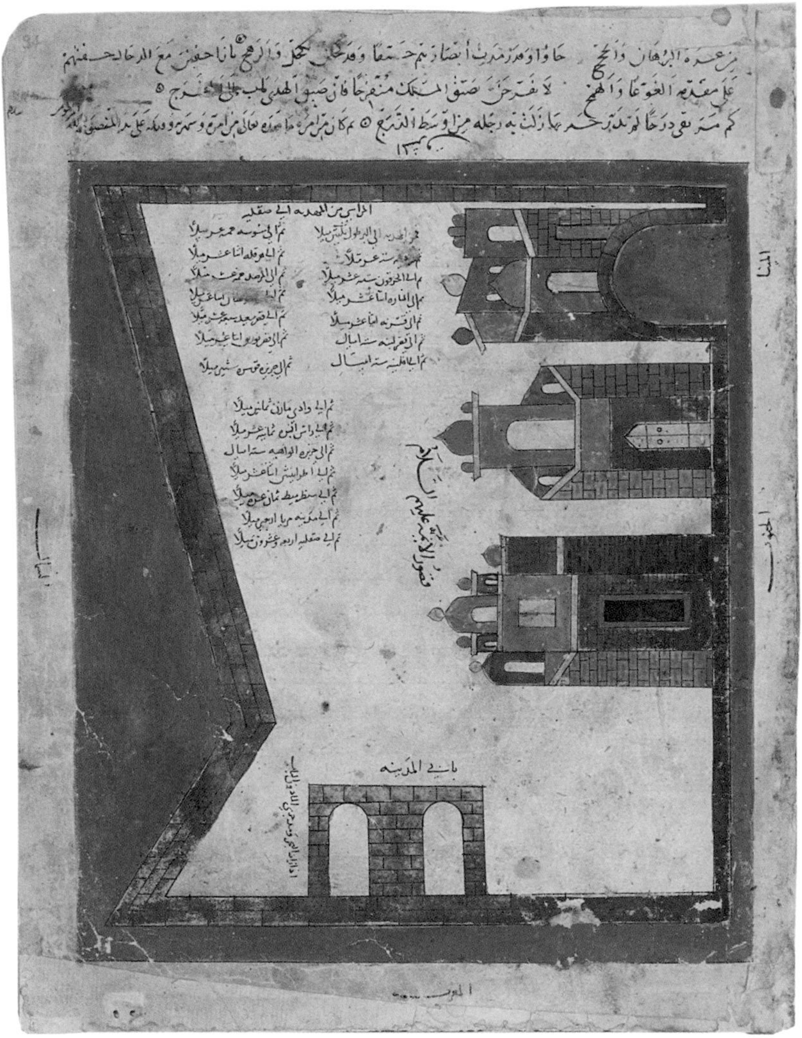

PLATE 9. Map of Mahdia in the Bodleian copy of the *Book of Curiosities*. Oxford, Bodleian Library, MS Arab. c. 90, fol. 34a, copied ca. 1200.

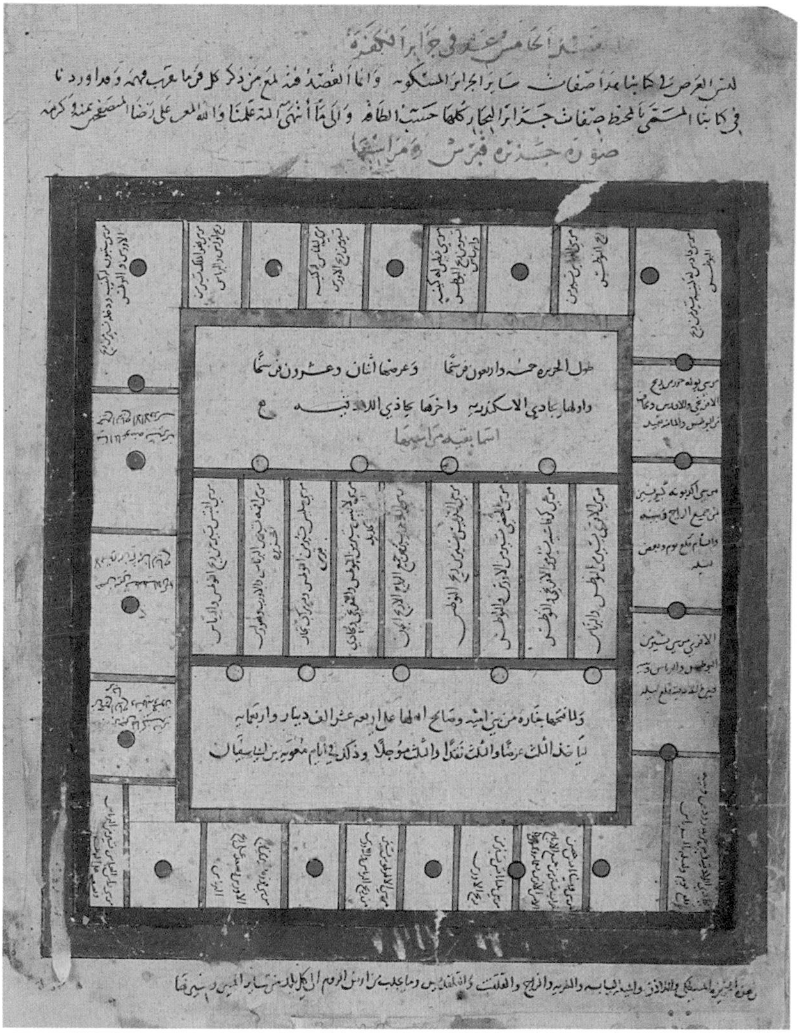

PLATE 10. Map of Cyprus from the *Book of Curiosities*. Oxford, Bodleian Library, MS Arab. c. 90, fol. 36b, copied ca. 1200.

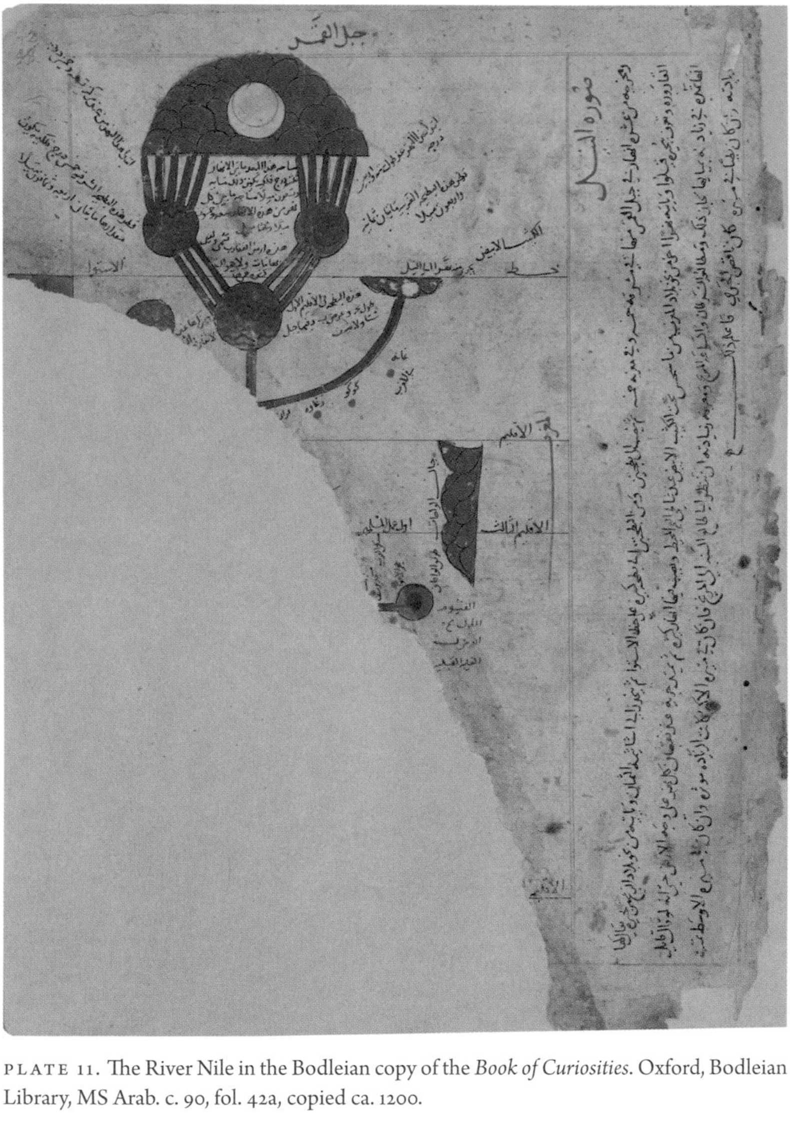

PLATE 11. The River Nile in the Bodleian copy of the *Book of Curiosities*. Oxford, Bodleian Library, MS Arab. c. 90, fol. 42a, copied ca. 1200.

PLATE 12. Map of the Nile by al-Khwārazmī. Bibliothèque Nationale et Universitaire de Strasbourg, MS 4247, fols. 30b–31a, copied 1037/429H. Reproduced courtesy of Coll. et photo BNU Strasbourg.

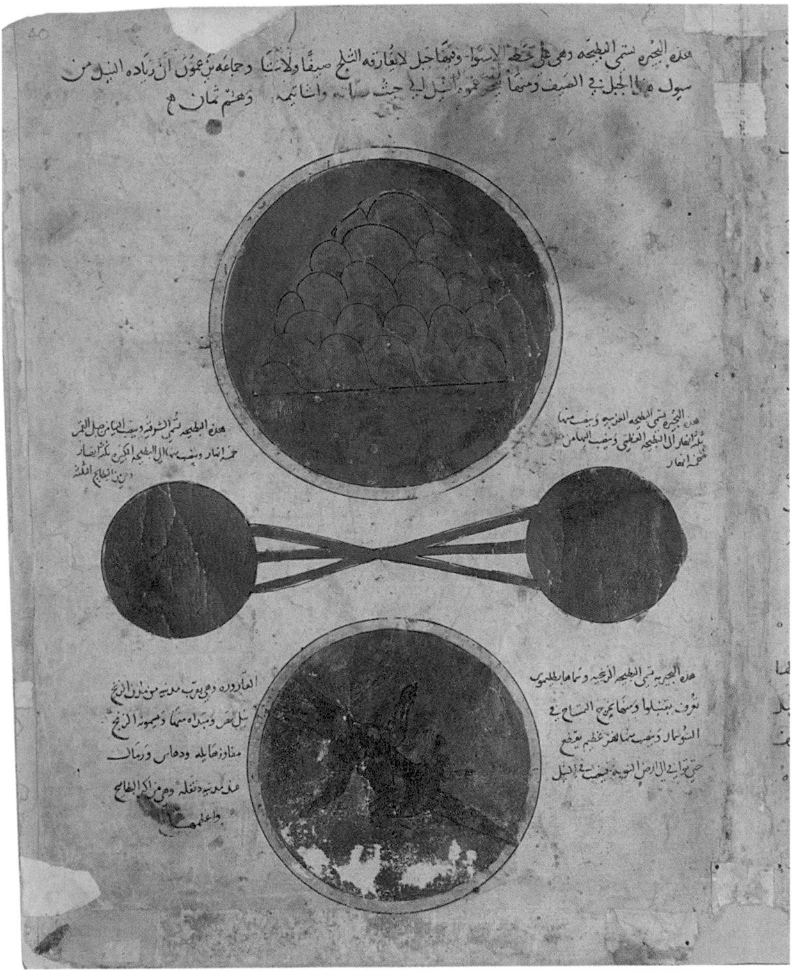

PLATE 13. Diagram of the Sources of the Nile from the *Book of Curiosities*. Oxford, Bodleian Library, MS Arab. c. 90, fol. 40a, copied ca. 1200.

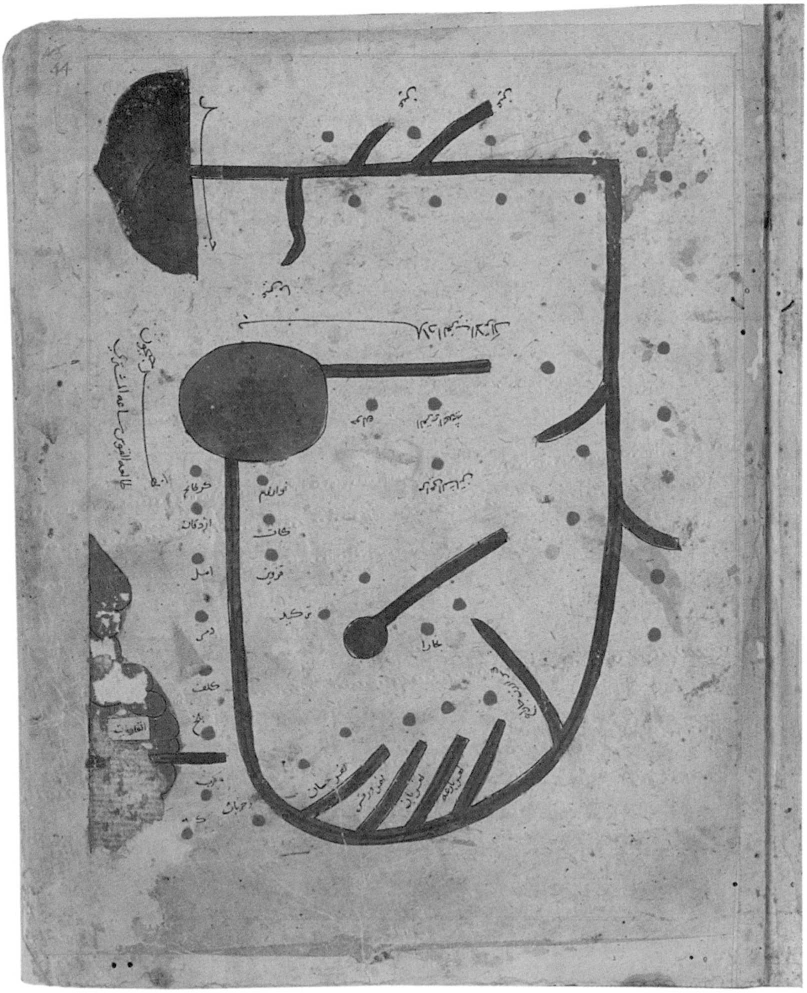

PLATE 14. The River Oxus. Oxford, Bodleian Library, MS Arab. c. 90, fol. 44a, copied ca. 1200.

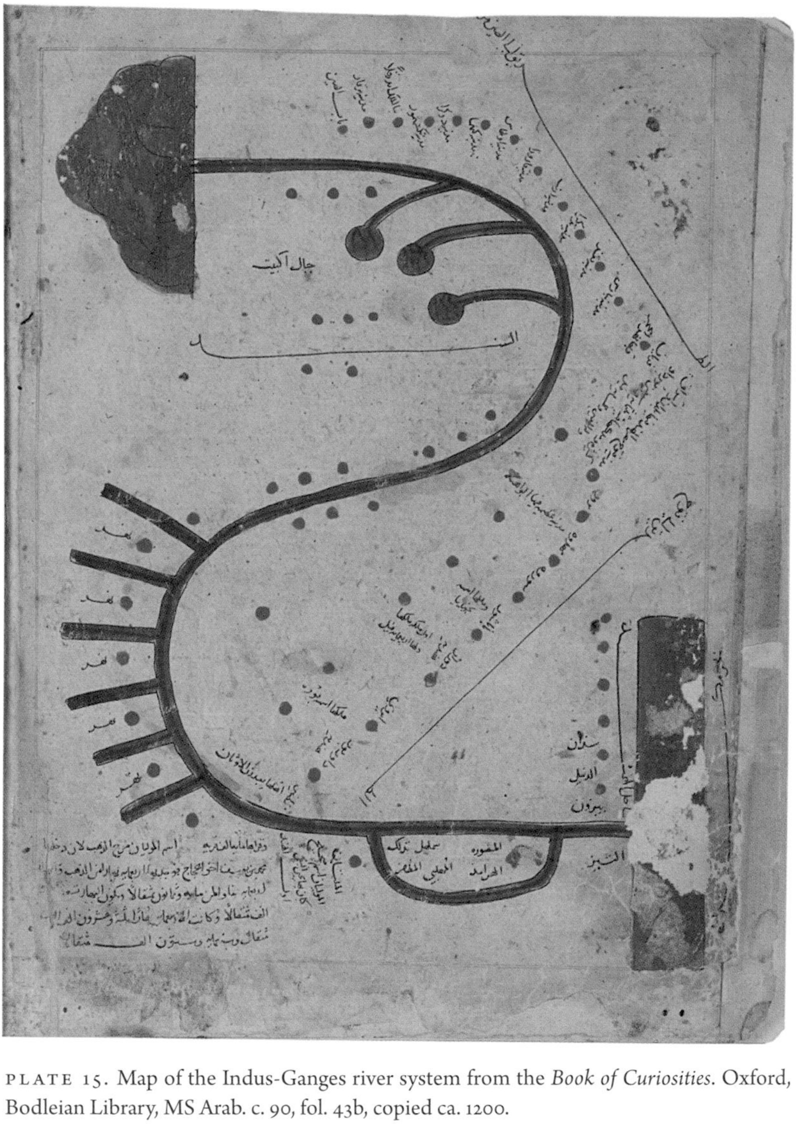

PLATE 15. Map of the Indus-Ganges river system from the *Book of Curiosities*. Oxford, Bodleian Library, MS Arab. c. 90, fol. 43b, copied ca. 1200.

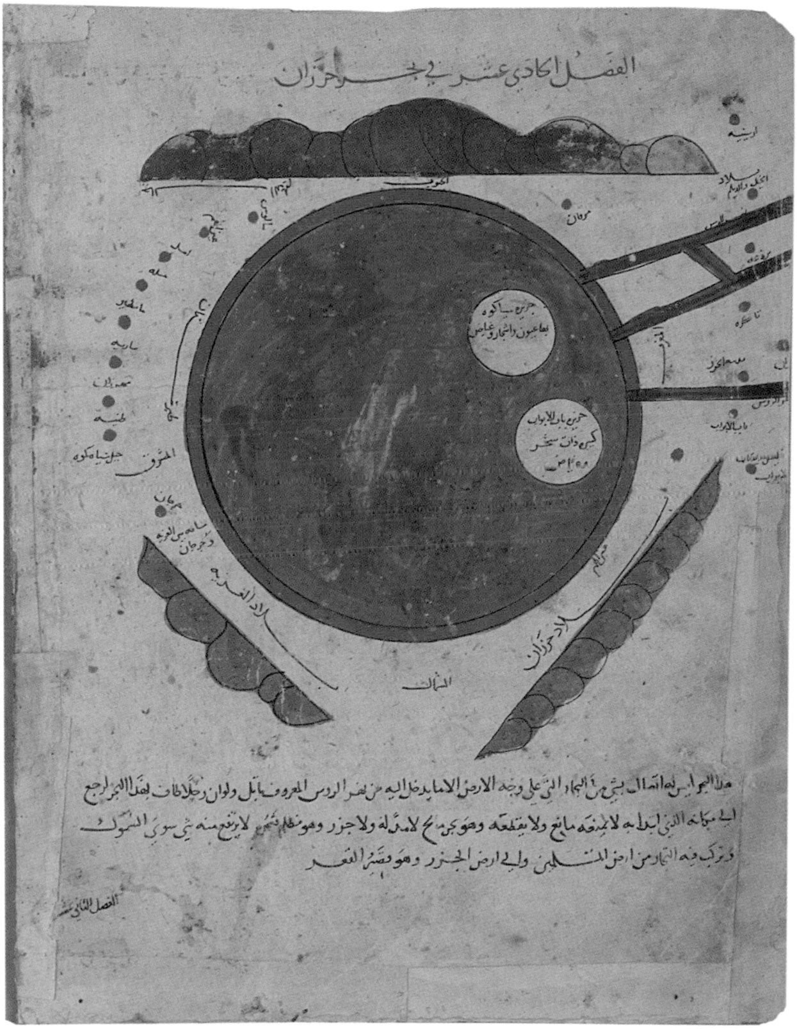

PLATE 16. The Caspian Sea in the *Book of Curiosities*. Oxford, Bodleian Library, MS Arab. c. 90, fol. 31b, copied ca. 1200.

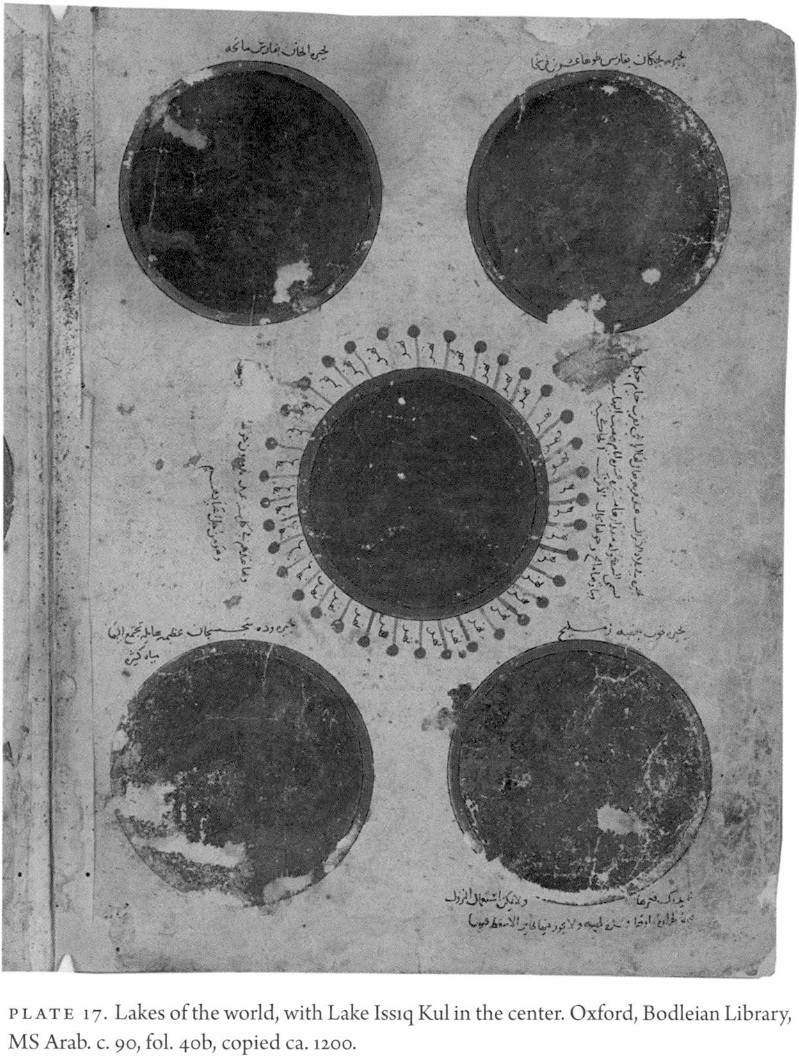

PLATE 17. Lakes of the world, with Lake Issıq Kul in the center. Oxford, Bodleian Library, MS Arab. c. 90, fol. 40b, copied ca. 1200.

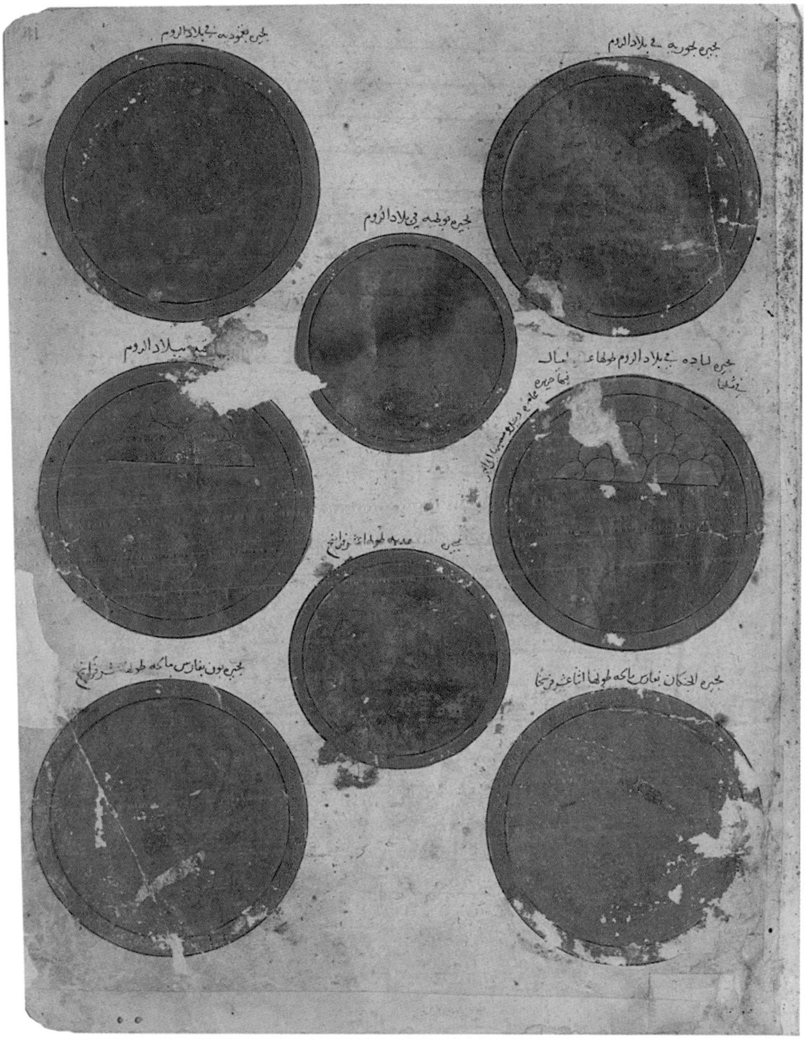

PLATE 18. Lakes of the world in the *Book of Curiosities*. Each lake is represented by a perfect circle. Blue represents sweet water, and green salt water. Short labels include name of the lake, its location and size. Oxford, Bodleian Library, MS Arab. c. 90, fol. 41a, copied ca. 1200.

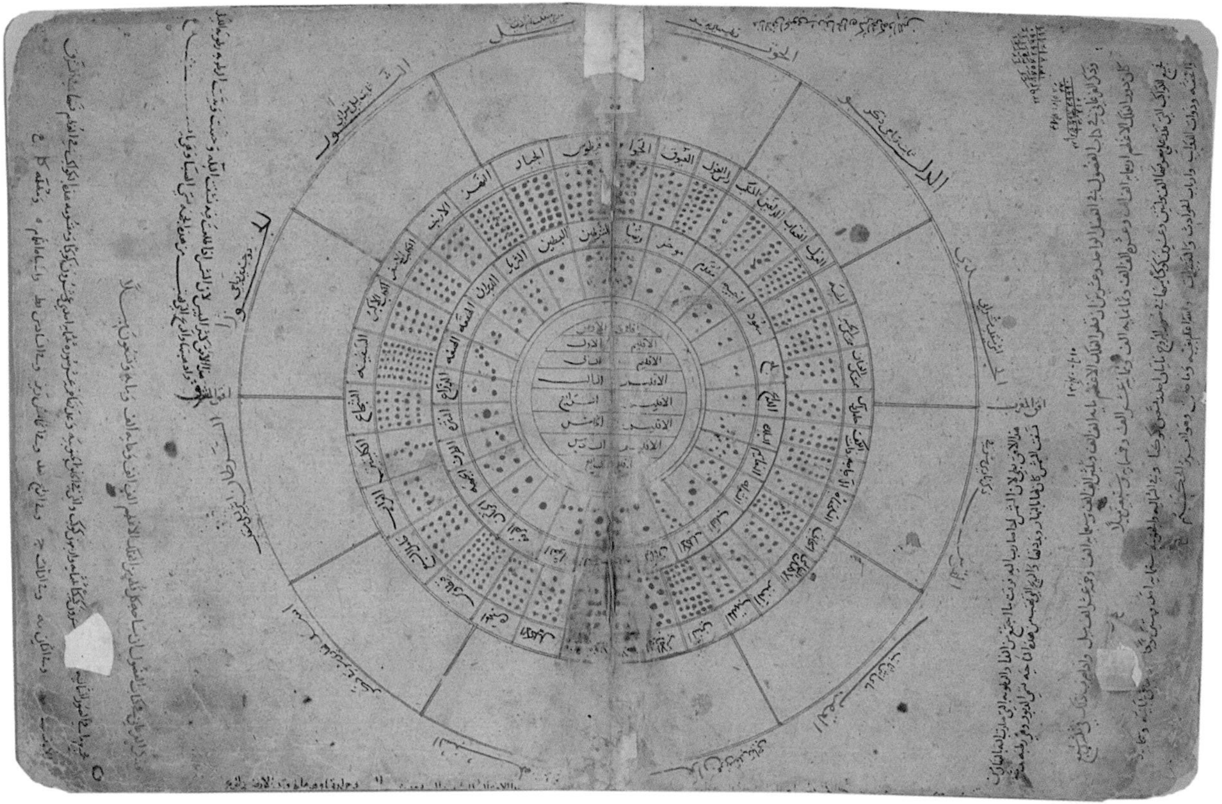

PLATE 19. The opening diagram of the *Book of Curiosities*, labeled "The Illustration of the Encompassing Sphere, and the Manner in which It Embraces All Existence, and Its Extent." Oxford, Bodleian Library, MS Arab. c. 90, fols. 2b–3a, copied ca. 1200.

PLATE 20. The "maps" of the lunar mansions as drawn in the *Book of Curiosities*. Oxford, Bodleian Library, MS Arab. c. 90, fol. 18b, copied ca. 1200.

1. *al-qāʾid* (the commander) / *al-rāmī* (the archer)

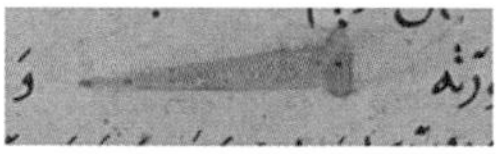

2. *kawkab al-dhanab* (the star of the tail)

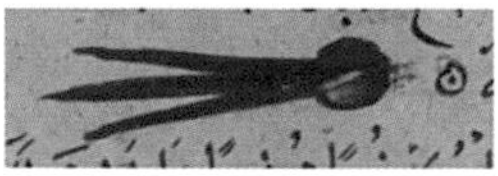

3. *al-waqqād* (the stoker)

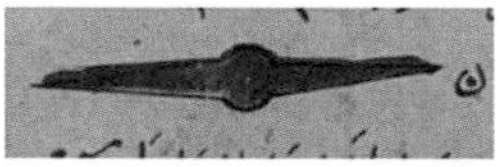

4. *al-muʿtaniqayn* (the embracing couple) / *alyat al-ḥamal* (the lamb's fat-tail)

5. *al-rāmiḥ* (the lancer)

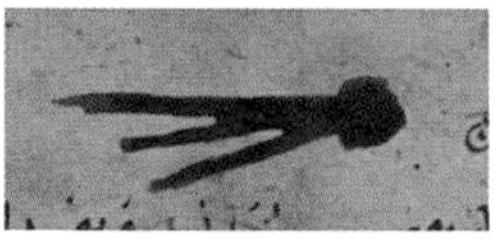

PLATE 23. Five "stars with faint lances" attributed to Hermes in the *Book of Curiosities*. Oxford, Bodleian Library, MS Arab. c. 90, fols. 15b and 16a, copied ca. 1200.

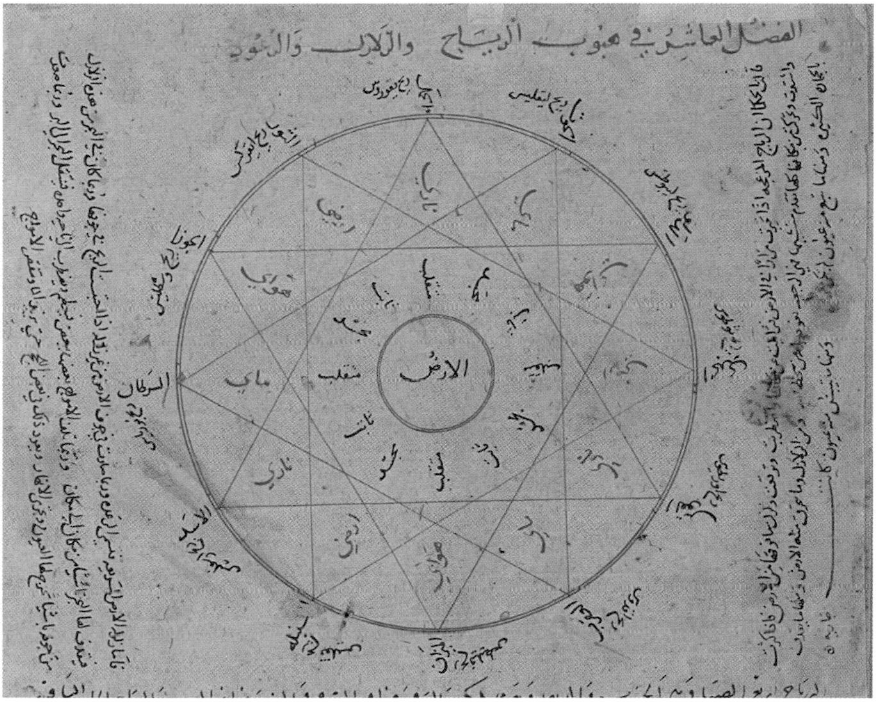

PLATE 24. A diagram of the winds. Oxford, Bodleian Library, MS Arab. c. 90, fol. 21b, copied ca. 1200.

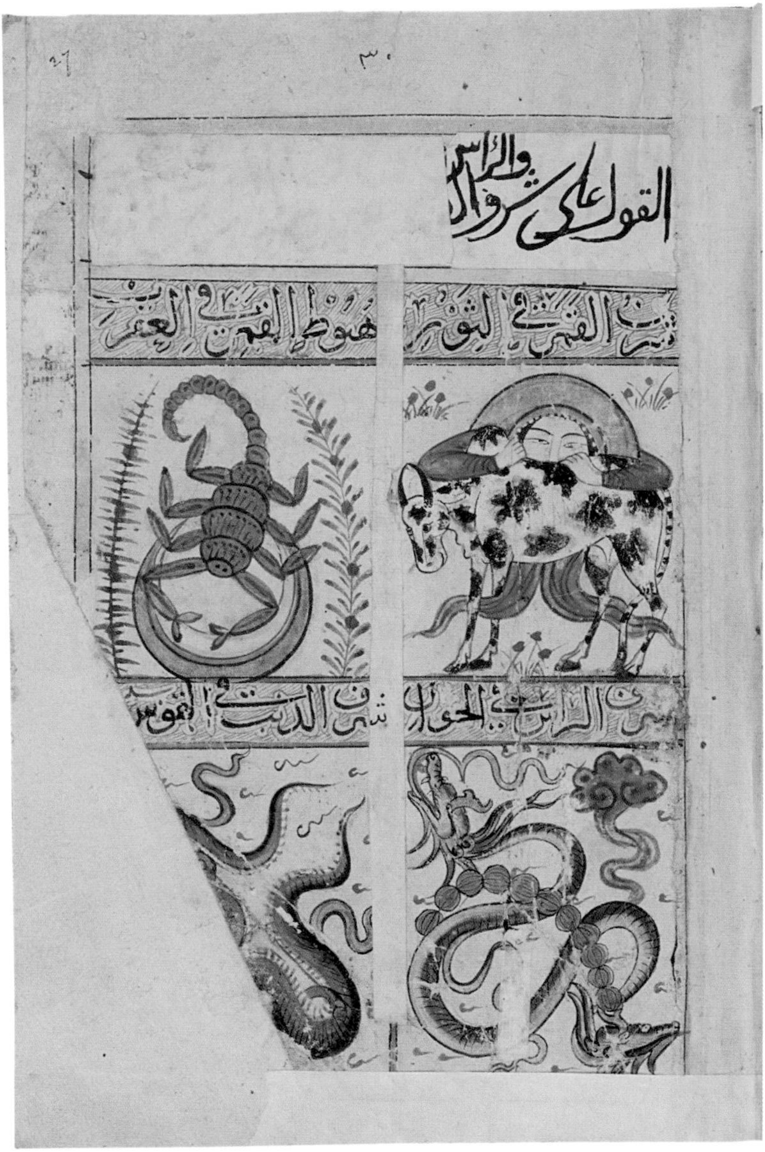

PLATE 25. Exaltation of the Moon in Aries (top right) and the fall of the Moon in Scorpio (top left). The exaltation of the Dragon's Head in Pisces (lower right) and the fall of the Dragon's Tail in Sagittarius (lower left). From a copy of *Kitāb al-Bulhān* produced in Baghdad during the reign of Sulṭān Aḥmad (reg. 1382–1410). Oxford, Bodleian Library, MS Bodl Or. 133, fol. 27a.

CHAPTER SIX

Ports, Gates, Palaces

DRAWING FATIMID POWER ON THE ISLAND-CITY MAPS

The logic of the *Book of Curiosities* is a logic of water. The maps of the three great seas—the Mediterranean, the Indian Ocean, and the Caspian—are followed by maps of Mediterranean islands and peninsulas, which are then followed by the maps of bays, lakes of the world and finally maps of great five rivers. In the previous chapter, we used the maps of the Mediterranean, of Cyprus, and of the Aegean Bays as windows onto the rudimentary charting of the Mediterranean coasts from the perspective of seamen and naval commanders. The three maps to be discussed in this chapter—the maps of the island of Sicily and the cities of Mahdia and Tinnīs—tell the story of Fatimid power in the Mediterranean from the perspective of a loyal Fatimid court official.

The choice of these three—Sicily, Mahdia, Tinnīs—is not coincidental. Apart from Alexandria, these were the three most important Fatimid holdings in the Mediterranean: the ports that made the Fatimids a Mediterranean power. Mahdia, the first Fatimid capital in modern Tunisia, and Sicily were the gateways to the eastern Mediterranean, and the most important stopping points for the lucrative trade coming in from the ports of the Maghreb and the Iberian Peninsula. Tinnīs, an island lying within a brackish lake formed by the meeting of the Mediterranean and an eastern arm of the Nile delta near modern Port Said, monopolised Egyptian maritime trade with Syria and Palestine. The maps of these three island-cities aim to project the tenacity of Fatimid control over the eastern Mediterranean through a visualization of political sovereignty and military might.

Modern scholarship has so far rarely discussed political messages of medieval Islamic maps. This is partly because the maps of the Balkhī School, which are the most common in the manuscript corpus, promote the power of Islam as a worldwide community, but are not explicitly aligned with any one dynasty. Their maps are a mosaic of ethnic groups and administrative divisions, while individual rulers are never identified.[1] They do not represent sovereignty in the way familiar to us from the iconography of European portolan charts, for example, where political authority is shown through an image of an enthroned king. The focus in Balkhī maps is on trade routes, and the Islamic world is envisaged as a chain of cities, a network of communication rather than a collection of potentates.

The Fatimid caliphs, however, did use maps to express political ambitions. We already mentioned the luxurious silk world map prepared for the caliph al-Muʿizz in 964, which carried the caliph's name and a statement of his longing to visit Mecca.[2] Its production was a declaration of the Fatimid goal of conquering the central Islamic lands, and the two Holy Shrines of Mecca and Medina in particular.[3] The map was made when al-Muʿizz was preparing for the conquest of Egypt, at the time seen only as a step toward Fatimid domination of the Islamic world as a whole. Specifically, the inscription at the bottom of the map linked the Fatimid caliph, who is "longing for the Sanctuary of God," with the sacred geography of the Prophet's life. This world map was more than a reflection of scientific inquiry. Rather, it expressed the aspirations of the Fatimid caliphs for universal hegemony, their hopes of conquering the Arabian Peninsula, and their claim to be descendants of the Prophet Muhammad.[4]

Similarly, the three Fatimid maps to be discussed here carry a message about political power, coupled with an exceptional aesthetic appeal. The three maps are joined together not only through a shared substance—that is, Mediterranean ports under Fatimid control—but also in their visual language and message. The visual representations of these cities focus exclusively on the walls and gates of the cities, the defenses of the ports and the fortified palatial complexes, at the expense of all other urban institutions. The aim in all three maps is to convey the impregnability of the fortifications. Although these maps are appended to textual descriptions, they stand independent of them, and they add or omit data in order to achieve their desired effect.

In the treatise, they are called "maps of islands," but in effect these are city maps, with Palermo and its environs taking up most of the space on the map of Sicily. Together, they form the earliest set of city plans to have survived and are a crucial addition to the handful of other urban maps to have come to us from medieval Islam. Other city maps known to us, including another map of Tinnīs from the pen of the Iranian al-Qazwīnī (d. 1283), focus on topography or on commercial institutions. This chapter opens a window on the hardly explored field of urban cartography in medieval Islam, in order to further emphasize the distinctly political message of the city maps in the *Book of Curiosities.*

Together, the three maps form a unique set of images of Mediterranean islands and peninsulas, zooming in, so to speak, on some of the nodal points in the dense map of the Mediterranean described above. Unlike the maps that we have discussed so far, the representations of these port cities are more than mere abstractions. While not aiming for verisimilitude, the images do capture the distinctive features of each city's defenses. Because of that, the three maps are also the most complex images in the treatise as a whole. They have no precedent known to us in the Islamic cartographical tradition, and only limited parallels from later centuries. To a degree, they underscore the tone of the entire manuscript.

The Impregnable Walls of Palermo

Chapter 12, "On the Largest Island in These Seas," is devoted to a textual and visual account of the island of Sicily, which had been under Islamic rule since the early ninth century and under Fatimid control since the early tenth. The focus of the narrative text is very much the human and physical geography of the capital city, Palermo, adapted from an account by Ibn Ḥawqal, who visited Sicily in 972–73. The author disregarded, or cut to a minimum, much of the moralistic and anecdotal elements in Ibn Ḥawqal's account, while appending a section on the Muslim conquest of the island under the Umayyads and the early Abbasids. The strategic value of the island as a bridgehead for attacks on the Italian mainland is stressed right at the beginning of the chapter: "The island of Sicily is the largest of the Islamic islands and the most honourable on account of its continuous military expeditions against the enemy."[5]

The text of the chapter provides a detailed account of the walls of Palermo, its gates, suburbs, markets, and water sources.[6] As for the walls, the author begins by saying that the wall of the city "is strong, tall and impregnable." His discussion of the nine gates of the city is particularly detailed, and the gates are listed in sequence:

> The most famous is the Sea Gate (Bāb al-Baḥr), because of its proximity to the sea. Close to it lies the Gate of Aḥmad ibn Abī al-Ḥasan Aḥmad ibn Abī al-Ḥusayn.[7] Next is the Gate of St Agatha (Bāb Shantaghathāt), which is an ancient gate. [Then comes] a gate which was created by Aḥmad ibn [Abī] al-Ḥusayn, where there is an excellent spring [which powers] many mills.[8] [Then come:] the gate called Gate Ibn Qurhub, then the Gate of the Buildings (Bāb al-Abnāʾ), which is the oldest of [the city's] gates. Then the Gate of the Blacks (Bāb al-Sūdān) opposite the blacksmiths, followed by the Iron Gate (Bāb al-Ḥadīd), from which

is the exit to the Jewish Quarter. Then there is another gate nearby, renovated by Abū al-Ḥusayn. The total number of gates is nine.[9]

Following the gates, the author mentions the nearby palatial town of al-Khāliṣah, now la Kalsa, which lay to the south of the port. He reports that it too has a wall and four gates, but provides far less information on it than Ibn Ḥawqal.[10] Suburban quarters, or *ḥārāt,* are also described, including three named by Ibn Ḥawqal, as well as a more recent one called al-Jaʿfarīya, which has ten thousand houses. The author is also interested in the freshwater sources of the city, reproducing information provided by Ibn Ḥawqal. Some space is also given to the markets of the city and to the main commercial street. He notes that all the markets are outside the walls, and only the grain merchants, a group of butchers, and the sellers of vegetables and fruits are found inside the city (a statement that conflicts with the account of Ibn Ḥawqal).[11] The only reference to the religious geography of the city is to a Christian relic associated with prayers for rain.[12]

The map of Sicily follows overleaf and is drawn over two folios (plate 7 and fig. 6.1). It shows the island as a flattened sphere, with no attempt to reproduce coastal details, except for a V-shaped indentation for the port of Palermo at the top of the map. The city of Palermo is represented as a circular enclosure in red, broken by eleven gates. The harbor, which lies outside the walls, is flanked by two towers, each labeled "Castle of the Chain." On the eastern side of the harbor, the arsenal (*al-ṣināʿah*) is shown. The domed structure on the top right is labeled the "Ruler's Palace" (*qaṣr al-sulṭān*) and represents the palatial city of al-Khāliṣah. Suburban quarters are indicated by rectangular boxes with yellow borders. A series of isolated peaks is found along the coast, with the volcanic Mount Etna shown on the lower left of the map with a red cap. Labels along the coast indicate distances in miles. Inland only the peaks of the valley of Conca d'Oro near Palermo are indicated. All other localities, including villages, valleys, mosques, markets, baths, springs, and castles, are indicated by red dots, or merely as labels.

Visually, the landscape of Palermo is dominated by the walls and the gates. The mapmaker is faithful to the textual description of the city, which specifically informs us that the city had expanded from a rectangle to a circle. The gates are shown in the same sequence in which they were described, with the Gate of the Sea correctly located opposite the port. Compared to the nine gates listed in the text, two additional gates are shown: the "Gate of the Chick Market" and the "Gate of the Well." The walls dwarf four labels referring to localities within the city walls, including the markets of the herb sellers and the flour merchants, which are mentioned in the preceding text as one of the few intramural markets. The other two labels are

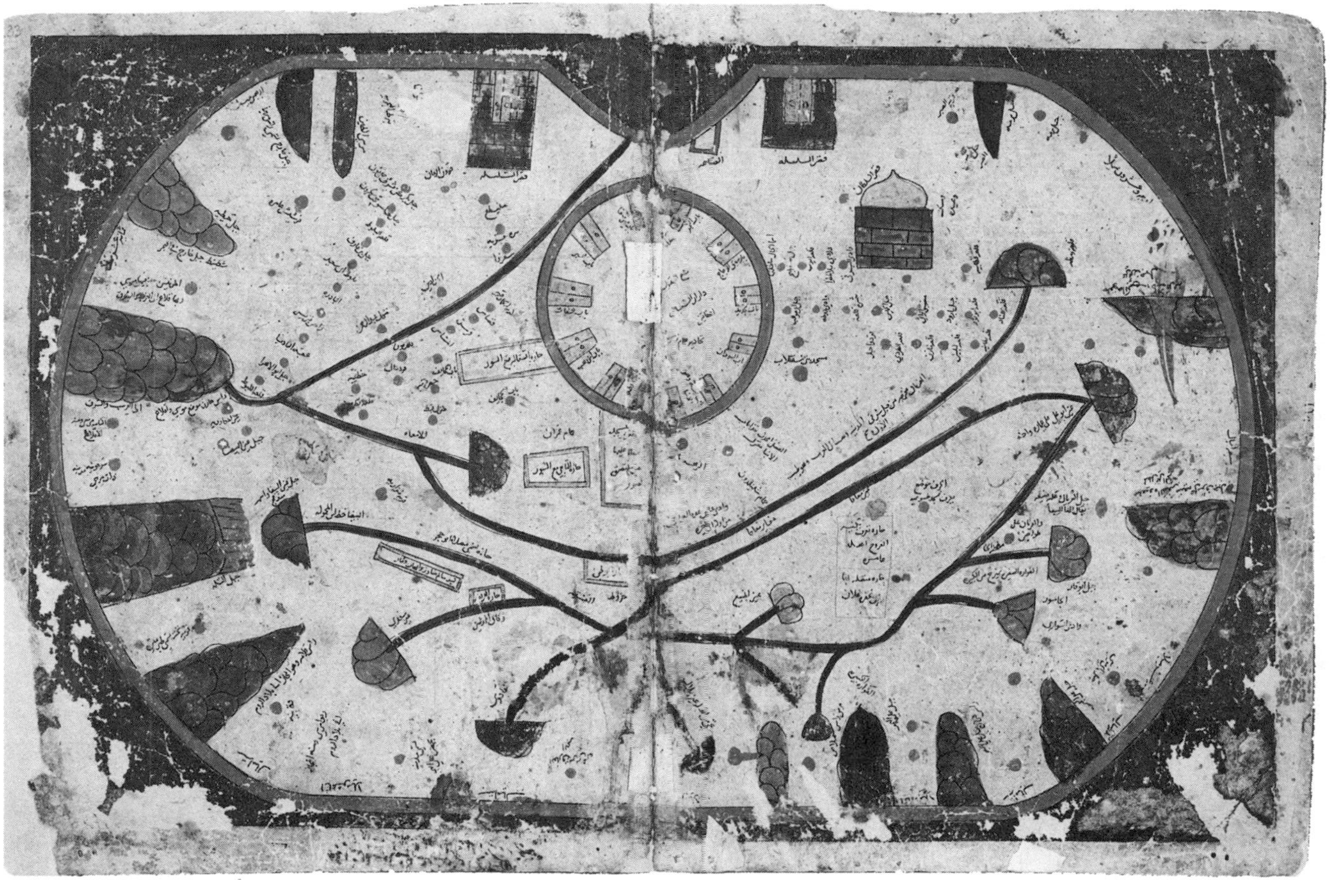

FIG. 6.1. Map of Sicily from the *Book of Curiosities*. Oxford, Bodleian Library, MS Arab. c. 90, fols. 32b–33a, copied ca. 1200.

enigmatic references to the "House of Ibn al-Shaybānī" and to the "Baths of Nizār."

The interest in walls extends to the eight suburban quarters, which are indicated by rectangular boxes with thin yellow borders. The thinness of these yellow borders is a visual reminder that their defenses are not as solid as those of the city itself. Two adjoining quarters on the bottom right ("Quarter of the Church of the Joyful" and "Quarter of the Ditch of Ghullān") are marked only with single-lined boxes, which may indicate an even lower level of defense, or merely an oversight on the part of the copyist.[13] Labels for three suburban quarters located nearest to the city—the Slav Quarter on the left, the "Quarter of the Mosque of Ibn Siqlāb," and the "Quarter of al-Tājī"—specifically mention the presence of walls. In fact, the fortification of Palermo's suburban quarters was a fairly recent development. We know from the author that the Slav Quarter had acquired walls only forty years before the composition of the *Book of Curiosities.* In 972–73, when Ibn Ḥawqal was writing his account of Sicily, none of the suburban quarters was fortified, so the map visualizes this rapid transformation of suburban sprawl into gated neighborhoods.

The map is in line with the expansion of the inhabited area of Palermo at the end of the tenth century, attested by narrative sources and archaeological excavations (fig. 6.2). The main city walls were reinforced in 967, and new gates opened by the emir Aḥmad ibn al-Ḥasan, who ruled Sicily between 954 and 969. This was followed by the building of the Jaʿfariyya suburban quarter and its walls. The Jaʿfariyya was almost certainly named after the emir Tāj al-Dawla Jaʿfar ibn Yūsuf (ruled 998–1019) and probably corresponds to the Tājī Quarter shown on the map.[14] Excavations confirmed the extension of extramural quarters in this period. In the areas which would have been part of the Quarter of the Slavs, structures show high technical quality and a regular street system, suggesting a planned growth.[15]

The map of the *Book of Curiosities* also gives a sense of the commercial activities taking place outside Palermo. Mills, orchards, and gardens lined up the streams to the west of the city, in particular Wādī ʿAbbās.[16] These may have been sites of aristocratic villas, predating the better-known Norman structures outside the city. Accounts of the Norman Conquest mention the delight of the victors in stumbling upon these suburban gardens.[17]

In addition to the walls, the two complexes that dominate the map are the harbor and the ruler's palace. The harbor is not mentioned at all in the preceding chapter, or in the account of Ibn Ḥawqal. The mapmaker, however, correctly locates the harbor to the north of the city and outside its walls. He prominently represents the twin citadels that guarded its entrance,

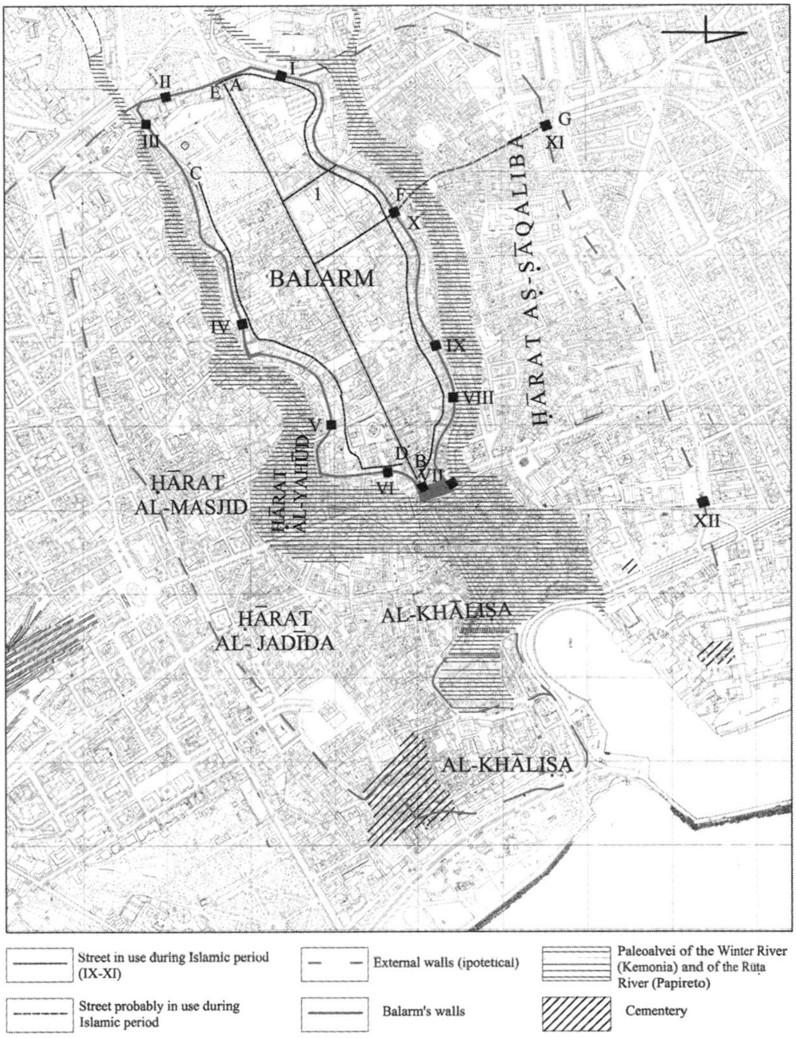

■ Gates

Fig. 1 Palermo in the late Islamic period (graphic by Dr. Maria Antonietta Parlapiano)

1 *jāmi'* mosque

Streets

AB. *Simāṭ, platea marmorea* ; CD. South *Shāri'*; EB. North *Shāri'*; FG. artery out of the Gate of Sant' Agata (via Beati Paoli, via Porta Carini)

***Balarm*'s Gates**

I. *Bāb Rūta*; II. *Bāb Ibn Qurhub?*; III. *Bāb al-abnā'*; IV. *Bāb al-Sūdān*; V. *Bāb al-ḥadīd*; VI. *Bāb sūq al-Dajāj*; VII. *Bāb al-baḥr*; VIII. *Bāb ash-shifā'*; IX. *Bāb al-Bi'r*; X. *Bāb Shantaghāt*

***Hārat al-Ṣaqāliba's* Gates**

XI *?*; XII *Bāb al-Ḥajjārīn*

FIG. 6.2. A plan of Islamic-era Palermo superimposed on a modern map, with south at the top. Prepared by Elena Pezzini and graphics by Antonietta Parlapiano, in Pezzini 2013, p. 228.

and from which a chain could be stretched in order to block access. Moreover, a red rectangle indicates the arsenal, or shipyard, of the harbor. This conflicts with the account of Ibn Ḥawqal, who reports that the arsenal was located within the palatial city of al-Khāliṣah. Modern scholarship locates the arsenal on the modern Piazza Marina, next to the harbor basin.[18]

The structure with the onion shaped dome, labeled the "Palace of the ruler (*sulṭān*), his residence and his slaves," is an emphatic visual representation of authority. This is supposed to represent the palatial city of al-Khāliṣah, established in 937–38 to house the representatives of the Fatimid caliph. The label on this map is derived from the account of Ibn Ḥawqal, who says that "it is there that the ruler and his followers reside."[19] Ibn Ḥawqal also notes that the walls of al-Khāliṣah are not as strong as those of Palermo itself. The author of the *Book of Curiosities*, on the other hand, only notes that the city has "a wall and four gates," and that the Palace, presumably meaning the same complex, is adjacent to the city walls. Yet, the visual message of the map accords this palatial complex a much more prominent place. It is located some distance away from the city walls, and its solid walls are indicated by brown bricks. The depiction of the palace and its walls is unlikely to be realistic. In one other medieval Islamic map, a schematic image of a domed rectangular structure represents Baghdad, the Abbasid capital.[20]

The map of Sicily is a very dense image that covers not only Palermo but the island as a whole. Yet its visual center is the capital city and its suburbs. It highlights the main aspects of the fortifications of a port city: gated walls, well-protected palace, and the defenses of the harbor. These three key elements are represented prominently and pictorially, and, although the designs are stylized, the mapmaker made an attempt to correctly represent actual distinctive features. Moreover, the coasts of the island are lined with peaks of mountains, which would make invasion more difficult. Each of the inhabited quarters around Palermo is encircled by a wall, and the city itself by a thicker wall. The harbor is well protected by imposing towers and chain, and the palace of the ruler is solidly built. The core of Muslim and Fatimid political power in Sicily appears secure—at least on the map.

Mahdia: The Palaces of the Imams

Strategically, Sicily was the key to power in the Mediterranean, and the jewel in the crown of the Fatimid caliphs. Yet the city of Mahdia, today no more than a small resort town on the Tunisian coast, was nearly as important. Its artificial harbor was as bustling with ships sailing westward from

Alexandria, and its significance for the Fatimid Empire was not only strategic, but historical too. Mahdia was the first capital built by the Fatimid caliphs when they were still a North African empire, and the city retained its symbolic importance to the dynasty even after its move to Egypt and the foundation of Cairo in 971. Fitting with this legacy, the map of Mahdia in the *Book of Curiosities* is an iconic image of Fatimid power. A millennium later, it still has distinctive charm. The Bodleian Library has elected to reproduce it on the mugs and scarves it sells, and Farhad Daftary has chosen it as the cover for his authoritative history of the Ismaʿili sect.[21]

The map of Mahdia is preceded by a textual account, devoted to two main topics. First, it tells the story of the foundation of the city by the Fatimid imam ʿUbayd Allāh al-Mahdī in 916–21. While this account has several parallels in medieval chroniclers, the version in the *Book of Curiosities* is one of the earliest to have reached us.[22] According to this narrative, al-Mahdī built his new capital at a site that could be fortified and defended against future attacks. When he laid the cornerstone of the city in the west, he predicted that an anti-Fatimid rebel called Abū Yazīd would lay siege to the city, but would be stopped at the public prayer grounds outside the city walls.[23] The second part of the chapter is devoted to Abū Yazīd's actual rebellion thirty years later.[24] The account dwells on the ruthlessness of the rebels and on the physical deformities of their leaders. It then recounts how the rebel forces were eventually held at the public prayer grounds outside the city walls, as foretold by al-Mahdī.

There is relatively little in this chapter about the layout of the city and its architectural landscape. It correctly describes the peninsula as being surrounded by the sea on all its sides, except the western side. It also states that "al-Mahdī fortified the city with a wall and sturdy iron gates."[25] It also notes in passing that al-Mahdī built "his palaces" in the city,[26] and that it had become the center of administration. It reports that when Abū Yazīd laid siege to the city, the Fatimid caliph ordered the digging of a ditch around the city, which halted the progress of Abū Yazīd's forces. This ditch was dug next to the public prayer grounds.[27]

In the map of Mahdia that follows (fig. 6.3 and plate 9), the focus is on the pictorial representation of three man-made complexes: the walls, the harbor, and the twin palaces of the imams. The stone walls surround the peninsula, pierced by two large gates in the west (bottom of map). These twin gates are labeled as "the two gates of the city," in the dual form. The walls form a rectangle, except for an indentation in the northern walls, which follows the contours of the coast. Second, an arch in the southeastern corner of the map, topped by another elaborate structure or structures,

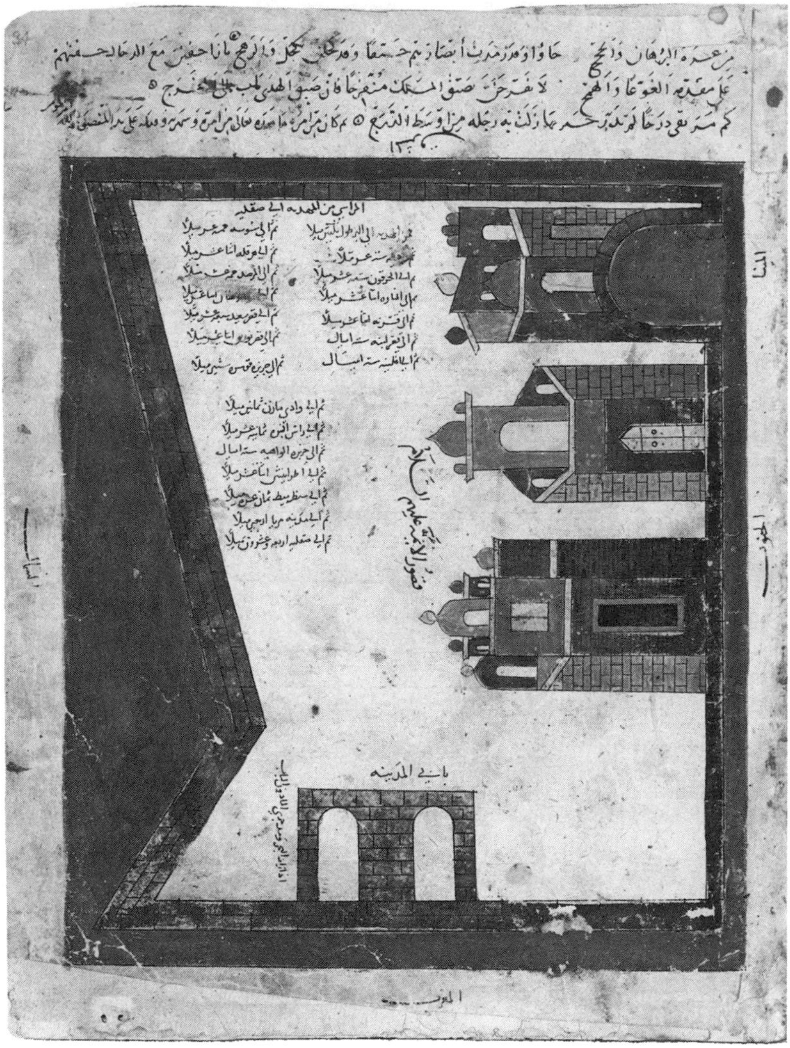

FIG. 6.3. Map of Mahdia in the Bodleian copy of the *Book of Curiosities*. Oxford, Bodleian Library, MS Arab. c. 90, fol. 34a, copied ca. 1200.

is labeled "the harbor." Thirdly, two isolated and elaborate buildings near the southern walls of the city are labeled "the palaces of the imams, may they rest in peace."

Apart from the three complexes of walls, palaces, and harbor, no other features are represented. The list in the top left, written in the open space within the walls, is a maritime itinerary between Mahdia and Sicily, and consists of the names of anchorages along the way and the distance

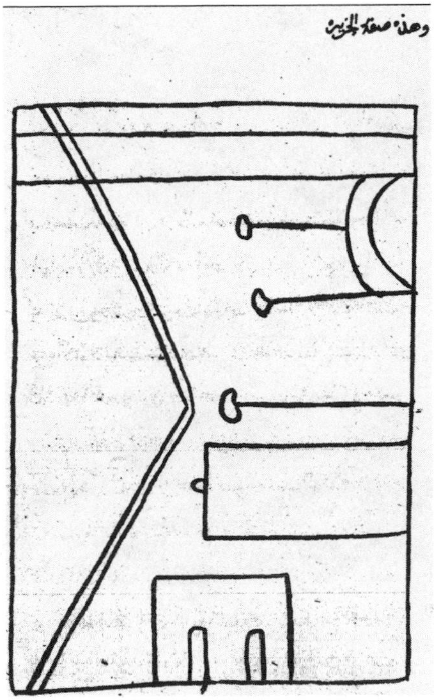

FIG. 6.4. Map of Mahdia in the Damascus copy of the *Book of Curiosities*. Damascus, Maktabat al-Assad al-Waṭāniyah, MS 16501, fol. 99a, copied 1564/972H.

between them in miles. Remarkably, the city is not represented as a peninsula but rather as an island. While the text of the chapter says that the city is linked to the mainland on its west side, the map shows the sea surrounding the city from all sides, and a label near the western gates reads: "When the sea rises and swells, the water reaches below the gate."

This plan of the city of Mahdia as preserved in the Bodleian manuscript is likely to be faithful to the original eleventh-century treatise. This is confirmed by the version of the map in the Damascus copy, which is not derived from the Bodleian copy (fig. 6.4). This abstraction corroborates the layout of the three complexes and confirms that all three were found in the original treatise. It shows an arch with two towers at the top right, a tower and another tall building on the right-hand side, and two gates at the bottom of the map. It also shows, as in the Bodleian map, a distinctive indentation in the shoreline on the left-hand side. The two lines at the top of the Damascus diagram were perhaps intended to accommodate the list of anchorages, which in the Bodleian copy are incorporated within the city walls.

The map of Mahdia in the Bodleian copy is far more detailed about the layout of the city than the text that precedes it. While the narrative has

FIG. 6.5. Satellite view of the peninsular town of Mahdia today. The rectangular shape of the Fatimid port is clearly visible in the southwest, as it is on the map of Mahdia in the *Book of Curiosities*. Courtesy of Google Maps.

no mention of the harbor of the city at all, the mapmaker is able to correctly locate it in the southeast corner and to provide an elaborate image of it. Against the textual description of a peninsula projecting into the sea, the mapmaker drew the city as an island surrounded by the sea from all its sides, including the western walls. Further discrepancies are also intriguing. The indentation in the northern walls of the city is not mentioned in the text. The text refers to "gates" in the plural, while the map depicts two gates, and specifically labels them in the dual form. Finally, while the text refers to the palaces of the imam al-Mahdī, the label on the map refers to the palaces of "the imams," in the plural, and visually depicts two main structures.

The additional information provided by the map, though absent from the preceding textual description, is corroborated by archaeological evidence and external textual sources.[28] The general layout of the city in the map corresponds to what is known about the actual layout of eleventh-century Mahdia (fig. 6.5). The indentation in the northern walls is clearly an attempt to represent the curve of the northern coast of the peninsula. The location of the harbor in the map also corresponds to the actual location of the artificial harbor, dug in the bed-rock of the southeastern corner of the city. Finally, the palaces of the Fatimid imams are known to have been located near the southern walls of the city.

Archaeological and textual evidence even suggests that the mapmaker attempted to capture the design of each of the complexes depicted. The visual representation of the harbor tallies with European reports on the existence of an arch above the entrance, linking two towers that guarded the passageway. The entrance, which can still be seen, forms a passage of about fifteen meters in width. Medieval Arab geographers noted that it was blocked by a chain stretched between two towers placed at each side.[29] A European map of Mahdia, dated 1554, shows these two towers linked by an arch, under which the ships passed. Such an arch above the entrance to the port is also mentioned by Luys Marmol in an account published in Spain in 1573.[30] The existence of this arch in Fatimid times has been questioned by the archaeologist A. Lézine, who pointed out the asymmetrical positioning of the remains of the two towers.[31] Yet, the map of Mahdia in the *Book of Curiosities* strongly supports the existence of such an arch during the Fatimid period. The map also shows structures on top or behind the arch—several of them in the Bodleian copy, two towers in the Damascus manuscript.

The twin palaces of the imams are located correctly near the southern walls of the city, and, at least in the Bodleian copy, are also made to look as if they are facing each other. The palace on the right has its gate facing south, while the left-hand palace is depicted with a large rectangular blind niche, suggesting that it is represented from the side or from the back. Such a presentation corresponds to contemporary descriptions of the twin Fatimid palaces that dominated the landscape of the city. According to the Andalusian geographer al-Bakrī (d. 1094), the large palace of ʿUbayd Allah al-Mahdī was distinguished by the magnificence of its main building. He also says that its gate turned toward the west. Facing it across a large square was the palace of his son Abū al-Qāsim, with its gate turned toward the east.[32] Archaeological excavations found very few remains of these Fatimid palaces, although Slimane Zbiss identified a north-facing (rather than east-facing) structure as the palace of the Abū al-Qāsim.[33] It seems clear that the mapmaker attempted to capture the position of the two palaces in relation to each other. It is also notable that some attention has been given to individuating each of the two buildings. The right-hand structure consists of a tower surrounded by high walls, while the left-hand structure shows a wider central building.[34]

The imposing twin gates at the western walls of the city were also noted by most medieval travelers to the city. Ibn Ḥawqal, visiting the city during the second half of the tenth century, states that the walls of Mahdia have two doors (*bābān*), "the likes of which I have not seen on the face of the Earth," except in Raqqa in Syria.[35] Al-Bakrī reports that "the city has two iron doors, without any wood. Each door weighs 1000 kintar and is 30 spans tall. Each of its nails weighs 6 pounds. Images of animals decorate these two doors."[36] A century later, the North African geographer al-Idrīsī (d. 1165) lauds the two iron gates as two of the world's wonders.[37] Modern scholars have debated the meaning of these references, with some arguing that there was only one gate of entry to the city, at the location of the current *al-saqīfah al-kaḥlāʾ*, fitted with two iron doors.[38] Other scholars have argued that there was a second entry gate, which either faced the sea, or guarded the entrance to an advance rampart that protected the main wall.[39] The plan of Mahdia in the *Book of Curiosities* is a valuable addition to this debate, strongly suggesting the existence of two points of entry to the city through its western walls. The depiction of the two gates on the map is evidently not incidental or ornamental, as it fits so neatly with the accounts of medieval travelers. Like the depiction of the port and of the palaces, this image aims to capture the distinctive features of the actual walls.

The omissions from this map of Mahdia are also telling. For example, given the centrality of the extramural public prayer grounds to the narrative of the rebellion of Abū Yazīd, one would expect to find them depicted on the map, but they are not. Two other notable omissions are of the arsenal, constructed by al-Mahdī,[40] and the congregational mosque of Mahdia, with its renowned portal. The mosque is widely discussed by medieval travelers and chroniclers, and to this day is a striking feature of the urban landscape. But the *Book of Curiosities* does not mention it, either in the text or in the map. The civic institutions of the city, such as mosques or markets, are of no interest. It is the political and military aspects which are the exclusive visual focus.

Part of the intramural space is taken up with the maritime itinerary from Mahdia to Palermo. The first thirteen stops are along the coasts of Ifrīqiyah, modern Tunisia, then from Kelibia to the island of Pantelleria, and from there to Mazara del Vallo on the southwestern coasts of Sicily and onward to Palermo. The function of this maritime itinerary, located as it is in the blank intramural space of Mahdia, is to underline the strategic location of this port city. Zayde Antrim makes the interesting argument that the itinerary, by its mere connectivity, "complicates or belies" the impenetrability of the city walls.[41] But it is also possible to read the message of this maritime itinerary differently. Tracing the maritime highway from Mahdia to Palermo, that other key Fatimid Mediterranean port, could be seen as reinforcing the visual message of the map about Fatimid power.

In representing the city as an island rather than a peninsula, the map is at odds with both the preceding text and the actual topography of the city.

We know that the mapmaker intended to draw a body of water below the western walls of the city, as a label informs us that the sea can rise up to just below the westerly gates. It is possible that the mapmaker intended here to illustrate the ditch dug by the Fatimid caliph "around the city" during the siege of Abū Yazīd, as described in the preceding text. A moat filled with rainwater near the western walls is also mentioned by al-Idrīsī.[42] In any case, even if the mapmaker has intended to represent a ditch or a moat, he did not label it as such, creating the false visual impression of Mahdia as an island.

The map of Mahdia stands out as an independent artifact, only loosely connected to the text that precedes it, and one that carries a distinct visual message. That message is expressed through the three Fatimid complexes depicted pictorially, which are the gated dark walls, the harbor, and the walled palaces of the imams. As with the walls of Palermo, here too the intended message is the sense of impregnability. Moreover, the representation of these key features is not abstract, but grounded in physical reality. It aims to capture the distinctive architecture of the city, such as the twin gates, the vaulted entrance to the port, and twin palaces facing each other. By doing so, it individuates them and allows the viewer to identify them immediately. Even the topographically erroneous depiction of the city as an island, rather than a peninsula, may be an attempt to underline its unique defensive features.

There are strong visual and thematic links between the maps of Mahdia and of Sicily/Palermo. Mahdia, the first Fatimid capital, served as the model for Fatimid building projects in Palermo. The same caliph, al-Qāʾim bi-Amr Allāh, who ordered the building of the palatial al-Khāliṣah compound in Palermo also constructed one of the twin palaces shown on the Mahdia map. More generally, both were seats of Fatimid government near to the major ports of the central Mediterranean. Both palatial areas were surrounded by larger suburban areas where commercial activities took place. While al-Khāliṣah abutted the old city of Palermo, palatial Mahdia depended on the unwalled residential town of Zawīlah, not shown on this map but mentioned in the text of the chapter.[43]

In the map of Sicily, the palatial city of al-Khāliṣah and the walls of Palermo take center stage, in the same way Manhattan is foregrounded in the famous tourist maps of New York. This is a nonconcrete representation of a round city and an onion-domed palace lying within a Sicily of an unfamiliar shape. Mahdia, on the other hand, is shown on its own, with its actual coastline, buildings and architecture. It almost gives the impression that it was drawn by someone looking at it from the outside, either from the residential town of Zawīlah further on the coast, or perhaps by a person on board a ship making its way to the city harbor. Even when we are on firm land, the *Book of Curiosities* keeps reminding us of the sea.

The Lost City of Tinnīs

Over the last millennium, Mahdia lost its medieval glory and strategic importance, and is today not more than a modest town. But at least it survived. The port of Tinnīs in the Nile delta, on the other hand, paid the ultimate price for the desirability of its location and vulnerability. Tinnīs was a major center for the production of textiles, as well as the major port on the northeastern branch of the Delta, a natural stopping point for trade between Egypt and the Syro-Palestinian coast. It was repeatedly attacked by Crusader fleets until Saladin ordered it to be evacuated from its civilian population in 1192–93.[44] The city was eventually completely abandoned during the course of the thirteenth century, and its remains are now entirely covered by sand. The map of Tinnīs in the *Book of Curiosities* is therefore like a map of a sunken Pompeii, a gift from Heaven for archaeologists in the early stages of trying to reconstruct the site. Owing to the salinity of the soil and the gradual nature of abandonment, Tinnīs will not yield the same treasures as Pompeii, but in its heyday this city of fifty thousand inhabitants and five thousand weaving looms was one of the most important Fatimid assets in the Mediterranean.

The textual account of the city of Tinnīs in the *Book of Curiosities* is not original to the author, but is taken verbatim from "The Companion Guide to the History of Tinnīs" (*Kitāb Anīs al-jalīs fī akhbār tinnīs*), written by a certain Muḥammad ibn Aḥmad ibn Bassām, who was the market inspector of the city at the beginning of the eleventh century.[45] In medieval Islamic cities, which had neither guilds nor corporate city councils, the market inspector was the key official responsible for regulation of weights and trades as well as supervision of public morality. Ibn Bassām's account of Tinnīs, preserved in the *Book of Curiosities* as well as in independent copies, is an exceptionally rich account of the economic and social life in this industrial town. He lists weaving looms and merchant inns, provides the precise measurements of the congregational mosque, and even gives an estimate of the city's population based on its bread consumption. Ibn Bassām specifies the layout of four extramural suburbs of the city, including the arsenal and the governor's residence (*dār al-imāra*), public prayer grounds in the northern suburb, and fishermen's storehouses in the southern.

The author of the *Book of Curiosities* follows up Ibn Bassām's account of Tinnīs with his own map of the city. In this map, remarkably, it is only the walls and the ports of the city that are pictorially represented (fig. 6.6 and plate 8).

The diagram of Tinnīs extends over two folios and shows the city with the green Mediterranean at the top of the map and, on the other three sides, the blue deltaic lake that encircled the island-city. The rectangular walls of the city have twelve gates, which are indicated by simple squares, without doors or arches. Each of the western and eastern sides has two gates, and four more gates are located in the northern side. The southern walls have six gates, two of which regulate water channels that flow into the city. The labels near these twin gates indicate that this is the entry to the harbor. The one on the left reads, "This is a harbor for ships, on which there is a gate"; and the right-hand label, "This is a harbor into which ships enter."

This diagram of the city walls and its port partly follows the preceding text. In the textual account of the city, the walls are said to have "nineteen gates for entry and exit, one of them plated with copper and the rest plated with iron. There are also two archways leading to two ports, each locked by an iron-plated gate preventing anyone from entering or leaving without permission."[46] The mapmaker chose to depict fewer gates (twelve), and made no special attempt to visualize their shape or strength. But he did choose to represent the two gates that regulated access to the two ports. These are located at the bottom left, on the southern side of the city (this detail is not mentioned in the preceding text). It is a schematic representation, especially when compared with the more elaborate depiction of the entrance to the ports of Mahdia and Palermo. Nonetheless, the image captures the distinctive features of the defenses of the harbor.

Results of a recent geophysical survey suggest that the map, schematic as it is, makes a cumbersome attempt to represent actual topography.[47] The contours of the city walls covered much, but not all, of the island.[48] Geophysical investigations also located a large canal, preserved as a low-lying topographical feature running into the town from the south. A somewhat wider basin was located immediately within the gates, and the canal then narrowed as it approached the center of the city. This canal probably served as the harbor of the city, very much like the depiction in the *Book of Curiosities* map. The location of this channel depression in the site is also a good fit with the location of the harbor on the map.

The survey of the site also found two other channel depressions farther to the north and west, a close parallel with the inlets drawn at the bottom left of the map in the *Book of Curiosities*. Alison Gascoigne, the archaeologist leading the survey, suggested that the twin inlets depicted on the map are the two westerly channels found in the survey, and that both were used for water intake.[49] As Ibn Bassām reports, during the flooding of the Nile,

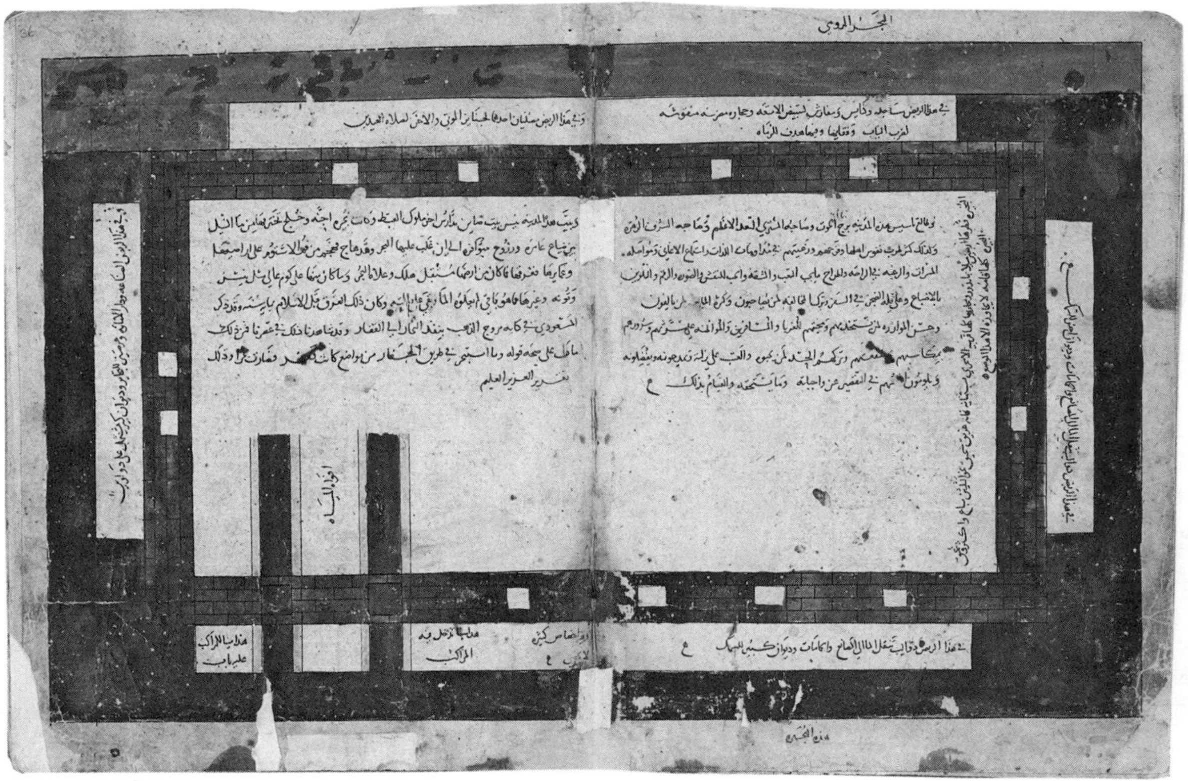

FIG. 6.6. Map of Tinnīs in the *Book of Curiosities*. Oxford, Bodleian Library, MS Arab. c. 90, fols. 35b–36a, copied ca. 1200.

when the water of surrounding Lake Manzala became drinkable, channels from the lake were opened and water was allowed to run into cisterns and water installations in the town. However, the map explicitly identifies the two inlets as harbors for the entry of ships. It is very likely that at least one of them is the large southern channel found in the geophysical survey.[50]

In the *Book of Curiosities*' map, the intramural area of the city is void of any topographical detail or iconography. The markets and textile workshops, the mosques and inns, all of which are so vividly described in the preceding text, do not merit a place on the map. The vast space inside the city walls is used for the text of two final paragraphs by Ibn Bassām devoted to the favorable traits of the city inhabitants and its pre-Islamic history. We know that these passages were an integral part of the map in the original treatise, since the independent Damascus copy of the treatise, which retains the preceding account of Tinnīs, omits these lines together with the map. Also within the city walls is the long label written vertically on the right-hand side, along the eastern walls of the city. It describes the size and depth of Lake Manzala, the deltaic lake surrounding the city. The information on Lake Manzala, and specifically with regard to the shallowness of its waters, is again of navigational significance, and therefore fitted with the overall message of the map.

As in the maps of Mahdia and Sicily/Palermo, the map of Tinnīs is meant to convey a message about the ability of this important Fatimid port city to withstand attacks, especially naval attacks. The mapmaker was not uninterested in other features of the urban topography. The labels on the four sides of the walls describe the four suburbs of the city, and are taken verbatim from the preceding account. But none of these extramural complexes is represented pictorially, not even the arsenal and the governor's residence on the western suburb. The reason for the omission of the government complex from the map of Tinnīs may partly be political. Depicting a local governor's residence would put him on a visual par with the glorious Fatimid imams or the autonomous Kalbid emirs of Palermo. It is also possible be that in Tinnīs the governor's residence was not fortified, and did not have any significant role to play in the defense of the city.

City Plans in Medieval Islam

In all three maps discussed, the most prominent pictorial representations are of the city walls and its gates. When deciding how to visualize urban space, the mapmaker made a deliberate decision to focus on fortifications. As far as we can tell, he was not following any cartographic convention. When other medieval mapmakers wanted to indicate a city on a regional map, they usually used icons of squares, circles, rosettes, or four-pointed stars, as witnessed in medieval copies of the maps by al-Iṣṭakhrī, Ibn Ḥawqal, or al-Idrīsī. Sometimes, especially if the localities are stopping places on a straight route, they resemble small tents or perhaps doors to caravanserais. Such representation of cities as connectives fits in with the main purpose of the Balkhī maps, which is to represent the network of communication across the provinces.[51] Representing cities by a wall or a tower is rarely found in the cartographic tradition. One of the few examples comes from the rectangular world map *Book of Curiosities* itself, where the city of Constantinople is flanked by a large brick wall on its western side.[52] All in all, the use of walls or towers to represent cities is surprisingly uncommon in medieval Islamic cartography.

The maps of Sicily/Palermo, Mahdiya, and Tinnīs stand in sharp contrast to the few other extant city maps or island maps from the medieval Islamic period. Jeremy Johns and Emilie Savage-Smith have already pointed out the differences between the map of Sicily in the *Book of Curiosities* and that made by al-Idrīsī a century later.[53] In al-Idrīsī's map, the island is shown with characteristic triangular shape in the midst of its archipelago, with the toe of Calabria to its left and Sardinia to the right (fig. 6.7). The principal rivers and dominant relief are indicated in stylized but recognizable form. Mount Etna is clearly shown in the northeast corner, a chain of mountains is illustrated just inland from the north coast, and the hills of the south and the interior are represented schematically but accurately in relation to the rivers of the island. Approximately twenty-five or so towns and cities—all on the coast, except for two or three in the interior—are marked on this map, invariably as rosettes. Palermo is indicated, but is one city among many, and no attention is given to its fortifications or to the fortifications of any other city. While al-Idrīsī aims to capture the topography of the island as a whole, as well as its distinctive shape, the mapmaker in the *Book of Curiosities* is primarily interested in Palermo's defenses, and is ready to distort the familiar shape of the island in order to make his point.

Another telling comparison is between the map of Tinnīs in the *Book of Curiosities* and another map of the same island, found in a 729/1329 copy of "The Monuments of Nations" (*Āthār al-Bilād*) by al-Qazwīnī (d. 1283); see fig. 6.8.[54] The chapter on Tinnīs in al-Qazwīnī's work was written after the city had been abandoned, but it reproduces some passages from the account by Ibn Bassām, in particular a long list of local fish and birds. The beginning of al-Qazwīnī's account is concerned with the layout of Lake Manzala, and its unique position between the Mediterranean and the Nile. The map that follows illustrates the text and explains it. It depicts the opening of Lake Manzala toward the Nile at the bottom right and two openings

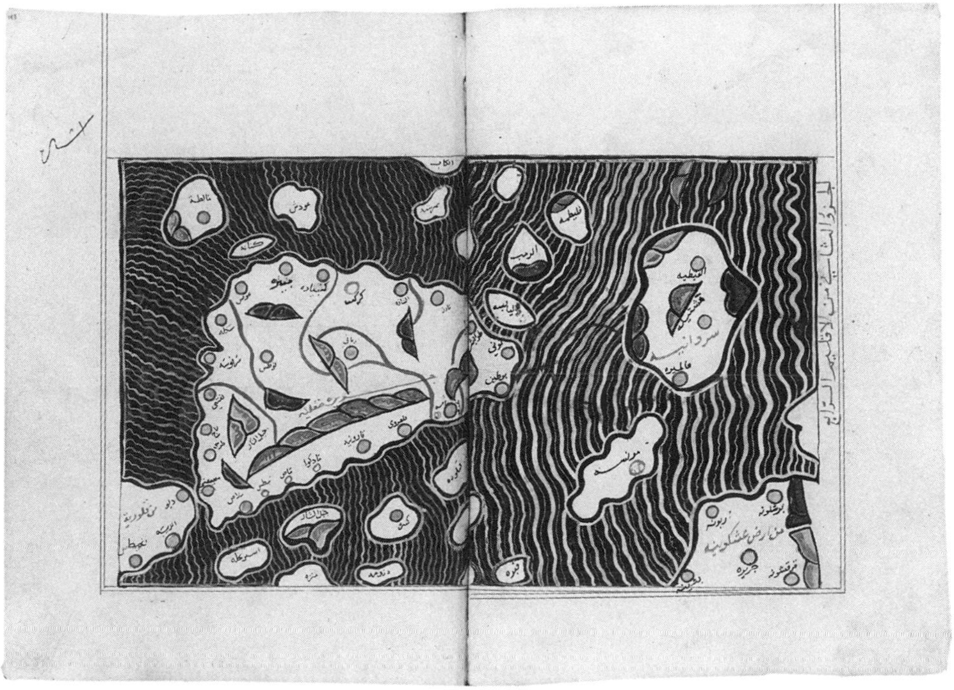

FIG. 6.7. Map of Sicily by al-Idrīsī. Oxford, Bodleian Library. MS Pococke 375, fols. 187b–188a. Copy completed July 25, 1553/13, Shaʿbān 960H, by ʿAlī ibn Ḥasan al-Hūfī al-Qāsimī.

toward the Mediterranean on both sides. The stretch of land that separates Lake Manzala from the Mediterranean is depicted as a semicircle at the top of the diagram, and the island of Tinnīs itself as a circle within the Lake. The mapmaker in al-Qazwīnī's treatise is interested in the unusual location of Tinnīs on an island within a deltaic lake. The walls of the city and its port are neither mentioned nor represented.

The same copy of al-Qazwīnī's work also contains a plan of his hometown of Qazvin in Iran. The plan is highly stylized and consists of four concentric circles (fig. 6.9).[55] The two central circles represent the old city and the "larger city," that is the city as it existed in the thirteenth century. The third and fourth bands represent, respectively, the surrounding orchards (or gardens) and cultivated fields. On this symmetrical diagram, two named rivers are shown to cut through the city and its suburbs. The diagram is directly linked to the preceding textual account, where the old city is said to be enclosed within the larger one. The concentric bands of gardens and fields, as well as the two rivers, are also mentioned in the preceding lines. In the text, the author specifically notes that both the smaller old city and the larger modern city have "walls and gates." But this detail is omitted from

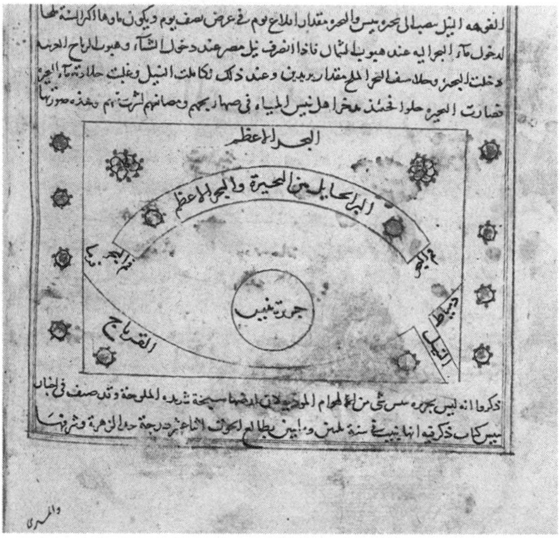

FIG. 6.8. Town plan of the city of Tinnīs by al-Qazwīnī. British Library, MS Or. 3623, fol. 49b, copied in 1329/729H. Reproduced with permission of the British Library.

the expository plan that follows, and the walls and the gates are not represented. The mapmaker, most probably al-Qazwīnī himself, was surely acquainted with the features of the walls, but had no interest in depicting them. The map highlights the typical concentric layout of a medieval town, disregarding any strategic or military aspects.

The largest set of town plans from the medieval Islamic tradition comes from the work of the traveler and merchant Ibn al-Mujāwir, written in the 620s/1220s, and betrays this author's primary interest in commerce and sacred geography. This work is a geographical and historical compilation on the Arabian Peninsula, combined with firsthand accounts. It includes thirteen schematic town plans or maps, depicting, among others, Mecca, Zabīd, and Taʿizz.[56] The most complex map drawn by Ibn al-Mujāwir is that of the port of Aden, which appears to be surrounded by the sea on the south and by the Lake of the Persians on the northwest. As in the map of Mahdia, the port town is made to look like an island, surrounded by water on all sides, and the message in both cases is that of a well-protected domain. But in Ibn al-Mujāwir's Aden, the emphasis is undoubtedly commercial. The two main features at the center of the map are the circle labeled "balance," probably the weighing house, and the label "custom house," which is prominently placed above the straight horizontal line that runs across the diagram. To the left, a long label indicates the place from which one can spot the ships of the Kārimi long-distance merchants arriving from Egypt.[57] The defenses of the port are not depicted. The map is drawn, as we

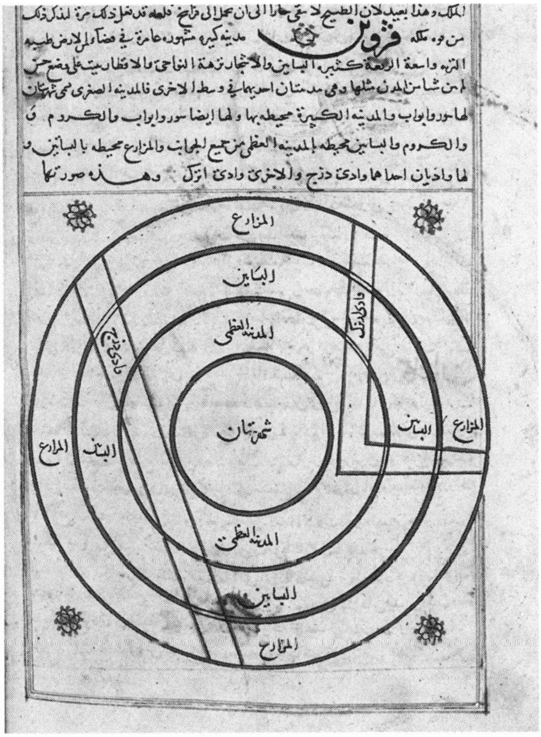

FIG. 6.9. Town plan of the city of Qazvin by al-Qazwīnī. British Library, MS Or. 3623, fol. 119b, copied in 1329/729H. Reproduced with permission of the British Library.

are explicitly told, from the fortress of Sīrah, an islet facing the main port. This means that a key feature of the fortifications of the port is omitted by the very design of the map.

One final comparison that underlines the visual message of *Book of Curiosities* is with the Late Antique mosaic maps of Madaba. The vivid images of urban agglomerations in mosaics dating from the fifth to the eighth centuries, mostly found in what is today modern Jordan, formed an important cartographic tradition in the Late Antique and early Islamic Near East.[58] In his study of these maps, Glen Bowersock dispels earlier interpretations of the mosaics as biblical commentary divorced from reality, and rather views them as the visualization of a union of the Late Antique Hellenized centers of the Near East, a cultural space that endured into the Umayyad period.[59] Specifically, a close analysis of the images of the various towns represented shows that the buildings carry distinctive and realistic forms. On the Madaba mosaic, Jerusalem famously shows recognizable features of the city, such as the Damascus Gate and the Church of the Anastasis (fig. 6.10).

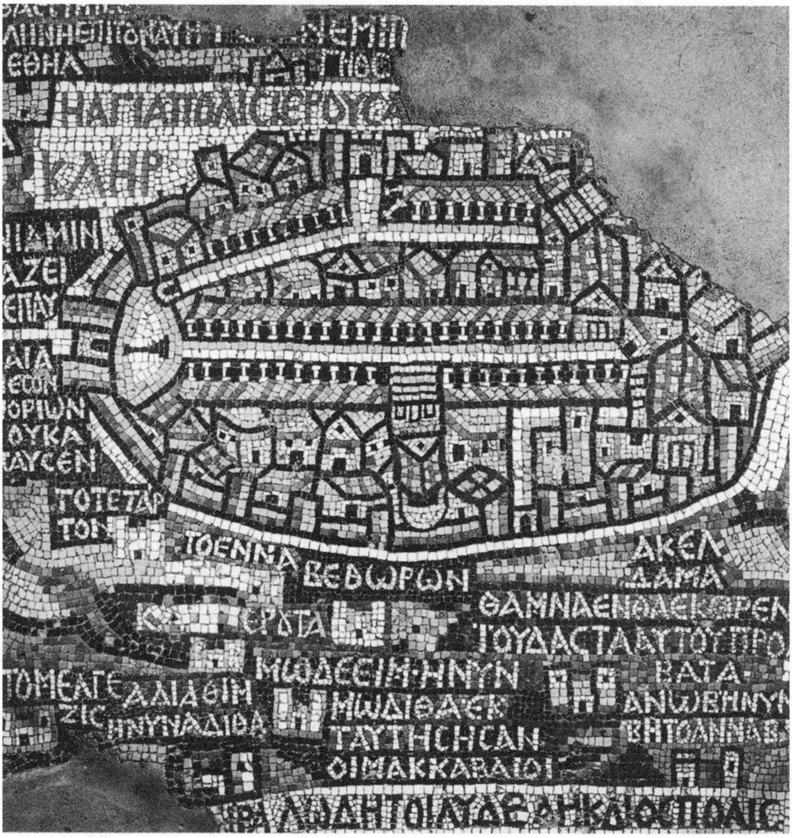

FIG. 6.10. Detail from the Madaba map. Reproduced courtesy of ACOR, the American Center of Oriental Research, Amman, from M. Piccicillot and E. Alliatta (eds.), *The Madaba Map Centenary 1897–1997: Travelling through the Byzantine Umayyad Period* (Jerusalem: Studium Biblicum Francscanum, 1998).

This tradition of mosaic representations does not aim at verisimilitude, but tries to capture some distinctive architectural features so that the viewer would be able to recognize the individual city.[60] A series of city mosaics at Maʿin, also in modern Jordan, shows highly individual buildings in Alexandria, Chracmoba (Kerak), Areopolis (Rabba), Gadoron, and Esobus. The recently discovered grand mosaic in the Church of St. Stephen at Umm al-Rasas, with an eighth-century inscription, has also architecturally detailed images and identifiable buildings, such as the Temple of Zeus in Nablus. In this respect, these mosaic maps are akin to the port city maps in the *Book of Curiosities*, which also select distinctive architectural features in order to individuate the depicted structures. But the mosaic maps are also very different. In the Madaba map cities are identified by means of key religious and civic institutions, densely cramped within the walls. The maps of the *Book of Curiosities*, on the other hand, show empty spaces. Whereas the Late Antique mosaic tradition represents churches and temples, none of the maps of the *Book of Curiosities* has a visual depiction of a mosque.

Cities, Ports, and Power

As Anna Contadini reminds us in her discussion of miniatures in medieval Arabic manuscripts, images and text need to be read in conjunction with each other, as two parts of a whole.[61] This applies to the *Book of Curiosities* as well. The maps and diagrams in the treatise not only need to be interpreted in light of the text to which they are attached. Rather, they should also be seen as conveying a visual message that interacts, reinforces and even conflicts with individual passages and with the overarching framework of the treatise.

The three city island maps of Sicily, Mahdia, and Tinnīs are embedded within the text of the treatise and autonomous of it. The common elements of the visual representation of the cities bridge the marked differences between the three textual accounts. The account of Mahdia mainly consists in historical narrative, with little geographical detail. The account of Palermo is based on Ibn Ḥawqal and addresses the urban topography of Palermo and its environs. The description of Tinnīs, taken verbatim from the account of its market inspector Ibn Bassām, is mainly concerned with social and economic life.

It is the maps that tie these three very different chapters together, by conveying a similar message about the impregnability of the fortifications of the three cities. That is not to say that there are no differences between the three maps. Palermo is depicted within the context of the entire island, surrounded by springs, mountains, and extramural walled quarters, while Mahdia is depicted as if seen by a visitor approaching the city. The map of Tinnīs is more schematic than the other two, with all extramural suburbs indicated by textual labels only. And yet, the three maps share an interest in fortifications and gates, visualization of ports, and, at least in the case of Palermo and Mahdia, a foregrounding of political authority. Comparable visualization groups these key Fatimid Mediterranean strongholds together.

These maps are intended to convey a message related to the submission of these city islands to Islamic, and Fatimid, authority. They stand in contrast to the map of Cyprus in the next chapter, where there are no indications of fortifications or visual representations of the ports. If you compare the map of Cyprus to that of Sicily, a few folios earlier in the treatise, the visual difference is striking. Both maps represent large Mediterranean islands, but in the map of Cyprus there is no pictorial representation of mountains or rivers, and no walls, either around cities or at the entrance to the harbors. While the Fatimid island of Sicily appears well protected through natural topography and man-made fortifications, the Byzantine island of Cyprus appears flat and open to attack.

The textual and visual accounts of the three island-cities invoke the Fatimid struggle to hold onto their Mediterranean empire, but also their hopes to expand it. Some of the passages in the *Book of Curiosities* even suggest Fatimid aspirations for recapturing some of the Byzantine possessions in the eastern Mediterranean. The chapter on Sicily opens with a reference to Muslim raids against the Christian enemy and ends with an account of the Muslim conquest of the island. There are similar accounts of the conquests of Cyprus, Crete, and Rhodes in the early Islamic period, three islands which subsequently fell under Byzantine control.[62] The three maps discussed in this chapter convey the ongoing military outlook of the Fatimids in the Mediterranean, aggressive as well as defensive. Port cities were always on the frontier because they were vulnerable to naval attacks even if located deep in Muslim territory.[63] The visual focus on the fortifications of the cities, and specifically the fortifications at the entry to the harbors or ports, has a defensive message, but also suggests that they could serve as launch pads for a Fatimid wave of conquest in the Mediterranean.[64]

Ironically, this grandiose depiction of walls and palaces could not prevent the imminent demise of the Fatimid caliphate as a Mediterranean power. These maps may have been useful as pieces of propaganda, but they were no substitute for military might. Drawing thick walls does not stop a Norman onslaught. Within decades, this trio of Fatimid Mediterranean ports fell in quick succession. The Fatimids lost Mahdia in 1057, when it fell to the hands of the Zīrid ruler al-Muʿizz ibn Bādis. A force of Pisans and Genoese raided the city in 1087, and it then came under repeated Norman attacks, leading to its seizure by Roger II in 1148 for a period of twelve years. Palermo was lost to the Normans in 1072, and the last Muslim stronghold in Sicily capitulated in 1091. The island was never to be recaptured by Muslim armies. As for Tinnīs, in 1227 the Ayyubid sultan al-Malik al-Kāmil ordered the destruction of the remaining fortifications so that they would not serve as a base for an invading Crusader army. They are now buried under the sand, with the map of Tinnīs in the *Book of Curiosities* a rare testament to its medieval brilliance.

CHAPTER SEVEN

The Fatimid Mediterranean

Medieval itinerary maps—like the road maps of today—highlight the routes most commonly taken, the most important networks of communication at a given point in time. As such, they are invaluable for the emerging field of global history, which seeks to understand how different parts of the world interacted with each other. The most famous example of an itinerary map is the Peutinger Table, which shows the road network of the Roman Empire in the fourth or fifth century AD. From the Islamic world, the most prominent examples of itinerary maps are those of the Balkhi School of geographers, which show the trade networks of the Islamic world in the tenth century, represented as straight lines joining together stops on long distance caravan routes.

But when the author of the *Book of Curiosities* designed his maps, what he saw from the window of his study in landlocked Cairo were not the dusty roads leading to the Red Sea port of al-Qulzum or the canal network leading to Alexandria. He didn't think of drawing maps that would show the land route from Cairo to Palestine and Syria. Instead, in his cartographic imagination, he saw the sea. He imagined travel as occurring mainly within and through maritime spaces. And by imagining travel as primarily seaborne, he also largely dispensed with the need to depict movement along a linear line. On land, one has to follow the beaten track. But the sea is open, and when winds and current are favorable, travel could be at any direction.

Viewed from Fatimid Cairo, the most important, nearest sea was the Mediterranean. But what kind of Mediterranean did our author see in his cartographic imagination? What did it include and what did it exclude? Most importantly, did he think of the Mediterranean as one sea? Historians have been debating the unity of the Mediterranean since Fernand Braudel's landmark volumes, and the debate was reinvigorated over the past decade with the publication of Peregrine Horden and Nicholas Purcell's *The Corrupting Sea*, with its emphasis on fragmentation and micro ecologies.[1]

From the perspective of Mediterranean studies, the *Book of Curiosities* offers a rare example of a medieval Islamic treatise that has the Mediterranean maritime space as its center of attention and as a subject of detailed familiarity. The knowledge available to the author, the choices made about omission and inclusion, the terminology used, and the visual representation of Mediterranean space can allow us to reconstruct the Fatimid Mediterranean—not only read about it, but also *see* it as it was imagined by an eleventh-century Egyptian author.

This Fatimid Mediterranean of the *Book of Curiosities* was very much unlike the Mediterranean as we envisage it today. For one, it was effectively only the eastern half. The Mediterranean familiar to our eleventh-century Egyptian author was bounded in the west by an imaginary line drawn from Mahdia to Sicily. The western half—including not only the West European shores but also the North African coasts—belonged to a different maritime sphere, even if acknowledged to be part of the same great sea. The Adriatic Sea and the Black Sea were similarly unknown, beyond the limits of the Fatimid horizons.

On the other hand, the Mediterranean of the *Book of Curiosities* is as Byzantine as it is Muslim. The Byzantine southern coasts of Anatolia, the coasts of the Aegean and many of its islands, as well as Cyprus, are described with as much detail, and often with more specificity than the Egyptian or Syro-Palestinian coasts under Fatimid control. It is not only the geographical detail that ties the northern and southern coasts of the eastern Mediterranean together, but also shared maritime technology and terminology. While religious divisions would lead us to expect a north-south divide, the *Book of Curiosities* presents an east-west rift, with a surprisingly integrated Greek and Arabic eastern Mediterranean.

The oval map of the Mediterranean, consisting of coastal localities and islands, is very much a map of the eastern Mediterranean, with the information on the western half of the Mediterranean either nonspecific or inaccurate (fig. 7.1).

On the northern shores of the Mediterranean, the map refers in a generalized way to the anchorages of al-Andalus, the Galicians, the Franks, the Slavs, and the Lombards, in this sequence. These ethnic groups are confusingly lumped together. Note that Muslim Spain, al-Andalus, is no more familiar for our eleventh-century Cairene author than the Franks, the Lombards, and the Galicians.[2] The anchorages of the Franks were located in the southern coasts of modern France, and those of the Lomabrds in central and southern Italy.[3] The reference to the Slavs is probably to Slavonic communities in the Peloponnesus, explicitly mentioned elsewhere in the account of the bays of the Aegean. Here, confusingly, the Slavs are located

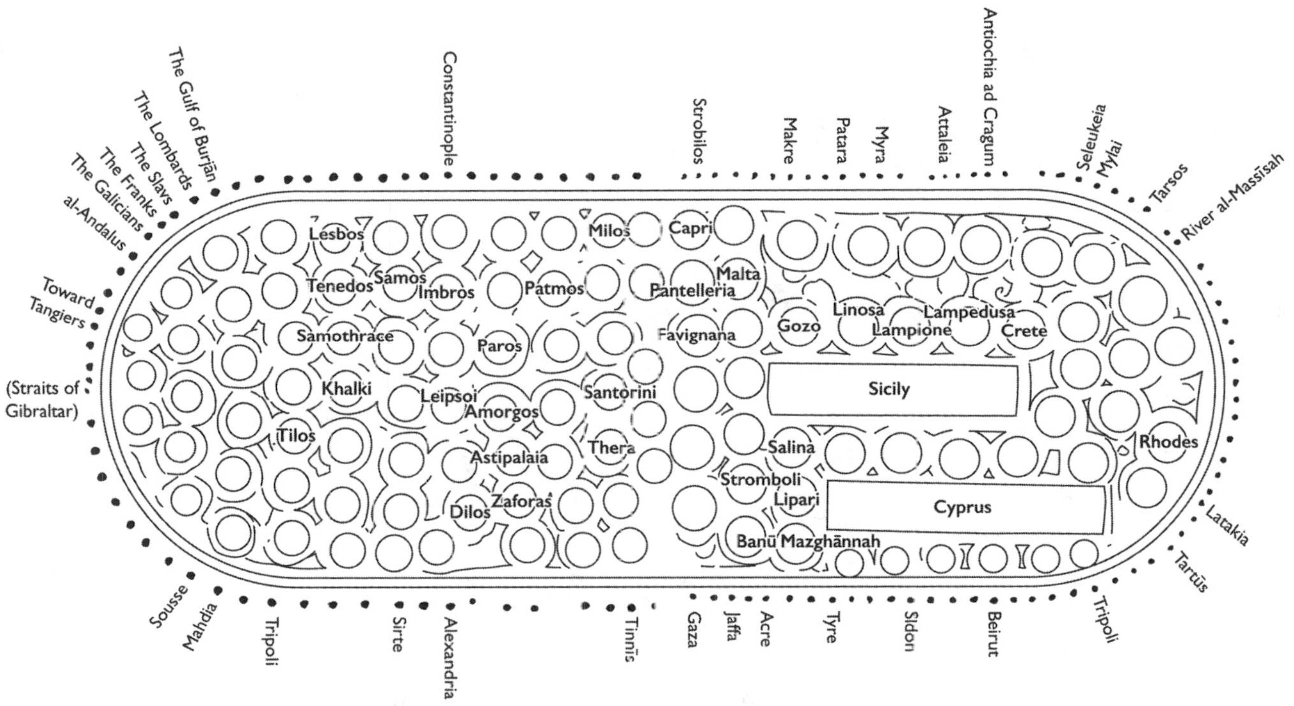

FIG. 7.1. Interpretive diagram of the map of the Mediterranean in the *Bock of Curiosities*.

between the Franks and the Lombards. The following label, "the Gulf of Burjān, with thirty anchorages for the skiffs of the Burjān," refers to the Black Sea, conceived by Islamic geographers as narrow straits that connect the Mediterranean and the Encompassing Ocean.[4] The Burjān, often mentioned in Arabic sources in connection with the Slavs, are the Bulghars who immigrated to the Balkans in the early medieval period.[5] The reference to the "Gulf of Burjān" is the only reference in this map to gulfs. The Aegean Sea and the Adriatic (known as the "Gulf of the Venetians" in Islamic geography), are completely omitted and not indicated either visually or textually on the perfect oval.

The information becomes more specific and accurate as we continue clockwise to the eastern Mediterranean. After the Black Sea, none of the next nine labels can be identified with certainty, but we know we are in Christian territory, with a couple of labels referring to churches. One is the "Church of Saint Badolo (or Padolo)." The names are definitely Greek, and possibly refer to localities along the coasts of the Hellespont, such as Sigeon, Sestos, Agios Georgios (San Georgi), and Gallipoli. The next label, "The fortress of *Qunsṭanṭīnīyah* on which there is a tower and an armoury," is almost certainly Constantinople itself, the Byzantine capital. After Constantinople, the next sequence of nine labels is again difficult to interpret, although the place-names are likely to be located on the western coasts of Anatolia. The difficulty in interpreting the place-names may be due to the limitations of transcribing Greek names in Arabic, copyist mistakes, or our meager knowledge regarding the names of Byzantine settlements in these areas at that period.

We pick up the sequence with a definite identification of the anchorage and arsenal of Strobilos, on the southwestern tip of the Anatolian coasts, ten kilometers southwest of modern Bodrum. Strobilos was an important naval and military post in the middle Byzantine period, and is mentioned by al-Idrīsī a century later.[6] Proceeding clockwise from Strobilos we are now on firm ground. The next twenty-six labels refer to anchorages on the southern coasts of Anatolia, from Strobilos to the anchorages at the mouth of the al-Maṣṣīṣah River, the modern Ceyhan, in southeast Anatolia. The place-names include all the major coastal towns of the south Anatolian coasts in the Byzantine period, such as Makre, Patara, Myra, Antalya (Attaleia), Antiochia ad Cragum, Seleukeia, Mylai, and Tarsos. The next forty labels mark well-known harbors and anchorage points on the Syro-Palestinian coasts, from the Gulf of Iskenderun in the north to Gaza in southern Palestine. Again, every major harbor is indicated, including Latakia, Ṭarṭūs and the nearby island of Arwād (Arados), Tripoli, Beirut, Tyre, and Acre (fig. 7.2).

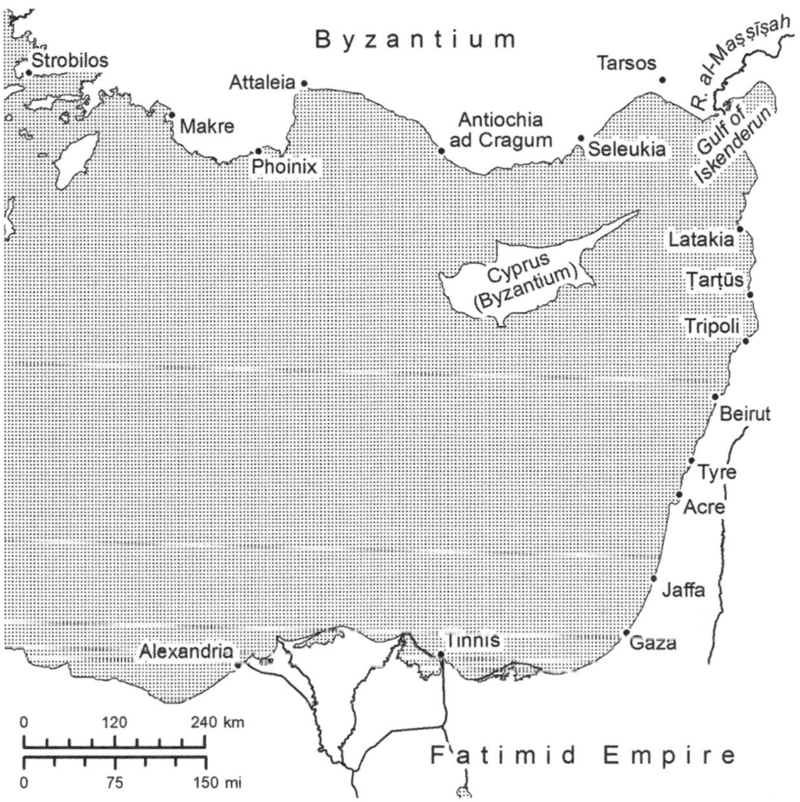

FIG. 7.2. The Eastern Mediterranean in the eleventh century.

Compared to the intensive coverage of the south Anatolian and the Syro-Palestinian coasts, the account of the North African harbors is considerably sparser. Seven labels indicate points on the Mediterranean coasts of Egypt, from Tinnīs in the east to Alexandria in the west. Another eleven labels mark localities on the coasts of modern Libya and Tunisia (medieval Ifrīqiyah), from Barka to Tunis, including Sirte, Tripoli, and Mahdia, the former Fatimid capital. The next segment, covering the Mediterranean coasts of modern Algeria and Morocco, is surprisingly empty. Only six labels separate Sousse in modern Tunisia from the Straits of Gibraltar. The dearth of information on the Mediterranean coasts of the Maghreb is followed by a reasonably rich account of the Atlantic coasts of North Africa, with seven extra labels listing anchorage points just west of the Straits, including Tangier and Asila.

The imbalanced knowledge of the Mediterranean coasts is replicated in the uneven knowledge of Mediterranean islands. There are 118 islands represented in the middle of the oval, of which the majority refer to specific named islands in the Aegean and around Sicily, while twenty-two circles on the far left of the oval are merely labeled with generic term "island." The picture is again of dense coverage for the eastern Mediterranean, with practically no island in the western half. Most of the circles left of the gutter can be safely identified with islands in the Aegean, from Samothrace and Tenedos in the north to the islet of Zaforas in the south. Pairs of neighboring islands such as Astipalaia and Amorgos, or Santorini and Thera, were placed next to each other, and it is very likely that the islands surrounding them were also in the Aegean, even if precise identification is often difficult. The other major cluster of identifiable islands is located around Sicily. It includes the islets that lie between Sicily and Ifrīqiyah, such as Lampedusa, Linosa, Lampione, Malta, Gozo, and Pantelleria. Also indicated are the Egadi group off the west coast of Sicily, including Favignana and Marettimo, and the Aeolian group off the north coast of Sicily, including Stromboli, Lipari, and Salina.

Very few islands are located north or west of Sicily. The northernmost island mentioned is Capri, south of the Gulf of Napoli. The Banū Mazghānnah island—in fact a peninsula on the North African coast—is the medieval name for Modern Algiers.[7] The "Island of Good Health" is possibly the islets of Bou Afia, off the modern Algerian coast between Jijel and Bougie (Bejaia), although the place-name would have been quite common along the North African coasts and may refer to other islets.[8] None of the large islands of the western Mediterranean—Sardinia, Corsica, or Mallorca—is even indicated.

The near exclusive focus on the islands of eastern half of the Mediterranean is reinforced in a chapter devoted to islands ruled by non-Muslims (chapter 15). This chapter includes the detailed map of Cyprus, discussed above, and accounts of history and topography of Crete, Rhodes, Halki, and Tilos (these last two are small islands north of Rhodes, also indicated on the map of the Mediterranean). Here there is a very brief note of Corsica and Sardinia and their size, followed by a reference to another treatise by the author. That other treatise, he explains, is richer in detail on the islands of the Mediterranean:

> All in all, there are 162 large inhabited and uninhabited islands in the Mediterranean [literally, "The Greek Sea," *al-baḥr al-rūmī*], but we have confined ourselves to a few so that the book would not be longer than intended. We have given a full list of the islands and detailed descriptions of their inhabitants in our other book, "The Comprehensive" (*al-Muḥīṭ*).[9]

That other treatise by the same author, "The Comprehensive" (*al-Muḥiṭ*, perhaps also a reference to *al-baḥr al-muḥīṭ*, "The Encompassing Ocean"), was apparently devoted exclusively to islands, not only in the Mediterranean. He also states elsewhere that this other treatise has "a description of all the islands of the seas, as many as possible and as much as is known to us." Unfortunately, this second treatise on the islands of the world appears to have been lost, and with it a unique perspective on global maritime history, and more information on the islands of the western Mediterranean.

In the *Book of Curiosities,* however, the focus is the eastern half of the Mediterranean. This is evident from the systematic manner in which the Mediterranean material is organized. After providing us with a map of the Mediterranean as a whole, the author zooms in on specific eastern Mediterranean islands, harbors, and coastal sections. He first describes Sicily, Mahdia, and Tinnīs, and then devotes space to a map of Cyprus and accounts of several other eastern Mediterranean islands (Rhodes, Crete). The account of the bays along the coasts of the Aegean Sea, both on the Anatolian side and on the Greek mainland, follows. Together, these Mediterranean harbors, islands, and bays form a maritime space bounded in the west by a line between Sicily and Mahdia in modern Tunisia, cutting off the western Mediterranean. It also excludes the Adriatic and the Black Sea. But this maritime space does include both the Fatimid coasts of Egypt, Palestine and Syria, as well as the Byzantine coasts of the Aegean, southern Anatolia, and the Byzantine islands of Cyprus, Crete, and Rhodes.

The visual prominence of Sicily, Mahdia, and Tinnīs in the *Book of Curiosities* is not much of a surprise. Along with Alexandria, they were the nodal points of Mediterranean commerce in the eleventh century. They are known to us from the records of the Jewish merchants preserved in the Geniza, a trove of hundreds of thousands of fragments recovered from the Ben Ezra Synagogue in Cairo, which mostly date from the tenth to the thirteenth centuries. A list of the origins of Geniza mercantile letters found in the Cairo Geniza shows that Alexandria was the point of origin of almost half of the mercantile letters found in the Geniza (46.6%). The second most important location was Mahdia (8.5%). Next on the list are the inland cities of Jerusalem, Qayrawān, Fustat (Old Cairo), and the flax production center of Busir in Middle Egypt. Then, the next prominent Mediterranean ports in the Geniza correspondence are Tinnīs (3.3%, twenty letters) and Palermo (2.2%, thirteen letters), followed by Tripoli in North Africa (twelve letters) and Tyre (ten letters).[10]

The economic historian Jessica Goldberg has used itineraries mentioned in Geniza letters to draw on a modern map the main trunk routes used by the eleventh-century merchants (fig. 7.3). To the east, Tinnīs was the end

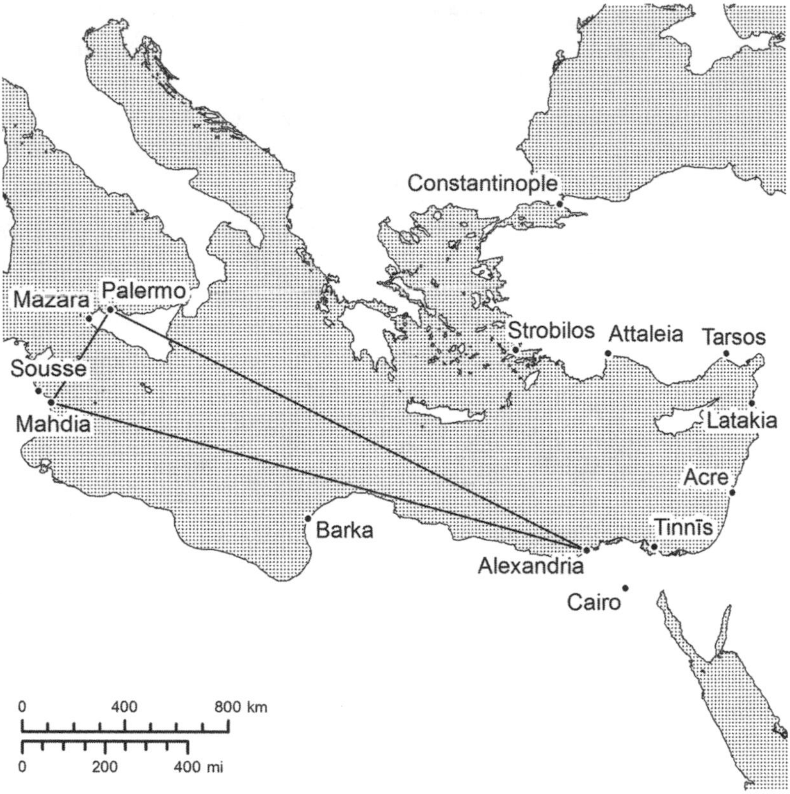

FIG. 7.3. Trunk routes used by the Geniza merchants in the eleventh century.

point of the coastal route from Ascalon and greater Palestine / Syria.[11] To the west, the trunk routes were Alexandria—Mahdia and Alexandria—Palermo.[12] The itineraries preserved in the Geniza show that sailings from Alexandria toward the west were made mostly in the spring, and merchants were able to complete a triangular route in a single season, going from Alexandria to Mahdia and then from Mahdia to Palermo, returning to Alexandria in the autumn. This is in complete accord with the information of the *Book of Curiosities,* where we have detailed maps of Palermo and Mahdia, as well as a maritime, coastal itinerary from Mahdia to Palermo incorporated within the map of Mahdia.

The relative absence of Alexandria from the *Book of Curiosities* is a surprise, especially given the prominence of Alexandria in Geniza correspondence. Alexandria was the Mediterranean gateway to Egypt since antiquity, and remained so under the Fatimids. But in the treatise it does not receive any detailed treatment of its history or fortifications, unlike Tinnīs, Mahdia, or Palermo. This does not mean that Alexandria was of lesser importance as a port. As Udovitch showed, Geniza merchants arriving in Alexandria treated it as a necessary stop. But the city was commercially dependent on Fustat, or Old Cairo.[13] The relative obscurity of Alexandria in the *Book of Curiosities* may therefore reflect its effective status as a suburb of Cairo. Remarkably, neither Fustat, nor even Cairo—the new Fatimid capital built in 971—is mentioned once in the *Book of Curiosities.* This is surely because Cairo was the vantage point from which the Mediterranean was observed. These maps are drawn from the perspective of a person residing in the Egyptian capital looking out. Perhaps, from this Cairene perspective, Alexandria was obscured precisely because it was so close.

Palermo, Mahdia, and Tinnīs were linked not only as nodes of the trunk routes of the eastern Mediterranean, but as examples of the constructed or man-made harbors that characterized this maritime space, from Syria to Ifrīqiyah, as opposed to the natural harbors of the Maghreb and al-Andalus.[14] On the Syrian coast, the Muslims inherited ports such as Acre and Tyre, which had since antiquity closed and fortified artificial basins for galleys.[15] These fortified harbors were blocked off by a chain between two towers, and sometimes also by an arch beneath which the ships entered. This type of artificial harbor was not found in the western Mediterranean at that time, where anchorages were natural, and arsenals not attached to a basin.[16]

The eastern Mediterranean artificial and internal harbor is fully visualized in the map of Mahdia, and at least partially in the maps of Palermo and Tinnīs. In Arabic sources, this is called a "harbor" (*mīnāʾ*, as opposed to mere *marsá*, "anchorage point"). The constructed port of the eastern Mediterranean is vividly described in 1047 by the traveler Nāser-e Khosraw in his account of Acre:

> To the south is a harbor. Most of these coastal towns have a harbor, which is like a stable for ships. Built right against the town, it has walls out into the water and an open space of fifty ells without a wall but with a chain stretched from one wall to the other. When a ship is about to enter the harbor, they loosen the chain so that it goes beneath the surface of the water, allowing the ship to pass over. Afterwards the chain is raised again lest strangers make untoward attempts on the ships.[17]

In the *Book of Curiosities,* the use of the term "harbor" binds together the northern and southern shores of the eastern Mediterranean. On the Syrian coasts of the map of the Mediterranean, the term is used to describe Arados and Tyre, the latter said to be a "harbor within a harbor." Barka, Tripoli, and Sousse are also described as harbors. The separate maps of both Tinnīs and

Mahdia also refer in labels and in iconography to the existence of artificial harbors (one in Mahdia and two in Tinnīs). There are also ten harbors on the Byzantine coasts of the Aegean and southern Anatolia, including Myra, Side, Attaleia, and Antiochia ad Cragum, and two harbors in the map of Cyprus—Ammochstos and Paphos.

The maps preserve other indications of shared maritime culture across the eastern Mediterranean. In our discussion of the navigation material we have already noted the extensive use of Greek nomenclature for prevailing winds. Harbors are said to be protected from the Boreas, the Notos, and the Euros, all Greek names—as well as the "Frankish" wind, meaning westerly winds. These names for winds are used for describing all eastern Mediterranean harbors, whether Fatimid or Byzantine, in the maps of the Mediterranean and of Cyprus.

The Greek names for winds are linked to classical meteorological theories, a source of cosomolgical knowledge in both Islam and Byzantium. The Boreas, Notos, and the Euros appear in book 1 within the context of an astro-meteorological diagram that links twelve named winds with the zodiacal signs. The use of these three Greek names here, on the maps of the Mediterranean and Cyprus, is on a much more practical level. Together with the "Frankish" wind, they constitue a more limited four-point wind-rose used to describe harbor conditions and access.

Another revealing aspect of a shared Greek-Arabic culture is the translation of Greek place-names into Arabic. Thus, the river Basileus in Cyprus is called here "River of the king" (*nahr al-malik*). On the map of the Mediterranean, "the Island of Thirst" (*jazīrat al-ʿaṭash*) is Leipsoi in the Aegean group of the Sporades, whose medieval Greek name was Dipsia or Dipsos (δίψα / δίψος), meaning "thirst."[18] The name of the nearby Island of the "She-goat" (*Jazīrat al-Māʿizah*) is likely to be a literal translation of Tragia (modern Agathonisi), also in the Sporades.[19] The Anchorage of the Monk in southwest Anatolia is also surely a translation of a Greek name. These examples show that the literal meaning of a Greek place-name was sometimes known, evidence of some bilingualism among mariners.

While the eastern Mediterranean is cut off from its western half, the *Book of Curiosities* draws no clear line of demarcation between the southern and northern shores. The map of the Mediterranean has no indication of a border separating the Fatimids and Byzantium. Rather, the northern and southern shores of the eastern Mediterranean appear remarkably integrated, with instructions for direct sailing including routes from Byzantine Cyprus to Fatimid Latakia, and from the Peloponnesus to Muslim Sicily.

This integration of the northern and southern coasts of the eastern Mediterranean in the *Book of Curiosities* is undoubtedly linked to an increase in the volume of commerce between Egypt and Byzantium around 1000 AD.[20] A series of peace treaties stipulated freedom of trade, and the presence of Byzantine merchants in Cairo and Alexandria, as well as Muslim merchants in Constantinople, is attested in literary sources.[21] In 1047, Nāser-e Khosraw depicts the bustling commerce in Tripoli, visited by ships from all over the Mediterranean, and the demographic and topographic expansion of Sidon and Tyre.[22] The archaeological evidence for Anatolia suggests a similar trend of recovery, with sites such as Smyrna and the abovementioned Strobilos showing considerable wealth.[23] The most impressive, tangible testimony of the expansion of Byzantine-Fatimid trade in the eleventh century is the exceptional shipwreck that sank around 1030 in Serçe Limanı, in southwest Anatolia opposite Rhodes. Excavations concluded that the ship carried a cargo of Syrian glass and glazed pottery destined to be sold in Constantinople, whence the ship had originally embarked.[24]

In the letters of the Jewish merchants found in the Geniza, the evidence regarding Fatimid-Byzantine trade in this period is conflicting. While there is a significant number of references to the activities of Greek (*Rūm*) merchants, many of whom were Byzantine (and not from Latin Europe, as was previously argued), and references to the importation of mastic and other medical plants and drugs from Anatolia and the Aegean to Egypt,[25] Jessica Goldberg found that Geniza merchants moved goods primarily on the axis Alexandria–Mahdia–Palermo. Only very rarely did they venture into the Byzantine realm, and never into Latin Europe.[26] She finds only a trickle of Greek or European merchants in the first half of the eleventh century, and their numbers were so small that local merchants made no attempt to distinguish between Amalfitans or Byzantines.[27] They became more important only from the 1060s, as the political instability in Mahdia and Palermo forced Italian merchants to travel directly to Alexandria. She finds even fewer indications of travel or transactions of Cairene merchants in Byzantium. In the Geniza documents, she says, Byzantium "is a blank," existing outside of mercantile knowledge and activity.[28]

This is clearly not the case with the *Book of Curiosities*, where the Byzantine harbors and islands are intimately familiar. The wealth of detail concerning the Byzantine coasts in the *Book of Curiosities* is unprecedented, utterly surpassing any earlier account of Byzantine coasts in Arabic geographical literature. One explanation for the discrepancy between the Geniza mercantile correspondence and the *Book of Curiosities* may be that Anatolia and the Aegean were not part of the specific trade network of the Jewish merchants of the Geniza, but were open to other merchants, such as those who traveled on the Serçe Limanı shipwreck. The Jewish Geniza merchants were not involved in the trade in glass, for example, or in the bulky trade in grains. Given the archaeological evidence and the *Book of Curiosities* itself, it seems quite certain that Fatimid-Byzantine trade was on the rise in the first half of the eleventh century, and that Geniza merchants were, for some reason, simply not part of that emerging commercial network.

But we should also bear in mind that much of the material on the maps of the Mediterranean has a distinct military utility, such as capacity of harbors and location of arsenals. This naval aspect of the *Book of Curiosities*' maps demonstrates the degree to which Muslims and Byzantines contested, as well as shared, the eastern Mediterranean, as they had done in previous centuries. In the early Islamic period, continuous raids by Muslim navies meant that their commanders were the ones best informed about the affairs of the Mediterranean. This is reflected in a passage by al-Masʿūdī, who states that while Omani and Sirafi captains (*nawākhida*) of commercial ships are the best authorities on the Indian Ocean, it is the naval commanders who are the most knowledgeable on the bays and coastlines of the Mediterranean. The most famous naval commander of his time was the military governor of Jabala, a town along the coast of modern Syria:

> ʿAbd Allāh ibn Wazīr, the governor of the town of Jabala on the coast of Homs in Syria, has remained today—in the year 332 AH (943–44 AD)—the most knowledgeable and experienced person with regard to the Mediterranean. Whenever any of the ship captains, whether military or commercial wishes to ride the Mediterranean Sea, they refer to his opinion and attest to his knowledge, wisdom, piety and long experience of Holy War at sea.[29]

Whereas al-Masʿūdī relied on the Abbasid governor of Jabala for his account of the Mediterranean written in the 940s, the political landscape had changed dramatically soon afterward. During the 960s the Byzantines recaptured territories they had lost in previous centuries, swiftly conquering Crete, Cyprus, and the southeast Anatolian major port of Tarsos. Even the frontier port of Jabala itself was lost, with the new border passing around Latakia. Byzantine advances in the Mediterranean coincided with the Fatimid conquest of Egypt in 969, and the Fatimids now replaced the Abbasids as the dominant Muslim power in the eastern Mediterranean.[30] From Egypt, the Fatimids turned to consolidate their rule on the Syro-Palestinian littoral. Between 970 and 975 they sent naval expeditions that helped secure Ascalon, Jaffa, Tyre, Beirut, and Tripoli.[31]

The final decades of the tenth century were characterized by political and military competition between the two eastern Mediterranean empires.

Armed conflict mainly revolved around the control of Aleppo, which Byzantium held as a protectorate. In the 990s Fatimid naval forces attempted to secure provisions for land armies operating deep in Byzantine territory.[32] The peak of Fatimid naval ambitions was under the Fatimid caliph al-ʿAzīz, who ordered in 996 the construction of a fleet (*usṭūl*) at the Cairo arsenal in preparation for a major naval expedition. After a fire in the dockyards destroyed sixteen ships and their armaments, a new fleet of twenty-four galleys was built and sent to Syria. When it arrived in Ṭarṭūs, near Tripoli, it was scattered by strong winds, and the crews, together with some serviceable ships, were taken captive.[33] As far as we know, this ill-fated naval expedition of 996 was the largest of its kind undertaken by the Fatimids. It is tempting to speculate that the intelligence gathered in preparation for this expedition was later used by author of the *Book of Curiosities* for compiling his maps of the Byzantine coasts.

Despite its naval orientation and visual and textual messages about Fatimid military power, the *Book of Curiosities* was actually written in a period of relative peace in the eastern Mediterranean. By the turn of the eleventh century, military confrontation was overshadowed by a marked increase in commercial relations. Fatimid naval expeditions in the direction of the Syrian coasts were few and far between. One major naval expedition took place in 1021–22, with galleys laden with horses and men headed to Tripoli to fight against the Byzantines. In the summer of 1024, newly constructed galleys were sent to several Syro-Palestinian coastal towns to quell local rebellions and Bedouin incursions. Then there is no further mention of Fatimid naval expedition until 1054, a result of the repeated truces with Byzantium and the expansion of trade.[34]

This means that the detailed knowledge of the Byzantine coasts in the *Book of Curiosities* could not have been gained through the direct experience of Fatimid naval commanders. Following Byzantium's capture of Cyprus, Crete, and Tarsos in the 960s, the Byzantine navy prevented Fatimid galleys from approaching the southern coasts of Anatolia and the Aegean. There are reports of occasional raids by North African pirates, the most successful leading to the sack of Myra on the southern coasts of Anatolia in 1035. Geniza letters also attest to the activity of pirates capturing Jewish merchants from Attaleia and Strobilos.[35] But these raiding fleets were not under the control of the Fatimids. Al-Muqaddasī, writing at the end of the tenth century, notes that the men with most experience of the Mediterranean are the Byzantines and the Muslims of Sicily and al-Andalus.[36] Unlike al-Masʿūdī in the 940s, who singled out the governor of Jabala as the foremost authority on the Mediterranean, al-Muqaddasī could no longer rely on the maritime knowledge of Syrian naval commanders.

How, then, did our Fatimid author collect such detailed information on the coasts of Byzantium? Without direct experience of the Byzantine coasts, Fatimid naval commanders could have employed indirect methods, such as interrogating prisoners of war, relying on the experience of North African corsairs, or translating Byzantine navigation guides—although we are not certain whether such guides even existed. It is also possible that the Fatimids relied on the experience of the surviving commanders of the navy of Muslim Crete, who fled to Egypt after the island was captured by the Byzantines in 961.[37] The emir of Crete himself was taken captive and sent to Constantinople, joining a large number of Muslim captives and slaves living in the city. Captivity and enslavement created a community of Muslims or ex-Muslims in the heart of Byzantium, facilitating contact across the porous borders. According to Byzantine sources, the successful Byzantine campaigns of the 960s brought back tens of thousands of Muslim prisoners. These were mostly enslaved and sold on to the leading imperial households. This large captive community was able to keep communication channels with the lands of Islam, sometimes securing their release through contact with Muslim rulers. Others made careers in the Byzantine army: the son of the Muslim emir of Crete became a high-ranking officer.[38]

Naval information could also have come from trade. The Byzantine merchant as a potential spy is the subject of a diatribe by the tenth-century geographer Ibn Ḥawqal, who reports that Byzantine commerce has largely replaced Byzantine piracy:

> They [the Byzantine commanders] send their ships to trade with the land of Islam. The men of the ships spy, observe and explore it inside out, and when they go back they bring first-hand information to the [Byzantine commanders] with regard to its affairs, and advise them on how to harm it. In this way they reach the interior, the plains and the rugged terrain of the lands of Islam, in full view and knowledge of the Muslim rulers. Many of these rulers even assist them [the merchants] in what they seek, reinforcing the enemy with a wonderful weapon and an effective tool. They do so out of desire for a few objects of vanity to come from the trade with Byzantium, a trade which gives them only a meager profit.[39]

Ibn Ḥawqal here accuses the Byzantine governors of Attaleia of using merchants to gain knowledge of the lands of Islam, manipulating Muslim rulers' insatiable thirst for luxury items. But the merchant-cum-spy was an accusation that had a long history and cut both ways. An anecdote in Byzantine chronicles tells of an Arab Syrian spy who came to Byzantium in the

880s to assess the state of Byzantine naval forces and to evaluate the success of a putative attack by an Arab fleet.[40] The military treatise of Emperor Nikephoros Phokas (963–69) recommends the dispatch of merchants to Muslim territory in times of tension, so that they will gather information.[41] Thus, the intimate knowledge of the Byzantine waters in the *Book of Curiosities* could well have been a side effect of the expansion of trade between the southern and northern shores of the eastern Mediterranean. It surpasses by far any other surviving account from preceding centuries of raids and conflict. Although the material has a clear military orientation, it comes at a time of relative peace, and it is precisely the increased contacts between the Fatimids and Byzantium that would have allowed for such familiarity with Byzantine coasts.

In their *Corrupting Sea*, Peregrine Horden and Nicholas Purcell reject the narrative of Mediterranean history as a sequence of maritime empires (the term they use is thalassocracy).[42] The *Book of Curiosities* visually demonstrates that one cannot speak of an Islamic Mediterranean versus a Christian one. The horizons of the Mediterranean here are not limited by the boundaries of religious communities. The knowledge of the coasts of Anatolia and even the Aegean would have come from the crews of ships like the one that sank in Serçe Limanı, or from captives and slaves. On the other hand, neither can we see here a unity of the Mediterranean. This treatise and the Geniza records both suggest a lack of integration between east and west, again in line with Purcell and Horden's contention that Mediterranean trade was often limited to distinct zones.[43] The Mediterranean is recognized as a single entity, but it is only the eastern half that really matters.

That unique moment of shared maritime culture ended with the arrival of the Crusaders and their Italian associates in the eastern Mediterranean. For all their rivalry, Byzantines and Muslims had built similar galleys and commercial vessels, followed comparable tactics of naval warfare, and influenced each other in the development of maritime law.[44] As the *Book of Curiosities* uniquely attests, Muslims also used the Greek nomenclature of wind navigation. However, this shared maritime culture, given such striking visual representation in the *Book of Curiosities*, was soon superseded by the dominance of the Italian fleets. The subsequent disappearance of Muslim shipping toward the Aegean must have led to a decline in the interest of Muslim navies in the coasts of Byzantium. The detailed maps of the Anatolian and Aegean coasts in the *Book of Curiosities* had, by a matter of less than a century, become an archaic relic of a maritime culture that had lost the day.

CHAPTER EIGHT

A Musk Road to China

If the Eastern Mediterranean was Fatimid Egypt's backyard, the maps of the *Book of Curiosities* also show us that Fatimid ambitions and Fatimid networks of communications were those of a truly global power. We might expect from our author certain knowledge of the Mediterranean, perhaps even of its Byzantine coasts. The *Book of Curiosities* allows us to understand better the dynamics of that shared Fatimid-Byzantine maritime space, and to visualize the nodal points of Fatimid commercial and political power. The Fatimid interest in the Mediterranean is natural—this is an empire that was bred in North Africa and matured in Egypt, creating what S. D. Goitein has famously called, in his study of the Geniza documents from Fatimid Cairo, "a Mediterranean Society."

Yet, far beyond the physical borders of their Mediterranean empire, the Fatimids believed they had a universal mission for the salvation of mankind. The caliph imams were, according to the Ismaᶜili messianic vision, divinely invested with spiritual authority, and it was their destiny to conquer the "eastern and western lands of the Earth" in order to fill it with justice. The conquest of Egypt in 969 was seen as a first step toward the toppling of the Sunni Abbasid caliphs, and a victory on the banks of the Euphrates would revenge the martyrdom of Ḥusayn, son of ᶜAlī, slain by the Umayyads in Karbala. The capture of the Holy Sites in Mecca and Medina would be a return home for the descendant of the Prophet.[1] Baghdad, the Abbasid capital, was the ultimate prize. It was fleetingly achieved, then lost, in 1058, wheh the Friday prayer was declared in the name of the Fatimid caliphs as a result of an alliance with the former Abbasid general, al-Basāsīrī.

If Fatimid poetry shows an obsession with taking Baghdad, the maps of the *Book of Curiosities* demonstrate the strategic importance of the Indian Ocean, and of the Ismaᶜili penetration in the Yemen and in India. Ismaᶜili missionaries were sent to Sind (Modern Sindh in Pakistan) from the Yemen already by the ninth century.[2] In the 960s, in tandem with their conquest of

Egypt, an Ismaʿili missionary captured Multān, on the banks of the Indus. By the late tenth century, the two capitals of Sind, al-Manṣūrah and Multān, were effectively under Fatimid control. It was during this period of expansion that the poet Ibn Hāniʾ fantasized about Fatimid forces moving even beyond Baghdad, leading Indian kings to lose sleep and their elephants, symbols of their military might, "lowing like a young camel."[3] India figured prominently in the Fatimid imagination as a worthy enemy, a fertile land for conversion as well as a font of cosmological wisdom.

The maps of the *Book of Curiosities* are thus of significant importance for the history of global communications. The representation of northern India and other East Asian and Central Asian localities is surprisingly detailed. India, Indian Ocean islands, Central Asia and China are depicted visually in three maps: a map of the Indian Ocean, a map of the River Oxus, and a map of the Indus, which also shows localities along the Ganges. It is this third map in particular that will merit most of our attention, as it the most important contribution of the *Book of Curiosities* to the history of medieval trade networks. This map of the Indus-Ganges river systems uniquely depicts an overland itinerary from Muslim Sind, then under Fatimid control, through northern India and then probably through Tibet, to China. There were of course other ways to reach China, either by sea through the Straits of Malacca, or through the Central Asian Silk Road. But the Silk Road is not mentioned at all in the *Book of Curiosities*, and information on the sea route to China is drawn entirely from much earlier ninth-century sources. Instead, the maps represent a time when the road to China passed through northern India and Tibet, a route that eclipsed its more famous alternatives.

Let us look first at the sea route to China, as illustrated in the map of the Indian Ocean, which constitutes the seventh chapter of the treatise (see plate 6 and fig. 8.1 and fig. 8.2). The Indian Ocean is not actually named, but the localities on the map all lie around or within the Indian Ocean as it was known to Islamic geographers. This map depicts the Indian Ocean as an enclosed oval sea, a form that parallels the form of the Mediterranean in the preceding map in the volume. The map is made up of two halves, an Asian one and an East African one (on which more in the next chapter). The Asian half of the Indian Ocean, occupying the right-hand side of the map, shows Indian and Chinese localities along the shores of the Indian Ocean, with the sea route to China indicated by a volcano and several islands in the Bay of Bengal and the Sea of China.

The sea route from the Persian Gulf to China is indicated by the islands depicted in the center of this Indian Ocean map. Most of the names here are familiar from the ninth-century work of the Abbasid geographer Ibn Khurradādhbih and, in particular, from *The Account of China and India*

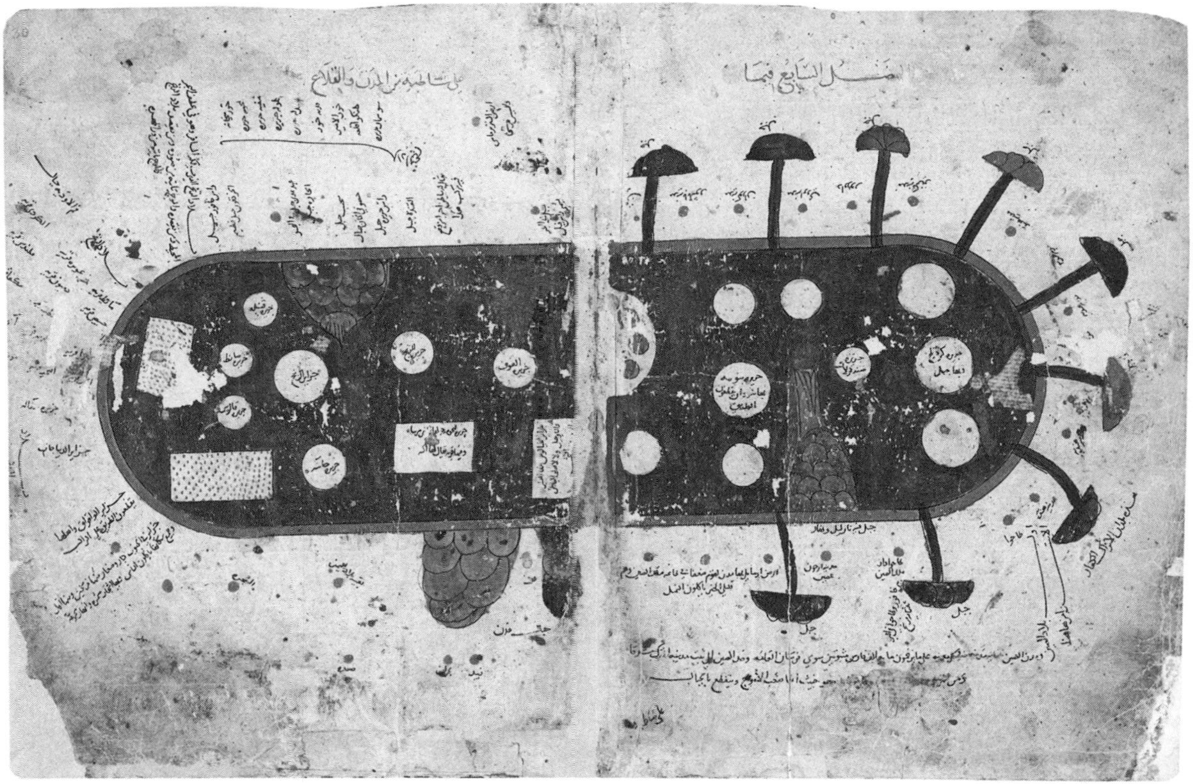

FIG. 8.1. Map of the Indian Ocean from the *Book of Curiosities*. Oxford, Bodleian Library, MS Arab. c. 90, fols. 29b–30a, copied ca. 1200.

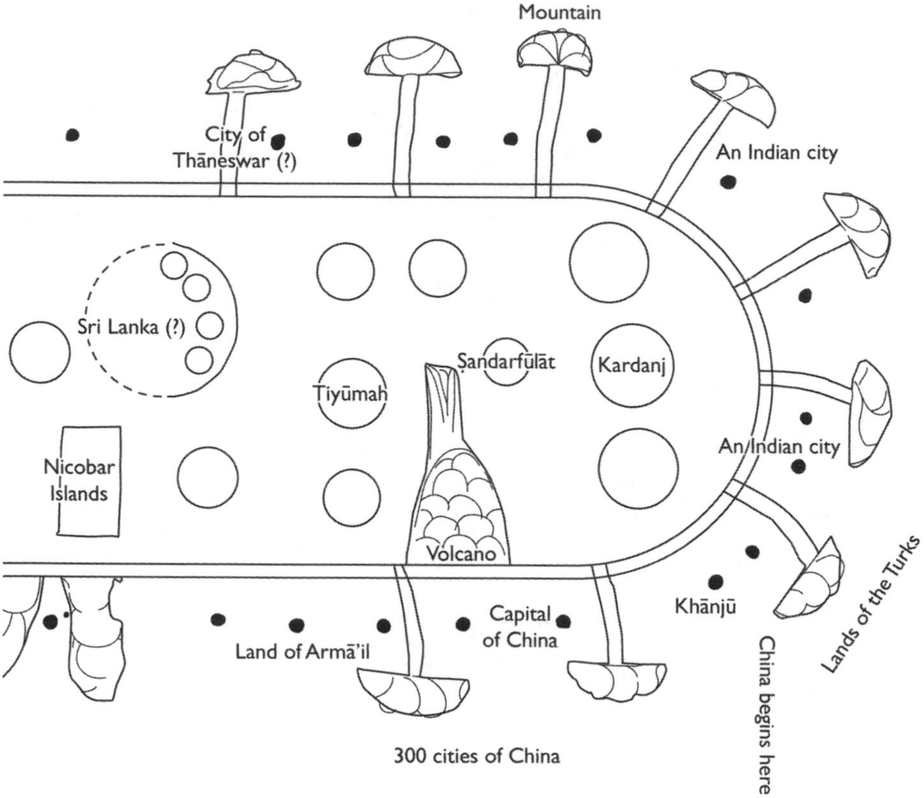

FIG. 8.2. Interpretitive diagram for the right half (India and China) of the map of the Indian Ocean from the *Book of Curiosities*.

by Sulayman the Merchant, composed in 851. They include the Nicobar Islands, the island of Tiyūmah (identified as Pulau Tioman off the coast of Malaya), and the island of Ṣandarfūlāt, one of the last stopping points on the route to Canton. There is also a large island, mostly lost in the gutter, which is almost certainly Sri Lanka. Other material on the sea route to China is found in the text of a separate chapter on the islands of the infidels, where we find entries for some clove-producing islands called "al-Azl" (possibly the Moluccas), the Andaman and Nicobar islands (again), and al-Rāmnī (Sumatra), as well as a lengthy account of religious customs in Sri Lanka, all taken verbatim from Sulayman's ninth-century account.[4]

China itself is located along the rim, at the bottom right. It is separated from India by the "Lands of the infidel Turks." A long label at the bottom of the map, partly illegible, states that there are three hundred cities in China, a number familiar from the account of Ibn Khurradādhbih, but adding that each has "100,000 standing cavalry, not counting the horsemen of the common people." Several Chinese place-names are indicated, but are badly deformed. Coming from the direction of India, the first label is *Ṭāḥū,* which may be a corruption of *Khānjū,* likely Ch'üan-chou (Quanzhou), later known to the Arabs as the port Zaytūn.[5] Alternatively, it may be read Anḥū or Anjū, the name by which earlier Arab geographers knew the Chinese capital of Chang'an (Xi'an) in central China.[6] Indeed, another label for Ṭāḥū adds that this city is the seat of the ruler of China. *Ṭāḥū* is also said to be three hundred farsakh from *Khāfūr,* which is possibly a mistake for *Khānfū* (Guangzhou), the port of Canton. Ibn Khurradādhbih gives an eight days' journey between Khānfū and Khānjū.

At the bottom of the map we find localities in southeast Asia, including what may be the first depiction of the Philippines. The mountain with a red cap protruding into the sea on the bottom right is a volcano in the Sea of China, which other Muslim sources locate near Java and not in China.[7] Beyond the volcano, one encounters another Chinese city, called *Arḥūn,* not otherwise mentioned in the geographical literature. Then, at the very edge of the Indian Ocean and the bottom of the map, a long label refers to the *Land of Armā'il,* a land with "cities of a weak nation in submission to the ruler of China, who have few good qualities and eat ants." The label is derived from Ibn Khurradādhbih, who mentions *Armā'il* as two months' sailing from China.[8] Hubert Daunicht plausibly suggested that *Armā'il* could be a mistake for *Baru-Manil,* the bay of Manila in the Philippines.[9] In this map of the *Book of Curiosities,* the *Land of Armā'il* is clearly located in Southeast Asia, confirming its identification as the Philippines. Although Chinese records show that Arab ships frequented the Philippines by the ninth century, the islands are not otherwise shown on medieval maps.[10]

The historical context of this material is that of the ninth-century sea route to China, when Arab and Persian boats made the entire journey from Basra to Canton.[11] Some of the information on China in this map is unique, such as the size of the Chinese army. But the image of China is based on acquaintance with the China of the Tang dynasty (618–906). All the place-names appear to be derived from the mid-ninth century accounts of Ibn Khurradādhbih and Sulayman, and reflect the popularity of the maritime route between India and China during the ninth century. This route, through the Straits of Malacca, was facilitated by advancement in navigational techniques and better understanding of the monsoon seasons, as well as by political instability in the overland trade routes in central Asia.[12]

This sea route to China came to a dramatic halt at the end of the ninth century. Arab sources see the root of its decline in the Huang Chao uprising against the Tang, leading to the massacre of thousands of foreigners in Guangzhou (Canton) around 879, including Muslims, Christians, Jews, and Zoroastrians.[13] The merchant Abū Zayd al-Sīrafī, who wrote a supplement to the *Account of China and India* in 916, says that this rebellion was a severe blow to Cantonese trade. It may well have been a watershed in the history of trade networks across the Indian Ocean. Al-Masʿūdī, writing in the mid-tenth century, reports that direct travel from Iraq to China is no longer possible, due to the decline in security in Chinese ports. In his time, Chinese and Arab ships met each other in Galla (Point de Galle) in India, a convenient halfway point.[14] By the tenth century, the long voyages across the breadth of the Indian Ocean, like those reported by the Sulayman, were replaced by shorter, segmented trips, from the Red Sea or Persian Gulf to Gujarat or Malabar, and from the Indian coasts to the Indonesian archipelago. While Muslims continued to play a major role in every circuit of Indian Ocean commerce, even in the trade between Chinese and Indian coasts, these circuits were not part of a direct trade with the Middle East.[15]

As the Arab and Persian merchants stopped making the long sea journeys to China, the ports and islands of the Indian Ocean began to be perceived as exotic, the material of marvels.[16] Indeed, many of the marvels in the *Book of Curiosities* are set in the context of the Indian Ocean. A chapter "on marvellous aquatic creatures" is for the most part devoted to marvels encountered on the sea route to China, and includes an Indian Crab which turns into stone as soon as it comes out of the sea, a fish that develops wings when it is thrown on land, and "a fish that, when you slice open its body, you find another fish within it, and when you slice open that fish you find another one, and so on without end." In a chapter on deformed humans there is an entry for Indian Ocean midget islanders who have their feet turned backward. While some other marvels in the *Book of Curiosities* were closer to home, in the Middle East itself, there is little doubt that the author of the *Book of Curiosities* and his Fatimid audience did not expect to set foot on a ship traveling directly to the Bay of Bengal or to China. For people living in the central lands of the Middle East, the maritime route to China, so important for the Abbasid economy in the ninth century, had become a wonder of the past.

If the knowledge of the sea route to China is outdated, the famed Central Asian Silk Road is not represented at all in the *Book of Curiosities*. Central Asia itself is depicted in the map of the Oxus (Amū Darya), shown in plate 14 and fig. 8.3, where the design is taken wholesale from Ibn Ḥawqal's regional map of Khorasan. The map shows the river emerging from a brown mountain at the top, then forming a nearly complete loop before emptying into the green Aral Sea in the center of the page. Other rivers are also shown, including the Syr Darya, also flowing to the Aral Sea, and the Ab-i

FIG. 8.3. Map of the River Oxus from the *Book of Curiosities*. Oxford, Bodleian Library, MS Arab. c. 90, fol. 44a, copied ca. 1200.

Qaysar that emerges in the red mountains on the bottom left. This map indicates some of the major cities on the Silk Road, such as Balkh and Faryāb. But they are not part of an itinerary toward China, which is not mentioned at all on this map, whose firm focus is the route of the river Oxus.

Knowledge of Central Asia is also demonstrated in the distinctive diagram of the enormous Lake Issıq Kul, in modern Kyrgyzstan, included in a chapter on the lakes of the world (fig. 8.4 and plate 17). Lake Issıq Kul would have been known to travelers on the Silk Road, and this map has been chosen as a cover of a recent book on the travel of images and maps along this famous route.[17] The design, which visualizes the numerous rivers flowing into the lake, is accompanied by a label describing the religious circumambulation of the lake by the local Turkish tribes. The information here is two centuries old, and derived from the account of a Muslim traveler to the Bēsh Baliq, the Uyghur capital, in 821. Again, there is no mention of

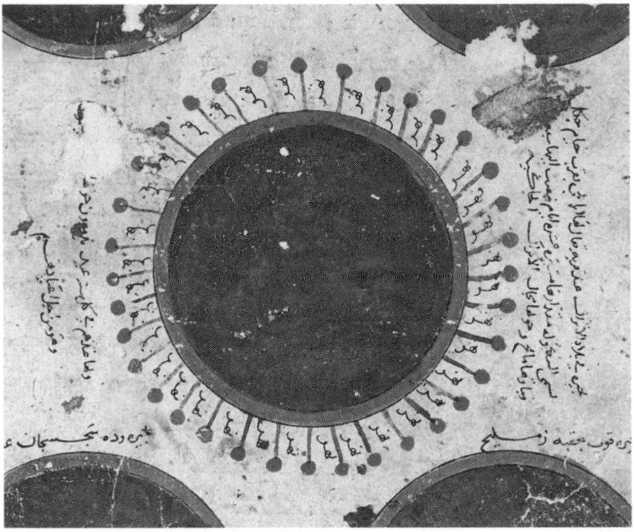

FIG. 8.4. Map of Lake Issıq Kul from the *Book of Curiosities*. Oxford, Bodleian Library, MS Arab. c. 90, fol. 40b, copied ca. 1200.

an itinerary toward China. This is typical of the scant information regarding the Silk Road in the Arabic geographical literature,[18] a reflection of the sharp decline in the importance of the Silk Road during the ninth and tenth centuries.[19]

From the perspective of Fatimid Cairo, however, the route to China and potential wealth that came with it passed not through the Silk Road or the Straits of Malacca, but rather through the Ismaʿili emirates of Sind. Indeed, the most important contribution of the *Book of Curiosities* to the history of global trade networks is the depiction of an itinerary that begins in Muslim Sind and the Indus Valley, traverses northern India, and then follows an overland route to China, almost certainly through Tibet. This itinerary is shown visually in a map of the Indus and the Ganges rivers, which is unique to this treatise. The focus on the river as the organizing principle of the map is in keeping with the maritime logic of the treatise, although the itinerary shown on the map veers away from the river and is entirely overland.

This untitled map of the Indian river systems curiously represents the major rivers of northern India—the Indus, the Ganges, and perhaps also the Brahmaputra—as one continuous river system that runs from east to west across the northern part of the Indian subcontinent (fig. 8.5 and plate 15). This single river originates in the mountains of Tibet[20], shown as red landmass at the top of the map, and then flows down what may possibly be taken to be the actual course of the Brahmaputra in eastern Bengal. The

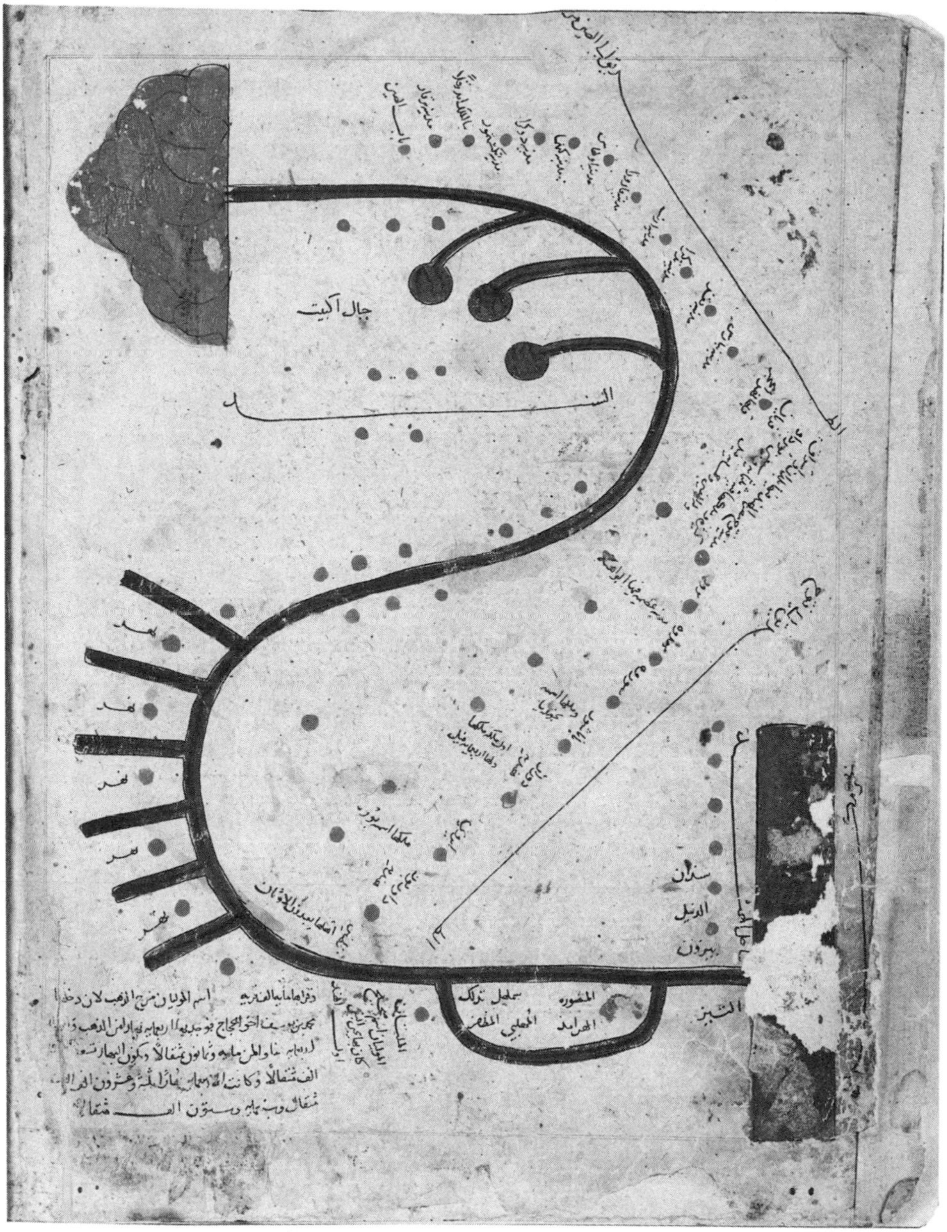

FIG. 8.5. Map of the Indus-Ganges river system from the *Book of Curiosities*. Oxford, Bodleian Library, MS Arab. c. 90, fol. 43b, copied ca. 1200.

labels at the center of the page refer to Hindu cities along the banks of the Ganges, such as Banares, Prayāg (*Allāhābād*), and Qannawj (Kannauj or Kannawj), the capital of Gurjara-Pratīhāra dynasty (836–1037). [21] The map then appears to show the Ganges running northward, in the opposite direction to its actual flow. After receiving six tributaries, the river bends southward, and then clearly represents the river Indus. The bottom part of the page shows Multān and al-Manṣūrah, the Muslim capitals of Sind, along the lower Indus. The river eventually empties into the Indian Ocean, marked by a green square.

The map is dominated by two itineraries, one from the Muslim city of Multān in Sind (in modern Pakistan) to the city of Kannauj in northern India, and a second itinerary from Kannauj to China. The itinerary from Multān to Kannauj begins with a label describing Multān itself and giving historical information about the city, including its association with the prophet Yaḥyā (John). A separate label describes the legendary amount of gold extracted from the city of Multān when it was first conquered by an Umayyad general at the beginning of the eighth century. The account is derived from earlier sources, but its inclusion on this map indicates a fascination with the riches of India.

There are eight place-names in the itinerary from Multān to Kannauj on the Indus-Ganges map, some with additional information on the population or the ruler of the city. Five of these place-names are also repeated on the Indian Ocean map, where they are depicted on the Indian shores of the ocean, even though in reality they must all have been inland. The place-names cannot be identified with certainty, but they all appear to be localities in northern India, ruled by non-Muslim, Indian kings. Coming from Multān, the first label reads *T-ṭ-y-z* (or *T-k-r-y-z* in the Indian Ocean map), where the inhabitants are said to be idol-worshippers. This may be the first city beyond the borders of Islamic-ruled Sind. The next label reads *D-a-w-r-b-w-r* (*Dāvarbūr ?*), which may be identified with modern Bahāwalpur, about one hundred kilometers south of Multān.[22] The label says that it is an Indian city, and that its ruler is called *būrah*, a name that is possibly a variant of the title of the king of Kannauj given by Arab geographers.[23] The fourth stop is *D-w-r-a-z* (*k-w-r-a-n* in Indian Ocean map). It is also described as an Indian city, and there is a mention of four hundred elephants, probably in the ruler's army. The fifth label, *Bānāshwar*, is possibly modern Benisar, in Rajasthan. Alternatively, assuming a northern route, it could be the major Hindu religious center of Thāneswar or Thānesar in the Punjab, mentioned in other Muslim sources.[24] The label also gives the name of its ruler. The seventh place-name on the itinerary is *M-h-d-w-a*, which is described as "a large city, in which the Brahmins are found." This is possibly Mahāʾūn,

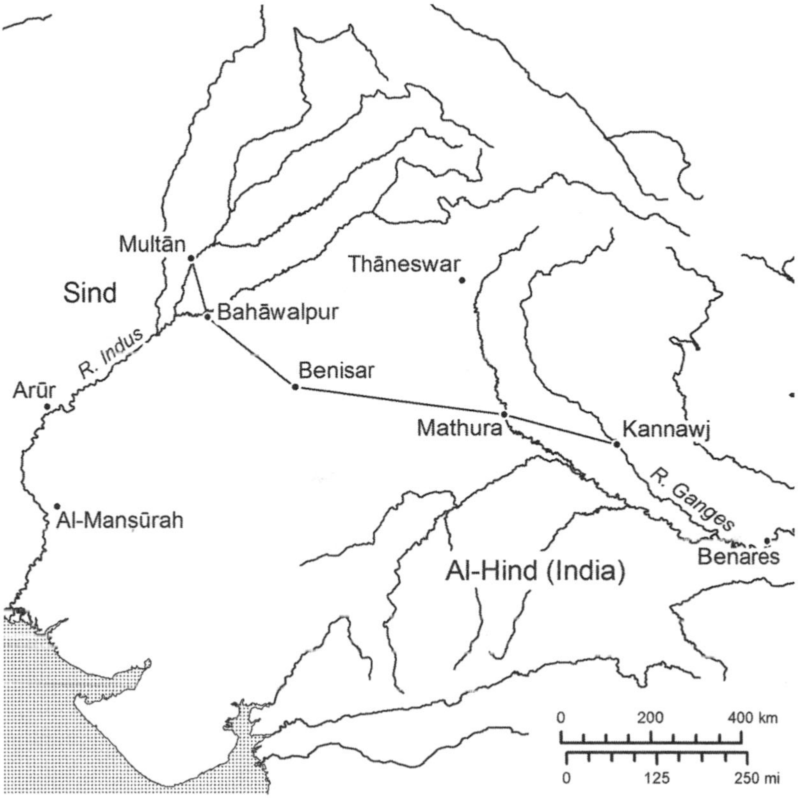

FIG. 8.6. Northern India in the tenth and eleventh centuries. The straight line indicates a possible route from Multān to Kannawj, as indicated on the Indus-Ganges map in the *Book of Curiosities*.

or Mahāvana, a sacred town in the vicinity of Mathura, on the Jumna river between Delhi and Agra.[25] The last stop on the itinerary is the city of Kannauj itself, which is the most prominent locality on the map, described as the capital of India and the seat of its ruler, whose army consists of 2,500 elephants.

Given the uncertainty in the identification of the place-names, it is difficult to trace the precise route of the itinerary from Multān to Kannauj, although both the visual depiction and the geographical literature of the time suggest a desert route, away from the river valleys (fig. 8.6). Most Islamic and non-Islamic sources relate that the route from Multān to central India cut through the Thar Desert, in what is the shortest direct line. Buddhist monks traveled via a route that proceeded away from the Indus delta, via Arūr (al-Rūr) and across the desert to Mathura, where it joined up with the main India trade route down the Ganges.[26] Early Islamic geographers attest to the continuous use of this desert route through Mathura, notwithstanding the Muslim conquest of Sind.[27] According to the anonymous Persian author of a tenth-century geography called *Ḥudūd al-ʿālam*, the trade of India was with Arūr and other small towns on the eastern banks of the Indus, all located south of Multān.[28] Al-Bīrūnī, who died around 1048, reports that Multān had become the western terminus point of the desert route. According to al-Bīrūnī, travelers between north India and Multān go through the city of Bazāna, also called Nārāyan, some 180 kilometers east of Mathura.[29]

The depiction of a detailed itinerary from Multān to Kannauj is a striking visual attestation to the religious, political and economic links between Egypt and Sind in the tenth and eleventh centuries. Even before the rise of the Fatimids, Sind had become a refuge for the dissidents of the Islamic world. By the late ninth century, Ismaʿili missionaries were sent to Sind from the Yemen, and one of them, known as al-Haytham, gained disciples in the principal trading outpost of Multān.[30] The city had become practically independent during this period, with a local dynasty of Arab tribal rulers residing in a nearby garrison and commanding a strong army supported by elephants.[31] Multān served merchants coming from the Gangetic valley and farther east. According to the ninth-century *Account of China and India*, aloewood was brought all the way from Kāmarūpa (Assam) to the temple of the sun god at Multān, where the merchants then bought it from the priests.[32] Multān also attracted substantial numbers of Hindu pilgrims who came to visit the sun god idol, first described by a Chinese traveler in the mid-seventh century.[33] In the Islamic period, the shrine of Multān became associated with the image of the prophet Ayyūb, or Job.[34] This was also a major source of revenue for Multān's rulers. According to al-Masʿūdī and Ibn Ḥawqal, the sun idol provided protection against the local Hindu powers—the rulers of Multān threatened to smash the idol in case of an Indian attack.[35]

Simultaneously with their conquest of Egypt, the Fatimid caliphs pursued a policy of aggressive intervention in Sind, and the map in the *Book of Curiosities* most likely was produced in that context. By the 960s—different dates are suggested for this event—Multān was captured by an Ismaʿili missionary called Jalam (or Ḥalam) ibn Shaybān. Jalam seized Multān, and then proceeded to destroy the sun idol and murder its priests, building an Ismaʿili mosque on the site and closing the Sunni Umayyad-era mosque.[36] Jalam was acting under the direct orders of the Fatimid caliphs. A letter written from al-Muʿizz to an agent in Multān criticizes a local missionary for his lax attitude toward the practices of the converts who joined the local Ismaʿili community.[37] A second letter congratulates Jalam for his capture of the city and destruction of the sun idol.[38] Al-Muqaddasī, who visited Sind in the late tenth century, reports Multān to be under the direct control of the Fatimid caliph, and that "envoys and presents go regularly from Multān to Egypt."[39] Direct Fatimid control is also attested by a series of coins minted in the Fatimid distinct concentric pattern.[40] There is also a record of an exotic gift from India received by al-ʿAzīz in 995.[41] The second capital of Muslim Sind, al-Manṣūrah, also came under Ismaʿili control circa 1010.[42] The two cities appear to have been under effective Fatimid control for a few decades, until they were captured by Maḥmūd of Ghazna in 1010 (Multān) and 1025 (al-Manṣūrah).[43]

Fatimid political and religious influence in Multān and al-Manṣūrah was intertwined with Indian Ocean trade, in particular trade with the Hindu powers of northern India. The Gurjara-Pratīhāra of Kannauj, who unified large areas of northern India between the 830s and 930s, were the main military threat to the cities of Muslim Sind.[44] But the relations between Hindus and Muslims also involved trade and pilgrimage. In the wake of the Arab conquest of the Sind valley, India was reincorporated into the global economy, with Hindus now actively involved in trade alongside Muslims, pushing aside the Buddhist merchant networks.[45] Kannauj in particular became the nodal point for trade routes crisscrossing northern India. Its visual prominence in the *Book of Curiosities* map is in complete accord with al-Bīrūnī's slightly later description of its central location: "The country around Kannauj is the middle of India, which they call *Madhyadeśa*, meaning 'the middle of the realm.' It is the middle or the centre from a geographical point of view, in so far as it lies halfway between the sea and the mountains, in the midst between the hot and cold provinces, and also between the eastern and western frontiers of India."[46] Al-Bīrūnī's geography of India is in effect a list of the routes connecting Kannauj with the different parts of India in all directions.[47] That included Multān and the Indus Valley, the Fatimid entry point into India.

For the author of the *Book of Curiosities*, this same Kannauj was not only the capital of a non-Muslim kingdom, but also a beacon of asronomical and atrological knowledge. As we have seen in chapter 2 above, the account of the origins of astronomy in India, found in book 1 of the treatise, identifies Kannauj as a city of wise men and the center of the learned in India, where, "to this day," the scientists prepare planetary equations that divulge the hidden truths—the esoteric influences of planetary conjunctions on events on Earth. At around the same time that al-Bīrūnī wrote his famous account of India, the *Book of Curiosities* reveals a similar fascination with the subcontinent. It is a place of idolatry and an enemy territory, but also a source of wisdom. In an Ismaʿili intellectual framework, where Heaven and Earth were closely intertwined in a web of symbols legible only to the initiate, Indian cosmology had a strong appeal.

The second section of the itinerary on the map, from Kannauj to China, shows that the horizons of the Fatimids extended even beyond central India. This second section of the itinerary is explicitly entitled "the road to China." It is a unique account of an overland passage from northern India to China, which is not found in any other Islamic source. At first, the itinerary follows the river—in reality, the Ganges, although it is not named—eastward from Kannauj, then turns away from the river and, after a dozen localities finally reaches to a place called "The Gate of China." Coming from Kannauj, the first label is *Frayān,* undoubtedly Prayāg (Allāhābād), which lies at the confluence of the Ganges and the Jumna. The label indeed adds that it lies on "the River of the Stone."[48] The next locality is written as *N-b-a-r-s,* certainly Banares. The third locality is written as *B-t-z,* which is likely to be *Putra,* a shortened form of Pāṭalīputra (Patna). From here on, however, the itinerary seems to be turning away from the river, although none of the following labels have been identified with certainty. The sixth label, "the city of *Qārūrā,*" may be a corruption of *Qāmarūpa,* that is Assam. It could also be *Qālūlā,* a locality mentioned by al-Idrīsī and tentatively identified with the present location of Dhaka.[49] The seventh label, "the city of *Awlhās*" can be Lhasa, in Tibet. The city of *T-k-sh-t-m-w-r* may possibly to be a corruption of *Lakshmibūr,* an unidentified kingdom in the vicinity of Bengal mentioned in Arabic geographical literature.[50] The eleventh label indicates the "Building of the King *A-m-d-r-f-l-a.*" The name could possibly refer to a king of the Buddhist Pala dynasty, who ruled Eastern India and Bengal since the middle of the eighth century.[51] The mountains of Tibet are indicated prominently at the top left of map.

Because the identification of the place-names is so speculative, it is difficult to ascertain the route this itinerary takes after it leaves the Gangetic valley. Two plausible options present themselves (fig. 8.7). One is that of an Assam-Burma route, through Bengal, leading to southern China. [52] The second is that of a route through Nepal and Tibet to Chang'an (Xi'an), the capital of central China. Muslim and non-Muslim sources report increased use of the Tibetan route in the tenth and eleventh centuries, and, given the prominent depiction of the Mountains of Tibet on the map, this is likely to be the route described here.

The Tibetan route is the shortest overland route linking India and China.[53] This route was inaugurated in the mid-seventh century, following the end of hostilities between Tang China and the emerging Tibetan empire. It quickly became popular among Buddhist monks and imperial embassies. When the Tibetans gained control of Dunhuang and the Gansu

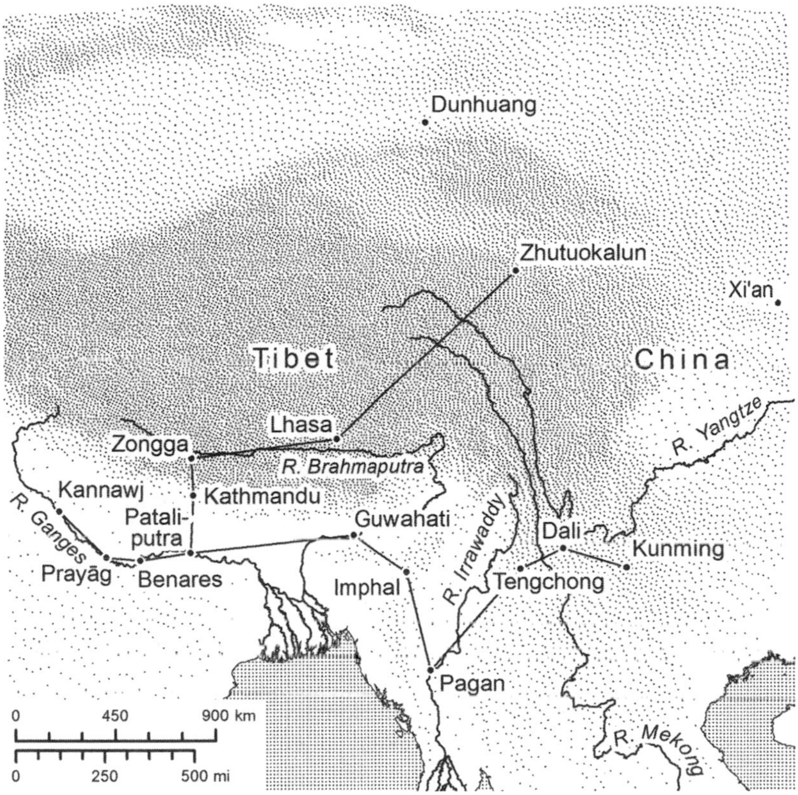

FIG. 8.7. The routes from Kannawj in northern India toward China in the tenth and eleventh centuries. Straight lines indicate possible routes through Tibet or through Burma.

corridor in 786, they ushered a period of intense traffic between Tibet, India, and China. Tibetan influence continued even after the Tang retook Dunhuang in 848. Among the thousands of documents found in Donhuang we find tenth-century letters of introduction, written in Tibetan, that accompanied a Buddhist monk traveling from Mount Wutai in China to the Buddhist center in Nalanda, on the banks of the Ganges.[54] According to Sen Tansen's study of Indian-Chinese Buddhist networks, travelers climbed the mountains of Nepal to Kathmandu, then passed through Lhasa and the Tibetan plateau, finally reaching the Chinese terminus at Chang'an (Xi'an).[55]

Apart from the map of the *Book of Curiosities* itself, there is some additional evidence for Muslims traveling along this Tibetan route. The author of the tenth-century *Ḥudūd al-ʿālam* reports the existence of a mosque in Lhasa.[56] He also describes a river crossing at the Sino-Tibetan border, a sign of some knowledge, even if not firsthand, of the overland trade route from Tibet.[57] Al-Bīrūnī, writing in the eleventh century, relates an account of a traveler who climbed up the mountains of Nepal, and from there reached Tibet and China.[58] Generally, medieval Muslim knowledge of Tibet was limited. Muslim historians commonly reported that Tibetan kings were of Yemenite origin, and the ninth-century historian al-Yaʿqūbī claimed that the king of Tibet converted to Islam and sent the Abbasid caliph al-Maʾmūn a golden image of the Buddha.[59] In the tenth-century maps of the Balkhī School Tibet is shown on the coasts of Indian Ocean, between India and China.[60] Elsewhere in the *Book of Curiosities,* we find a miraculous account of the Tibetan musk deer whose glands are the source for the lucrative fragrance of the same name. [61]

How informed and accurate was this map? The labels describe Kannauj as the capital of India and the seat of its ruler, and this seems to originate at a time when the Gurjara-Pratīhāra power was at its height, between 850 and 950. By the end of the tenth century the empire had shrunk to the territory immediately surrounding Kannauj, and it was then sacked by Maḥmūd of Ghazna in 1018. By al-Bīrūnī's days it was still commercially important, but no longer the capital of the kingdom.[62] The prominence of Multān on the map also appears to precede the Ghaznavid conquests during the first decades of the eleventh century. It seems that the map reflects the knowledge of India circulating in Egypt in the second half of the tenth century, perhaps during the period of the most intense Fatimid involvement and interest in Muslim Sind, during the 960s.

It is certain, however, that this *Book of Curiosities*' map was not based on firsthand experience. The misconception of Indian rivers demonstrates that the author of the *Book of Curiosities* did not travel through India himself. This confused depiction of the river systems is not attested in any other Islamic or Greek source. Ptolemy and Strabo clearly knew that the Ganges and the Indus were separate rivers. The Peutinger Map, reflecting Roman geographical knowledge of the fifth century AD, demonstrates this distinction visually.[63] Early Islamic geographers were likewise aware of the distinction between the Indus and the Ganges, even if confused about the course of the Indus.[64] At the same time, they sometimes show surprising lacunae in their knowledge. The ninth-century litterateur al-Jāḥiẓ suggested a connection between the Indus and the Nile, an idea ridiculed by al-Masʿūdī.[65] But al-Masʿūdī himself curiously notes elsewhere that the Indus (*Mihrān*), like the Nile, flows from south to north.[66] The author of the *Ḥudūd al-ʿālam,* otherwise well informed about India, unexpectedly fails to mention the Ganges at all.

Whatever the source of confusion about the Indian river system, the map must have relied on informants in Muslim Sind, who provided the itinerary to Kannauj, and from there, to China. The informant could have been a Hindu or other non-Muslim traveler. Multān attracted non-Muslim merchants and pilgrims from the far eastern reaches of India. It is possible that one of them recounted his itinerary to an Ismaʿili propagandist or a merchant, and the itinerary was then superimposed over what was perceived as the single large river of India. It is very unlikely that this Arabic map was an adaptation of an Indian one, as we have practically no evidence of indigenous Indian tradition of itinerary maps during this period.[67] A much later eighteenth-century Mughal scroll map indicates itineraries from Delhi to Kandahar in straight lines and adds brief information on each town. It is substantially different from the map of the *Book of Curiosities*, with place-names visualized as forts and towns, not merely as red dots.[68]

The Indus-Ganges river map visually shows how much Sind mattered to the Fatimids. Before the discovery of the *Book of Curiosities* we had only occasional glimpses of Fatimid involvement in Muslim Sind, mainly through the cryptic letters sent by the caliph al-Muʿizz to his missionaries in India. But here we see that from a Fatimid perspective, Multān was the gate to central India, and from there to China. Multān coming under Ismaʿili-Fatimid control was an opportunity for the Fatimids to rival the wealth of the great empires of the medieval world, much in the same way the Abbasids were made rich by their pole position in the global trade networks, and in particular through the direct maritime trade with China.

Surprising as it may seem for an Egyptian treatise, the *Book of Curiosities* constitutes an important contribution to our knowledge of the trade routes between India and China. Historians of the networks of the Indian Ocean have already shown that the overland and maritime routes were complementary, and we should not assume that there was only one option for pilgrims, merchants, and envoys.[69] However, the *Book of Curiosities* does suggest that for the tenth century at least, neither the sea route nor the Silk Road were the dominant or preferred routes. Instead, an alternative overland route, most likely through Tibet, appears more prominent. We may have underestimated the importance of this route for global communications: The *Book of Curiosities* shows a "Musk Road" through Tibet that overshadowed the more famous Silk Route to the north.

CHAPTER NINE

Down the African Coast, from Aden to the Island of the Crocodile

As much as Sind was the Fatimid gateway to China, the map of the Indian Ocean shows the Gulf of Aden as a gateway to the ports and islands of the East African coasts, known today as the Swahili coasts. Fatimid commercial relations with East Africa are very rarely documented, and recent scholarship has actually belittled Fatimid impact on the region during the formative period of its Islamization. But the detailed depiction of East Africa in the *Book of Curiosities* points to an unexpected level of familiarity, based on information gathered from navigation along the coasts of the Horn of Africa. We have here what may be the first recorded references in Arabic to the islands of Zanzibar (al-Unguja), Mafia, and several localities and capes along the coasts of modern Somalia. Prior to the late fifteenth century, no other medieval text describes the East African coasts in such detail. Viewed from Fatimid Cairo, the Indian Ocean was as much about East Africa as it was about India and China.

We have said above that the map of the Indian Ocean in the *Book of Curiosities* is made up of two halves, which are quite independent of each other and joined up in an arbitrary fashion, with the Yemen lying next to China in the bottom. The left-hand side of the oval depicts several distinct regions, or segments, along the Gulf of Aden and the Swahili coast (Fig . 9.1). The Yemen is located at the bottom of the map, with Aden and its mountains prominently indicated. Capes and landmarks along the African coasts of the Gulf of Aden, in what is today northern Somalia, lie at the top of the map, directly opposite from Aden and its mountains. A third segment, on the left-hand side, is a list of villages along the East African coasts, mostly on the coasts of Berbera but possibly also farther south, on the coasts of modern Kenya and Tanzania. The islands of Zanzibar (al-Unguja), Pemba (Qanbalū), and a generic "Island of the East Africans (Zanj)" are indicated inside the oval of the Indian Ocean. A list of islands and bays of the Zanj is also found at the top of the page, above the map itself. This list includes

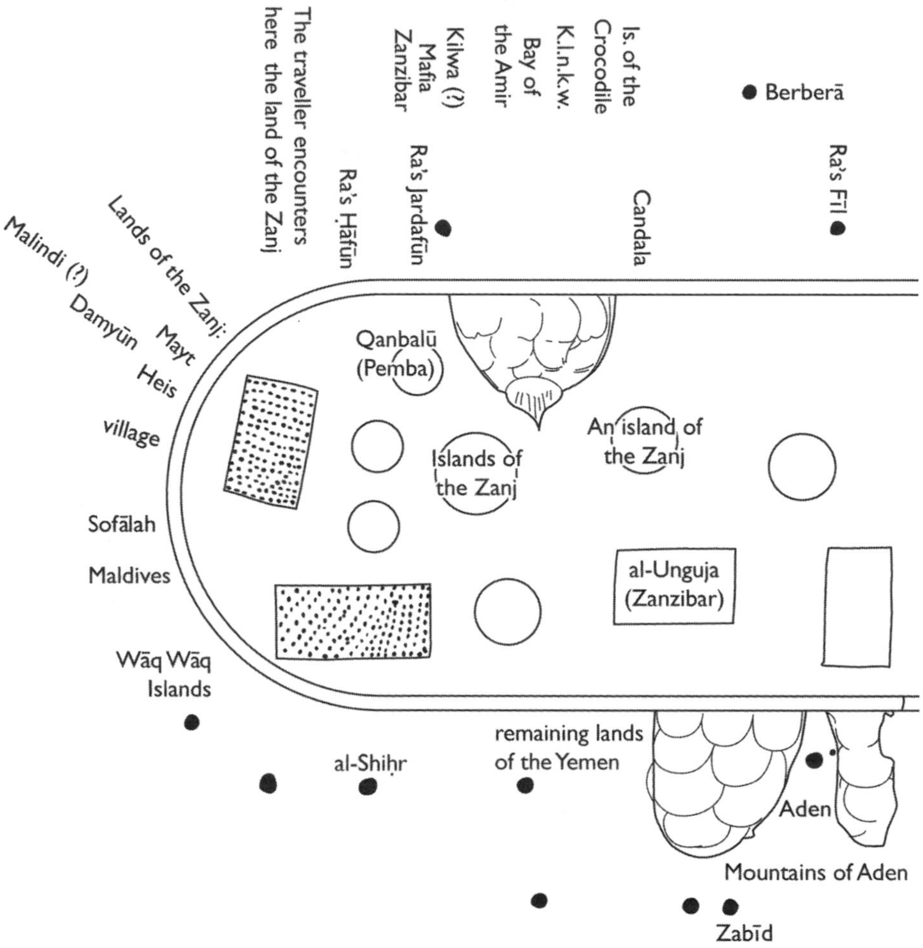

FIG. 9.1. Interpretive diagram for the left half (Gulf of Aden and East Africa) of the map of the Indian Ocean from the *Book of Curiosities*.

Zanzibar again, and eight other place-names, including the nearby island of Mafia. Finally, the Maldives (Dībājāt), the Wāq Wāq islands, and the island of Sofālah in what is today Mozambique are all found on the lower left-hand side, along the rim of the oval at the southernmost areas of the East African coast.

The port of Aden is the most prominent Yemeni locality, located on the coast between two prominent mountains, labeled "the mountains of Aden." Further from the gutter, the other Yemeni port indicated along the coasts is al-Shiḥr (incorrectly written as al-Shajarah), east of Aden. Below Aden and away from the sea, the inland cities indicated are Zabīd, established as

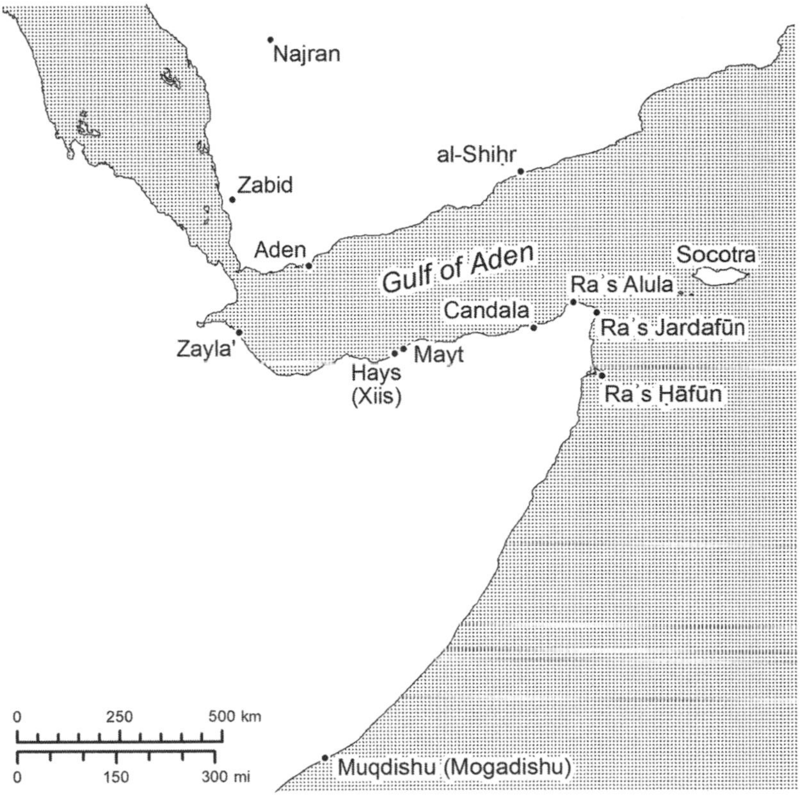

FIG. 9.2. The Gulf of Aden in the eleventh century, with reference to the map of the Indian Ocean in the *Book of Curiosities*.

the capital of the Sunni Ziyādid dynasty in 819, Najrān and Ṣaʿdah. An additional label refers to the "remaining lands of the Yemen."

The African coast of the Gulf of Aden is shown straight across from the Yemen, over the waters of the Indian Ocean, at the top of the map. Directly facing the port of Aden is a label located away from the rim, and therefore inland, which reads "The lands of the Berbers." It refers to the people known to Arab geographers as Berberā, in what is today Somalia.[1] Below this label, on the coasts of the Indian Ocean, the map has a sequence of capes and other localities on the African coasts of the Gulf of Aden. The easternmost cape is Raʾs Ḥāfūn, the prominent mountain protruding into the sea on the eastern Somali coast (fig. 9.2). The next cape is Raʾs Jardafūn, described as a very large mountain. This is the cape at the tip of the Horn of Africa, known today as Raʾs ʿAsir and in European literature as Guardafui.[2] The remaining labels in this sequence indicate mountains west of Cape Guardafui, on the northern coasts of modern Somalia. These include al-Qandalā, a name that has survived to this day as the name of the village of Candala, ninety-five miles west of Cape Guardafui.[3] The identification of the other mountains, or capes, is speculative. The label "Raʾs Fīl" may be a mistake for Raʾs Faylak, known today as Raʾs Alula, the last major cape west of Cape Guardafui, and the northernmost point of the Somali coast.[4] It is described as protruding into the sea, and is found very close to the gutter. The label which reads "Raʾs Ḥarīrah" may refer to Raʾs al-Khanzīrah, or Raʾs Anf al-Khanzīrah ("the cape of the Pig's Nose"), located between Berbera and Mayt and opposite Aden.[5] The sequence also includes "a fortress in the mountains of *A-n-kh-a-n*," a name which may be a corrupt form of Injār, modern Angar, on the African coast, just south of the Bab al-Mandeb straits.[6]

The sequence of capes suggests that this map was based on firsthand knowledge of sailing along the East African coasts of the Gulf of Aden. The prominent mountain depicted as protruding into the sea from the coasts of Berbera is either Raʾs Ḥāfūn or Cape Guardafui, even though both are listed somewhat to the left of it. Either way, the visual effect is that of the mountain as it appears from aboard a ship circumnavigating the tip of the Horn of Africa. Another label reports the prevailing currents and winds that affect southward sailing toward East Africa. The label states that "the traveller encounters here the land of the Zanj (East Africa) at the curve of the Encompassing Ocean. Whoever wants to go there [that is, to the Encompassing Ocean] is thrown back by the waves, but whoever seeks the land of the Zanj, the sea waves come from behind [and assist him]." In effect, the author advises that the open sea beyond the Horn of Africa does lead to the ocean, but that the prevailing winds draw ships toward the shores of East Africa. This seems to be a reference to the northeast monsoon, which allowed ships to sail with the wind and waves at their back, toward what is known today as the Swahili coasts. Sailing in the other direction, out into the Indian Ocean, is practically impossible.[7]

The account of capes along the African coasts of Berbera is complemented by a list of African localities found on the left-hand side of the map. This list begins with the label "Lands of the Zanj (East Africans)" and is arranged in several columns. Nearly all the place-names are described as "villages," and most, unfortunately, are illegible. The few labels that can be identified with some degree of certainty demonstrate that this is a sequence of anchorage points along the East African coasts, in the Gulf of Aden as well as farther south. The first familiar place is *Māyiṭ*, modern Mayt, on the northern coasts of modern Somalia, opposite Aden.[8] Next is *Hiīṣ*, modern spelling Heis or Hais, another village on the northern Somali coasts.[9] In the second column, away from the rim, we find *Damyūn*, which is likely

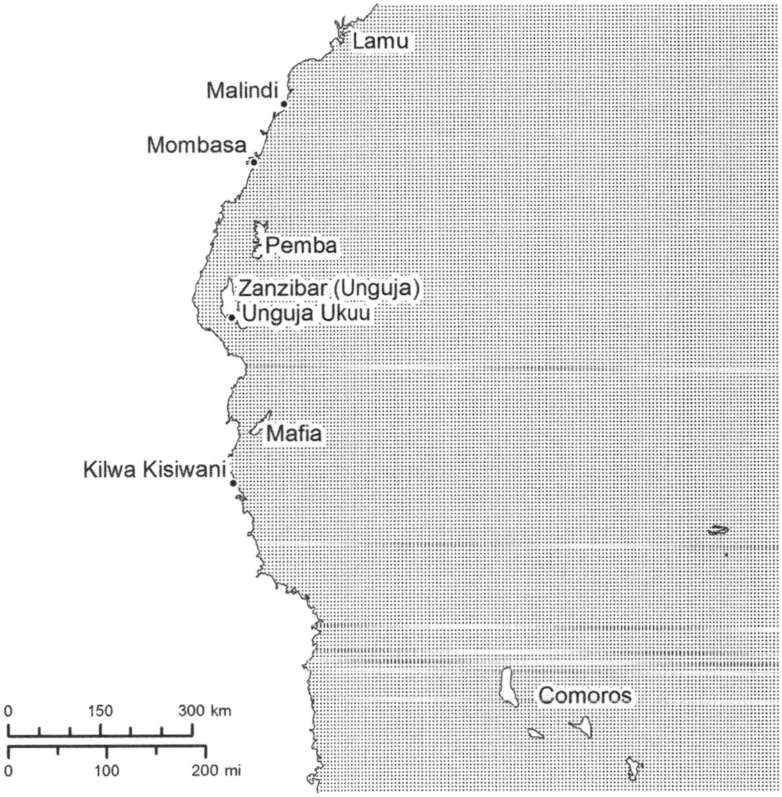

FIG. 9.3. The Swahili Coast of East Africa, with reference to the map of the Indian Ocean in the *Book of Curiosities.*

to be the same as *khaṭṭat Damyūn* or *Ḥaṭhat Damyūn,* an anchorage in the eastern coasts of modern Somalia mentioned in the navigational text of Sulayman al-Mahrī, written in 1511.[10] *M-l-n-d-s,* on the third column and further away from the rim, is likely to be the port of Malindi in modern Kenya, which is mentioned in 1060 by Ou-yang Hsiu as Ma-lin, and by al-Idrīsī in the twelfth century as *M-l-n-d-a.*[11] This list, together with the sequence of African capes in the Gulf of Aden, attests to routine navigation of these coasts and to a systematic recording of capes and anchorages along this route.

Islands are indicated in the Indian Ocean as circles or squares, and the named islands include al-Unguja (Zanzibar) and Qanbalū (Pemba) (see fig. 9.3). Zanzibar is given particular prominence, and shown as a square within the oval of the Indian Ocean, in the same way Sicily and Cyprus are accorded prominence in the map of the Mediterranean. A longer inscription reads: "The Island of Unjuwa (Zanzibar). There are [twenty ?][12] anchorages around it. It has a village called *A-k-h*."[13] Unjuwa is undoubtedly a corruption of Unguja, the Swahili name for Zanzibar.[14] The name of the village, *A-k-h,* is probably *Ukuu,* "greater," reference to Unguja Ukuu at the south of the island. As excavations have shown, Unguja Ukuu was a major trading site during the early Islamic period, probably since the ninth century.[15] The island of "Q-n-y-l-h" is certainly Pemba (Qanbalū), visited by al-Masʿūdī.[16] Two other circles in the oval refer to the generic "Islands of the Zanj."

Another extended list of East African localities is found at the top of the page and away from the rim of the oval. It is also entitled "the land of the Zanj" and includes further islands and bays along the modern Swahili coast. It may have been inserted there arbitrarily, due to lack of space elsewhere on the folio. The first locality is the bay (*khawr*) of *M-y-kh-a-n-h,* possibly a corruption of Mtwapa, between Malindi and Mombasa, recorded as *M-t-w-a-f-h* in the Arabic nautical guides of the fifteenth century.[17] The second locality is again Zanzibar, written this time as the island of *Lanjuwah.* It is followed by *M-n-f-y-h,* surely the island of Mafia, south of Zanzibar, which is not otherwise mentioned in medieval Arabic sources.[18] The next island, *K-l-w-l-h,* is likely to be Kilwa, known today as Kilwa Kisiwani, south of Mafia. In subsequent centuries, Kilwa would go on to become the wealthiest of the East African trading towns, controlling the gold trade with Sofālah to the south. Between 1000 and 1150, at the time of the *Book of Curiosities,* Zanzibar, Pemba, Mafia, and Kilwa all participated in long distance Indian Ocean trade, under the control of a closely related dynasty.[19]

The remaining East African localities of this sequence at the top of the map must refer to bays frequented by vessels traveling along the coast. The fortress of *K-l-n-k-w,* an indigenous place-name, may be one of the fortified sites in the Kilwa bay, such as the one excavated on the islet of Sanje ya Kati.[20] Two other place-names are Middle Eastern: One is the "Bay of the Amir," and the very last is the island of *Sūsmār,* Persian for crocodile. This island may be the same as Pemba (Qanbalū), which elsewhere in the treatise is associated with the Lake of the Crocodile, one of the sources of the Nile.[21] Intriguingly, the Roman *Periplus* also mentions crocodiles on the island of Menuthias, identified as either Pemba or Zanzibar.[22]

A final segment of the Indian Ocean map, located on the bottom left, between the villages of the East African coast and the Yemen, has islands thought to lie at the extreme southern end of the Indian Ocean: Sofālah, the Maldives, and the Wāq Wāq. The region of Sofālah in modern Mozambique was famed for its gold trade from at least the tenth century. Al-Masʿūdī claims that Sofālah was the capital of the Zanj, at the extreme limit of navigation for Omani vessels from Siraf in the Persian Gulf. Sofālah and its gold are also mentioned by other tenth-century Islamic sources.[23] Below Sofālah, a label indicates the islands of al-Dībājāt (the Maldives). A semilegendary account of these islands, complete with coconut trees and a naked queen, is found in a chapter on the islands of the infidels.[24] Further down we find the Wāq Wāq islands, whose inhabitants are said to be pirates. The Wāq Wāq Islands were the site of much of the lore of the Indian Ocean, and the author of the *Book of Curiosities* too has an account of the Wāq Wāq human-shaped fruits in his chapter devoted to wondrous plants. There he describes the islands as "bordering on Sofālah, one of the isles of the Zanj."[25] The author thus follows al-Masᶜūdī and his contemporary, the Persian ship captain Buzurg, who locate the Wāq Wāq next to Sofālah. This led modern scholars to identify the "African" Wāq Wāq with Madagascar.[26] The final label in this segment refers to islands even farther to the south and away from civilization, where "all the inhabitants are cannibals and the rivers are scorching hot." These islands seem to mark the southern, unexplored end of the Indian Ocean, beyond the seas familiar to Muslim sailors.

While Sofālah and the Wāq Wāq were at the exotic edge of the geographic horizons of Islam, the rest of the East African coast was familiar to the author of the *Book of Curiosities* or his informants. Taken as a whole, the geographical data suggests a maritime route that ran along the African coasts, from the Gulf of Aden and down to the Lamu Archipelago. The map is particularly detailed about the Berbera coasts of the Gulf of Aden, that is the northern coasts of modern Somalia. Mayt, Heis, Candala, and several major capes are indicated and even visualized. The information about landing points south of Raʾs Ḥāfūn is sparse, perhaps due to the coincidence of the legibility of labels. But we do have a reference to the landing point of Damyūn on the eastern Somali coast and uncertain reference to Malindi. Farther south we pick up more information, some of it significantly new, on the major trading islands of Zanzibar, Pemba, and Mafia.

The prominence of the Yemen on the Indian Ocean map is in line with its crucial strategic importance for the Fatimid dynasty, and for other Egyptian empires. Aden in particular lay at the natural gateway for the trade of the Indian Ocean. By the tenth century Aden emerged as a major emporium, a concomitant result of a shift in trade routes from the Persian Gulf to the Red Sea. Al-Muqaddasī described Aden as "the anteroom of China, entrepot of Yemen, treasury of the West."[27] The poet and missionary Nāser-e Khosraw, who traveled in Fatimid lands in the 1040s and 1050s, also tells us that the main trade routes of the Indian Ocean bifurcated from Aden, leading either to India in the east or to Zanzibar and Ethiopia in southwesterly direction.[28] The Fatimids would eventually consolidate control over the

Yemen through the Ṣulayḥid Ismaʿili dynasty, which rose to power in the 1040s and in 1087 established its capital in Dhū Jiblah. The depiction of the Yemen in the *Book of Curiosities* predates this direct Fatimid involvement in Yemeni affairs, but already highlights its strategic importance for the spread of the Ismaʿili message and the promotion of Fatimid influence in the Indian Ocean.

The *Book of Curiosities*' account of the Somali coasts of the Gulf of Aden far surpasses any other medieval text, including Yemeni sources. Ibn al-Mujāwir, writing in the early thirteenth century, reports that boats sailed directly from Aden to Mogadishu (Maqdishū), but does not give further details.[29] Under the Rasulid dynasty, which ruled the Yemen from 1229, the capital of Berbera was Zaylaʿ, a destination for Rasulid ships traveling there during the winter months. Mayt and the village of Berbera are also mentioned as embarkation points toward Aden in the Rasulid Almanac of Sultan al-Ashraf ʿUmar (died in 1296).[30] In the Geniza correspondence of Jewish merchants, dating from 1080 to 1160, Berbera is only mentioned when a ship bound to Aden is swept off course.[31] Thus, as with Egyptian—Byzantine trade, the *Book of Curiosities* again can shed light on commercial networks that lay outside the reach of the Geniza merchants.

In contrast to the detailed knowledge of the Gulf of Aden and the Swahili coast, the Red Sea is completely absent from this map and from the *Book of Curiosities* as a whole. The Red Sea is only indicated on the rectangular world map, and even there the labels for Suez (al-Qulzum) and Eilat / Aqabah (Aylah) are most probably taken from Ibn Ḥawqal's map of the Persian Gulf. Other than this secondhand and limited depiction, neither the text nor the maps of the *Book of Curiosities* refer to any ports or islands within the Red Sea. This is a strange omission, given that the any Fatimid involvement in the Indian Ocean had to pass through the Red Sea, and that Fatimid influence in the Red Sea began even before taking over Egypt. Ibn Ḥawqal notes that merchants in Suakin were calling the prayer in the name of the Fatimids, even before 969.[32] Chinese wares found in Red Sea sites dated to the tenth and eleventh centuries demonstrate increasing integration into the commercial network of the Indian Ocean.[33] So do funerary stelae found in the Dahlak archipelago, where inscriptions show that from the ninth to the late eleventh centuries the islands were a stopping point for merchants coming from as far away as Valencia and the Caspian.[34] Nāser-e Khosraw traveled extensively in the Red Sea in the mid-tenth century. He boarded a ship at al-Qulzum (Suez) and traveled down to al-Jār on the coasts of Hijaz, western Arabia. For his second pilgrimage, in 1050, he traveled down to ʿAydhāb, bypassing al-Qulzum, whose importance was on the wane from that period onward.[35] While waiting in ʿAydhāb for a suitable wind to sail to Jedda, Nāser-e Khosraw had the chance to note down the Fatimid customs levied on goods from Ethiopia, Zanzibar, and the Yemen.[36]

We can only speculate about the omission of the Red Sea from the *Book of Curiosities.* It may be that the Red Sea was actually described in more detail in the treatise, but that this description has not survived. In both the Bodleian and the Damascus copies of the treatise chapters 8 and 9 of the second book on the Earth are missing. These two chapters should have been located after the map of the Indian Ocean (chapter 7) and before the Mediterranean map (chapter 10). The internal logic of the treatise suggests they would have consisted of sea maps, perhaps of the Red Sea and the Persian Gulf, which is also absent.[37] There is no mention of Oman and Siraf in the Persian Gulf, the traditional embarkation points for the Islamic ships going to East Africa, such as the ones that carried al-Masʿūdī directly to Pemba in 916. Even the island of Socotra is not indicated, although it is described in the chapter on the islands of the infidels.[38] Since both the Red Sea and the Persian Gulf are not represented or mentioned—not even in the Indian Ocean map—it is possible that they were treated as distinct maritime spaces, worthy of separate maps.

Be that as it may, the Indian Ocean map demonstrates the importance of a trade route from Fatimid Egypt to Aden, and from there down the coasts of East Africa. The map has Aden, at the bottom of the map, as a point of embarkation, from where ships crossed the Gulf of Aden in a southerly direction—going toward the top of the map—seeking guidance from prominent capes on the African coasts. This explains the way the map visualizes mountains and capes on the African coasts of the Gulf of Aden as positioned directly across from the lands of Yemen. The capes that can be identified with certainty are near Cape Guardafui, and none close to the Straits of Bab al-Mandeb, surely because the vessels did not travel directly from the Red Sea but made first a stop at Aden. The list of settlements along the coasts of the Gulf of Aden (Mayt, Heis) and on the Swahili coast (Damyūn and Malindi) is shown separately from the sequence of capes. While the capes are the prominent landmarks during the perilous journey on the open seas from Aden across the gulf, the list of settlements, or "villages," provides a sequence of possible landing points. After turning Raʾs Ḥāfūn with the aid of current and waves, one can sail down the coast as far as the trading emporia of Pemba and Zanzibar. The coasts of East Africa are presented as a segmented itinerary, suggestive of a coastal route that begins east of Cape Guardafui and ends in Sofālah, in modern Mozambique.

This account of the African coasts in the *Book of Curiosities* has parallels with the Greco-Roman maritime itinerary of the Indian Ocean known as the *Periplus of the Eryhtraean Sea,* composed in the eastern Mediterranean a millennium earlier, circa 40–70 AD.[39] Along with mooring points and favorable winds, the *Periplus* provides information on emporia and commercial products, and is commonly thought to be redacted by an Egyptian-Greek merchant.[40] In the Gulf of Aden, it mentions a number of towns on the Somali coast, some of which are identifiable, such as Malao (Berbera) and Mundu (Heis).[41] The next section of the Greco-Roman itinerary begins at Opone, almost certainly Raʾs Ḥāfūn, and is given in stages of one day's sailing down the coast, finally arriving at the populous and wealthy island of Menouthias, which is likely to be Pemba or Zanzibar. The next major trading post was Rhapta, two days' sailing from Menouthias, and a place described as having close mercantile and political connections with the Arabian Peninsula.[42] The *Periplus* represents new knowledge originating from a discovery of a new route. Material evidence from first century AD East African sites confirms a brief period of direct trade with the Roman Mediterranean.[43]

Sailing down the East African coasts was a complex and long journey, usually fifteen months. For Greek or Roman ships sailing down the Red Sea, it was probably less lucrative than the journey to India, which, aided by the Monsoons, took less than a year to complete. East African trade was therefore eventually left to boats coming from the Arabian Peninsula or the Persian Gulf.[44] In later centuries the main axis of trade clearly shifted in favor of the Persian Gulf. As al-Masʿūdī reports, East African slaves, ivory, and timber, in the form of mangrove poles, were brought to the Gulf ports of Sohar and Siraf. Excavations at East African sites of the early Islamic period confirm the dominance of this route, with finds replicating those found in Siraf.[45] When Muslims first came to East Africa in the eighth century, it was almost certainly through contacts with the Gulf region, giving birth to the still current myth of Shirazi merchant-princes bringing Islam to the Swahili coast.[46]

But, as the *Book of Curiosities* shows, this pattern was challenged by the Fatimids. Like the Greek or Roman merchants who sailed down to East Africa a millennium earlier, the Fatimids moved down the African coasts and established direct contacts with East African trading emporia.[47] The most tangible evidence of this Fatimid connection to East Africa is the spectacular Mtambwe hoard, found near Pemba in 1984 (fig. 9.4). The hoard consists of about 2,100 silver coins, the vast majority of which were minted by local Muslim rulers. But the hoard also included thirteen gold coins, including eight Fatimid dinars and three imitation dinars of Fatimid type, dating from 969 to 1066.[48]

Another very tangible indication of Fatimid direct trade with East Africa is the sudden emergence of a market for African luxury materials in Cairo,

FIG. 9.4. The discovery of the Mtambwe hoard, near Pemba, in 1984. Photograph courtesy of Professor Mark Horton.

resulting in some of the best-known Fatimid artifacts. The extraordinary expansion of ivory carving in the Mediterranean world from the 960s, including in Fatimid Egypt, may be linked to increased level of commerce with East Africa.[49] The East African origin of Fatimid ivory is supported by the account of Nāser-e Khosraw, who reports seeing in Cairo's markets elephant tusks from Zanzibar.[50] The exquisite Fatimid rock crystal vessels, centerpiece of almost any catalogue of Fatimid art, also point to an East African connection. The surviving examples from Fatimid Egypt were carved from unusually large and pure pieces, suggesting that the Fatimid fascination with rock crystal artifacts was driven by the availability of a new source of raw material.[51] There seems little doubt that Fatimid rock crystal originated in East Africa, most probably in the river valleys of northern Madagascar.[52] Al-Bīrūnī also states that the finest crystal is brought to Basra from the islands of East Africa and from the Maldives.[53] Nāser-e Khosraw reports that in Cairo he has seen lamps made of Maghrebi rock crystal, but "it is said that near the Red Sea there is an even finer and more translucent crystal than the Maghrebi."[54]

The Indian Ocean map of the *Book of Curiosities* is a testimony to a resurgent trade between Egypt and East Africa. Admittedly, the Fatimid dinars found at the Mtambwe hoard are a tiny minority of the total pieces, and there are very few, if any, shards of Fatimid pottery unearthed in excavations.[55] East African imports predominantly originated from the Persian Gulf and India throughout the tenth and eleventh centuries.[56] And yet, while East Africa probably remained closer to the Persian Gulf than it was to Egypt, the *Book of Curiosities* demonstrates that an Egyptian Fatimid author was able to provide a detailed account of the capes and villages along the coasts of Berbera in the Gulf of Aden, and to provide a southward itinerary toward the trading emporia of Zanzibar, Pemba, and Kilwa. The information on the map appears to be based on routine sailing, a record of existing maritime activities.

Unlike the map of the Mediterranean, the Indian Ocean map does not have a military orientation. There is nothing here on the capacity of ports, availability of water, or references to fleets of galleys. The absence of a naval perspective fits what we know of the slow southward advance of the Fatimids. No Fatimid galleys traveled to Aden at the time the *Book of Curiosities* was composed. The first reference to the presence of Fatimid navy patrols in the Red Sea comes only in 1118, when five galleys (*ḥararīq*) were sent to ʿAydhāb to accompany merchants traveling to Suakin.[57] Convoys of ships traveling from Egypt, carrying the merchants known as Kārimī, did not appear in the Red Sea until well into the twelfth century.[58] It is therefore unlikely that the map of the Indian Ocean in the *Book of Curiosities*, dating a century earlier, was based on naval records, or was to serve a military purpose.

Likewise, however, there is no information on goods and commercial activities, unlike the Roman *Periplus of the Eryhtraean Sea.*[59] Instead, the treatise allows us to visualize the Fatimid Indian Ocean as a maritime space, with the Ismaʿili anchors of Sind and the Yemen as the two crucial nodes for further political, religious, and economic penetration.[60] In the Indus-Ganges map, as we have seen, Sind appears as a gateway to central India and China, following the establishment of the Ismaʿili principalities of Multān and al-Manṣūrah in the lower Indus during the tenth century. In the Yemen, long-standing Ismaʿili activity dating back to the ninth century culminated in recognition of Fatimid suzerainty in the 1060s, under the Ṣulayḥids.[61] While the Yemen was not yet under the suzerainty of the caliph in Cairo when the *Book of Curiosities* was written, it was already one of the key targets of the Ismaʿili *daʿwah.*

In fact, the most plausible source for the author's material on East Africa is Ismaʿili missionaries traveling to and from their headquarters in Fatimid Cairo. The Ismaʿili missionary network was a means for exchange of wide-ranging knowledge about the world. Indeed, one missionary *naqīb* called Aḥmad ibn al-Marzubān was the author's informant on the marvelous trees of Nubia. Each missionary was expected to have sufficient understanding of the exoteric and esoteric aspects of the faith, and, in particular, knowledge of the lands and peoples in their area of operation. In his manual for Ismaʿili missionaries, al-Nīsabūrī recommends that the missionary should travel and observe the various regions in his area of operation so that he be acquainted with the nature of the inhabitants and "the kind of knowledge they desire,"[62] as well as with their religion.[63] Lectures were written down in Cairo and copies given to the heads of the regional missions, who in turn sent back information about other countries, rulers, and creeds.

The two maps of the great maritime spaces—the Mediterranean and the Indian Ocean—should be understood in relation to each other. Both are enclosed, perfect ovals, with no indication of major gulfs and bays, a result of the cartographic method chosen by the author in his representation of maritime spaces. His method privileged clear display of textual information over accuracy in the depiction of actual coastlines and corresponded to the way coast-hugging mariners thought out their journeys as a sequence of landing points. As we have seen in all the other maps of maritime spaces, scale and orientation are removed not because they are unknown, but because they were of lesser importance as one traveled along the shore. The oval form is thus an intentional abstraction. The Indian Ocean is presented as an enclosed sea, but this representation is a result of cartographic choice, not a reflection of lack of geographical knowledge. While Ptolemy contended that the Indian Ocean was bounded by land on all sides, his theory was rejected by medieval Islamic geographers. Other Islamic maps, including the two world maps found in the *Book of Curiosities* itself, allow for an opening at the eastern edge of the Indian Ocean.[64]

It is not only that the Mediterranean and the Indian Ocean share a perfect oval. We find here too a thematic division of maritime knowledge into categories which are never integrated. As in the Mediterranean, information on harbors is treated separately from the information on bays and from the lists of islands. For the East African coasts the list of capes is separate from a sequence of coastal villages and lists of islands. The lists are not integrated.

In reality the villages of Heis and Mayt on the northern Somali coast lie quite near to the capes of Raʾs al-Khanzīra and Candala, yet there is no indication on the map of their physical proximity to each other.

But the representation of the Indian Ocean also marks the different strategic importance it held for the Fatimids. The Mediterranean was the realm of naval warfare against a renascent Byzantium, which recently captured key islands and long stretches of the eastern Mediterranean coasts. The visualization of Fatimid naval power in the Mediterranean is bolstered here by accounts of past Islamic conquests which the Fatimids hoped to emulate. Like the Abbasids before them, the Fatimids viewed the Mediterranean as the space of war and of frontiers, notwithstanding a surge in commercial activity. The Indian Ocean, on the other hand, is not a militarized space; the Fatimid ambitions in the Indian Ocean relate not to military conquest, but to the propagation of the Ismaʿili *daʿwah* mission and the extension of Fatimid suzerainty to key commercial nodes, such as Aden and Sind.[65]

On another level, the two maps also show different ways of imagining the sea. The map of the Mediterranean is solely a map of the maritime space. There is no hinterland, only ports and a dense network of islands, hundreds of dots and circles connected to each other only by means of the sea itself. As Horden and Purcell argue, the map of Mediterranean in the *Book of Curiosities* visualizes it as an essentially maritime space, "a remarkable type of sea, rather than a larger area."[66] The map of the Indian Ocean, on the other hand, has much more of the hinterland, with rivers and mountains indicated outside the rim to represent Yemeni, Indian, and Chinese localities which were, in reality, sometimes hundreds of miles inland. The maritime space, on the other hand, is relatively empty, and there are far fewer islands.[67] Along the coasts, the itineraries are segmented, with distinct visual and textual language for each of the regions surrounding the maritime space. Compared with the perfect symmetry of the Mediterranean, clearly conceived as a unity despite its militarization, the Indian Ocean is a cacophony of disparate segments, loosely connected and rather crudely put together.

CHAPTER TEN

The *Book of Curiosities* and the Islamic Geographical Tradition

Our journey through the *Book of Curiosities* has taken us from an erudite adaptation of Late Antique world maps to remote corners of Somalia, presented here in considerably more detail than they are on the most recent version of Google maps. Some of our maps relate to navigation, some are pieces of imperial propaganda, and others allow us to retrace global networks of communication prevalent a thousand years ago.

But what brought all these images together in one treatise? What is it that our anonymous author is trying to tell us about the world? So far, we have looked at individual maps in a piecemeal fashion, rarely inquiring about the context and the message of the treatise as a whole. This is not, of course, how the author intended us to approach his work. The maps come in a certain order. The organization of the material conveys a specific message about the division of the globe. Text and maps interact in certain ways, revealing new ways of thinking about spatial representation. Intertwined with all these is also an implicit discourse about the purpose of geography, the higher objective of studying the Earth and its wonders.

We should remind ourselves that the second section of the *Book of Curiosities*, the one on the Earth, complements the first book on the sky: the two are, as the author says, "the raised-up roof and the laid-down bed."[1] Yet the section on the Earth is also an independent and cohesive geographical treatise with its own internal logic, a work that stands out as a unique contribution to the already rich geographical literature of the early Islamic world. This contribution is twofold. In terms of substance, the author added significant geographical data, especially on the Eastern Mediterranean but also on Africa and East Asia. But more importantly, he also offered radically novel approaches to the organization and presentation of geographical material.

The *Book of Curiosities* was written on the back of the exceptionally rich Islamic geographical tradition, unparalleled in scope and size among pre-modern cultures.[2] This geographical literature, mostly in Arabic, originated with the administrative practices of the Abbasid Empire and its capital Baghdad. It was facilitated by access to the Greek and Persian geographical heritage, made available through the extensive translation movement of the eighth to tenth centuries, and by the expanding horizons of Arab and Muslim merchants, who dominated the Indian Ocean trade. To a lesser degree, it attempted to interpret the geographical material of the Qur'an and the Hadith, and was driven by the religious requirement of directing one's prayer toward Mecca.[3]

The different audiences of the geographical literature resulted in a variety of approaches.[4] The mathematical geography of Ptolemy was adapted by the astronomer and mathematician al-Khwārazmī in the mid-ninth century, and was the inspiration for lists of latitude and longitude and a division of the inhabited world into seven climes.[5] A second strand was primarily concerned with imperial communications and itineraries. It is usually known the "Routes and Kingdoms" (*masālik wa-mamālik*) genre, and its early representative was Ibn Khurradādhbih (died in 912), who himself was an official of the Abbasid postal service.[6] The third approach, of the Balkhī School geographers, developed later in the tenth century. This group's main interest lay in distinguishing regions of the Islamic world, which they depicted both textually and in separate maps. They also included sections on human and natural wonders, while at the same time placing emphasis on eyewitness accounts and deriding the armchair geographers of previous generations.[7]

The author of the *Book of Curiosities* set out to weave together these strands, adding a further emphasis on wonders. In his preface to the geographical section he says:

> The second book [on the Earth] consists of the seven climes, their length and their width, their seas and their islands. It describes the extent of the regions, highlands, lowlands, famous rivers and proverbial localities that are found in them. Then I added a description of the wonders and curiosities of the Earth, including those humans who are deformed as exemplary punishment, as well as those muted and abandoned; and also mentioning curious plants, stones and waters of every region and desert.[8]

In fact, there is much more in what follows. The geographical material does start with a section on the size of the Earth and its division into climes, distinguished by exclusive reliance on Greek material, and ends with a fascinating excursus on marvelous fish, plants, animals, and humans. But sandwiched between them is a highly original and systematic account of the major bodies of water on Earth, providing maps for three great seas

(Mediterranean, Indian Ocean, and Caspian), and cartographic and textual descriptions of major islands, peninsulas, lakes, and rivers. Most importantly, unlike earlier geographical literature, the focus of attention shifts firmly from dry land to bodies of water.

The author starts off the second book, the one on the Earth, with a citation of Ptolemy on calculating its circumference.[9] He follows with an account of the measurement of the length of one degree on the meridian by astronomers commissioned by the Abbasid caliph al-Maʾmūn in Baghdad, sometime around 830.[10] The rectangular map of the world is introduced at this point, followed by a thoroughly Hellenistic account of the seven climes and their inhabitants, as well as the legendary Amazons. The account of the climes in the *Book of Curiosities* is a reproduction of an unidentified Greek or Syriac source, and a slightly fuller version appears in the universal history of the Christian Bishop Agapius, written in Arabic in Mesopotamia around 940.[11]

After Ptolemy, our author turns to Hippocrates. He provides an abridged adaptation of Hippocrates' *Airs, Waters and Places* on the effect climatic conditions have on the health of local populations, a classical medical theory that carried great influence among early Islamic geographers and physicians.[12] All in all, the author of the *Book of Curiosities* is unabashed in his Hellenistic credentials. Unlike his contemporary al-Muqaddasī (d. ca. 1000), he spends no time on trying to reconcile the Greek material with the geographic data in the Qur'an and the Hadith.[13]

And then the author takes an explicit and radical break away from the Hellenistic material and from mathematical geography. The transition is introduced in a remarkable short chapter entitled "On the Depiction of the Seas, Their Islands and Their Havens." This is an introduction to the maps of the seas that follow, in which he spells out his mapmaking methods. He explains how he collected the data, and, crucially, why he chose such abstract, nonmathematical designs. First, the author states:

> We only mention here what we have heard from trustworthy sailors, from which I selected and made my own judgments; and what had reached my ears from the wise merchants who traverse the seas, and from any ship captain who leads his men at sea, from which I mention what I had knowledge of.[14]

Thus, unlike the previous chapters attributed to the Greek sages, the material that follows is up-to-date and based on credible reports by sailors and merchants. This is very reminiscent of the statements by al-Masʿūdī and Ibn Ḥawqal, who reject the armchair speculative geography of the past in favor of direct accounts of eyewitnesses.

As we have seen, the author then moves away from the mathematical geography of Ptolemy: "These sea maps (*al-ṣuwar al-baḥrīyah*) are not accurate representations," he says, acknowledging that he made no attempt to represent coastlines accurately. He gives two reasons. First, land can turn to sea and sea to land. Citing the work of al-Masʿūdī, he points out that Najaf, in southern Iraq, used to be covered by water but now is on dry land. Therefore, he argues, there is no need to bother with tracing the coasts in detail, as they change over time. The second reason is that the Ptolemaic method does not satisfy the needs of those using the maps of the seas. Thus, even were it possible to produce an accurate map of a sea in the manner described by Ptolemy by employing precise drawing instruments, the irregularity of the coastline would not leave room for labels.

In what follows, the *Book of Curiosities* shifts the emphasis from land routes to seaborne communication and from regions or provinces to bodies of water. This thoroughly maritime orientation has no direct precedent in the Islamic geographical tradition. It is very different from al-Khwārazmī's classification by climes, or the administrative framework of Ibn Khurradādhbih, who organized geographic data, fiscal revenues, and itineraries in accordance with the provinces of the Abbasid Empire. In that Abbasid framework, maritime space was attached to the lands ruled by the caliph in Baghdad, an uninhabited extension of imperial power.[15] The geographers of the Balkhī School, for the most part, delineated boundaries around the cultural and religious categories of different regions, or provinces, and these boundaries invariably cut through terrestrial, not maritime, spaces. Balkhī itineraries were always on land.[16]

The *Book of Curiosities*, on the other hand, offers systematic focus on bodies of water, organized around a clear hierarchy of seas, islands, bays, lakes, and rivers. First, there are maps of the three great seas of the world—the Indian Ocean, Mediterranean, and Caspian—comprising chapters 7, 10, and 11 (chapters 8 and 9 are missing from both the Bodleian and the later copies). Then come three chapters on Mediterranean islands—or peninsulas, as the word is the same in Arabic—held by the Fatimids. These are Sicily, the peninsular city of Mahdia, and the island-city of Tinnīs. The next chapter is devoted to Mediterranean and Indian Ocean islands held by non-Muslims, beginning with Cyprus. The chapters on islands are followed by the chapter on the bays along the Aegean coasts, and then by an account of the lakes of the world. The shift to inland bodies of water is completed by a chapter devoted to maps of five great rivers: Nile, Euphrates, Tigris, Oxus, and Indus. This systematic survey concludes with a brief note on water as the substance of life, reminding us, a bit belatedly, that water has other uses apart from being a conduit of travel from one place to another.

The focus on maritime spaces gives precedence to coasts and islands at the expense of inland routes, cities, or regions. This is most striking in the map of the Mediterranean, where nearly all localities are either coastal or islands. The distinct oval shape of the sea maps also brings the maritime space to the fore. A similar emphasis on the coastlines is evident in all the maps of the islands and peninsulas of the Mediterranean. In the map of Cyprus all the labels refer to anchorages, and no inland localities are depicted, and in the map of Sicily most labels refer to the port city of Palermo and its environs, or to prominent features on the coastline. The account of the bays of the Aegean Sea also excludes any discussion of inland features that cannot be seen from the coast. The diagrams of lakes (fig. 10.3, below, and plate 18) are remarkable for depicting only the bodies of water as a series of circles. They show no localities or human habitation on the shores. The five maps of the major rivers do include localities and itineraries at some distance from the river, but the curved, meandering water courses are undoubtedly visually dominant.

In developing this novel maritime framework, the author of the *Book of Curiosities* was drawing on certain elements in the works of earlier geographers, in particular the tenth-century works of Ibn Ḥawqal and al-Masʿūdī. Ibn Ḥawqal devotes separate chapters to the Mediterranean, the Indian Ocean, and the Caspian. His map of the Maghreb (see fig. 3.5, above), is in reality a map of the entire Mediterranean, bringing together sections concerned with separate regions (Muslim Spain, the Maghreb, and Sicily).[17] In the *Book of Curiosities* some of Ibn Ḥawqal's maps were adapted into a framework of maritime spaces. For example, the map of the Oxus River is a renamed copy of Ibn Ḥawqal's map of the province of Khorasan. The map of the Caspian Sea (fig. 10.1 and plate 16) is simply cannibalized from a map of the same area in Ibn Ḥawqal's treatise (fig. 10.2). The maps of the rivers Euphrates and Tigris also owe much to the maps of Iraq and Mesopotamia by Ibn Ḥawqal. But whereas Ibn Ḥawqal's maps draw regional boundaries that cut through the course of the rivers and divide them into separate maps, the *Book of Curiosities*' focus is the rivers themselves, from their origins to the sea.

Sections on seas and rivers in al-Masʿūdī's universal history the "Meadows of Gold" *(Murūj al-dhahab)* similarly anticipate the maritime focus of the *Book of Curiosities.* Al-Masʿūdī has a chapter on the changing shorelines of the seas, an account of major world rivers, and separate chapters devoted to the Indian Ocean, the Mediterranean, the Black Sea and the Sea of Azov (the last two grouped together), and the Caspian. In another work, "The Book of Notification and Verification" (*al-Tanbīh wa'l-Ishrāf*), al-Masʿūdī also devotes chapters to each of the great seas (Indian Ocean,

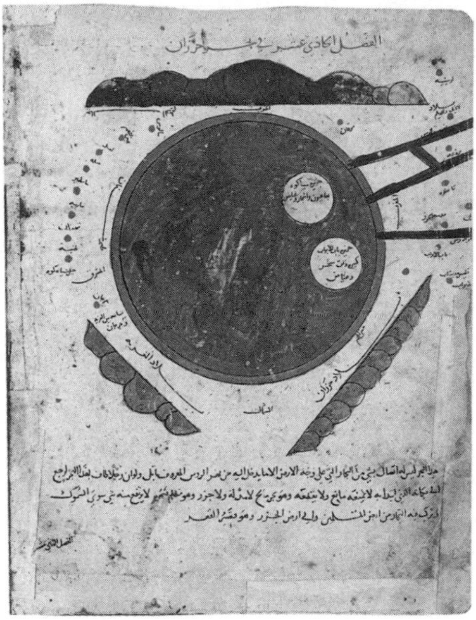

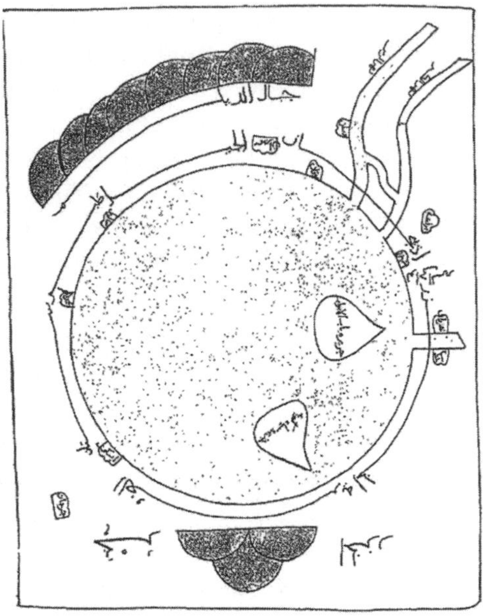

FIG. 10.1. The Caspian Sea in the *Book of Curiosities*. Oxford, Bodleian Library, MS Arab. c. 90, fol. 31b, copied ca. 1200.

FIG. 10.2. The Caspian Sea from Ibn Ḥawqal, *Opus geographorum auctore Ibn Ḥauḳal* (*Kitāb Ṣūrat al-arḍ*), ed. J. H. Kramers [Bibliotheca Geographorum Arabicorum, 2; 2nd ed.], 2 vols. (Leiden: Brill, 1938–1939), vol. 2, p. 387; based on Topkapı Sarayı Müzesi Kütüphanesi, Ahmet III MS 3346. The image has been rotated 180 degrees, to allow easier comparison with the map of the Caspian in the *Book of Curiosities*.

Mediterranean, Caspian Sea, Black Sea, and the Encompassing Ocean). His account of the rivers of the world is arranged according to the sea into which they flow.[18] Al-Masʿūdī's works lack the hierarchical organization of the *Book of Curiosities* into seas, islands, gulfs, and lakes, and are far less detailed, but were clearly a source of inspiration.

The *Book of Curiosities* aims to represent the maritime spaces of the entire known world. At the same time, it is evident that the Mediterranean, and specifically the eastern Mediterranean, is at the center of attention, and its depiction is far better informed than that of the Indian Ocean, the Caspian Sea, or the major rivers of Asia. The map of the Mediterranean, with more than two hundred labels for ports and islands, mainly in the eastern Mediterranean, is the most detailed of the sea maps in the treatise, and the information on it is original to the treatise and contemporary with its composition. The same is true for the maps which zoom in on specific Mediterranean localities. The account of Sicily is a much updated version of the account of Ibn Ḥawqal, who visited the island in 972–73, while the account of Mahdia is probably based on the contemporary narrative of the Zīrid secretary Ibn al-Raqīq (died after 1027).[19] The account of Tinnīs is a verbatim quotation of the work of the market inspector Ibn al-Bassām, composed sometime after 1014–15. The maps of Asian seas and rivers, however, are not as fresh. Much of the material on the map of the Indian Ocean is taken from the *Account of China and India* composed more than 150 years earlier, around 851. Neither was the author able to improve on the maps of the Caspian and the Oxus, and he copied them wholesale from Ibn Ḥawqal.

The *Book of Curiosities*' maritime logic and its emphasis on the eastern Mediterranean are directly related to the Fatimid context in which this treatise was produced. The focus on the sea reflects the political interests and orientation of the Fatimid caliphate as a Mediterranean power.[20] More than previous Islamic empires, the Fatimids relied on naval power and maritime commerce to pursue their political and religious ambitions. Their conquest of Egypt, in 969, was achieved through a combined naval and land attack, with galleys entering the Delta in Tinnīs and Damietta, then sailing down the Nile. This was followed in the 970s by Fatimid naval operations in Upper Egypt and in Jaffa on the Palestinian coast. A major naval campaign against Byzantium set out in 996, and a squadron of galleys suppressed a rebellion in Acre in 998. The regularity with which the Fatimids sent navies to the Syro-Palestinian coast, despite occasional disasters and scarcity of wood, shows their determination to have a military presence in the Mediterranean.[21]

The navy was an essential element of Fatimid military strategy, especially with regard to Fatimid control over Syria. A typical Fatimid military campaign involved army troops moving forward on land, accompanied by fleets carrying equipment and extra manpower.[22] The symbolic importance of the navy was pronounced by periodic processions and manoeuvres of the fleet in front of the Fatimid caliph imam seated on a Nile belvedere.[23] The Fatimid control over Syria relied on seaborne communication with the coastal towns. According to Ibn Shaddād (d. 1285), the Fatimids introduced a novel administrative structure for the port cities of Syria, such as Tripoli, Jubayl, Tyre, and Sidon. The Fatimids detached them from the hinterland and put in place governors appointed directly by the Fatimid caliph.[24] By the twelfth century at the latest, and possibly earlier, these coastal towns were placed under the financial control of a single government department, the Bureau of the Outposts (*dīwān al-thughūr*), separate from the Bureau administering the Syrian hinterland.[25]

The organization of geographic material according to maritime spaces could be seen as a Fatimid adaptation of the Abbasid "Routes and Kingdoms" genre. Abbasid geographers collected data from the imperial records, especially the records of the postal communication service, and their organization of geographical material replicated Abbasid provincial administration. The Fatimid caliphs in Cairo, on the other hand, did not even maintain a service of station posts for horses but favored the pigeon-post and merchant networks, which would often be seaborne.[26] One would not expect maps of land itineraries from an empire that preferred sailing (and flying birds). Instead, one could think of the diagrams of the Mediterranean and of Cyprus as its imperial maps of communications. Indeed, we have already alluded to the possibility that the military orientation of these maps, occasionally indicating safe harbors for navy fleets, may have been derived from the records of the Fatimid navy.

The *Book of Curiosities* reflects a mature Fatimid Empire, which had found its distinctive voice. Admittedly, we have little else to compare with. Only two other Fatimid geographies survive, and even those are only in fragments. Both date from the early years of the Cairo caliphate. The litterateur al-Muhallabī, who died in 990, wrote a treatise entitled "Routes and Kingdoms" for the caliph al-ʿAzīz. The work survives through quotations by later authors. The extant passages give accounts of Damascus and Jerusalem, with much Jewish and Christian lore, and a list of the Muslim rulers of Egypt.[27] It has been argued that the rest of al-Muhallabī's work consisted of itineraries, combined with longitude and latitude coordinates.[28] The other early Fatimid geographical treatise to have survived—but again only in excerpts—is the *History of Nubia* by Abū Sulaym al-Uswānī. Al-Uswānī was sent by the Fatimid general Jawhar to the King of the Nubians sometime between 969 and 973, and later wrote an account of travels there. The extracts that have come to us demonstrate that al-Uswānī's work was an informed account of Nubia, reminiscent in its format of Abbasid embassies to the edges of the world.[29] Neither of these early Fatimid geographical works was used by the author of the *Book of Curiosities.*

The shift from land-based geography to one rooted in seas and rivers could also be linked to a specifically Ismaʿili conception of space. The network of the secretive *daʿwah* mission divided the world into "islands," a terminology with intriguing maritime overtones. According to the idealized scheme mentioned in the Qur'an commentary of al-Qāḍī al-Nuʿmān (d. 974), who was both chief judge under the Fatimids and the head of the Ismaʿili mission, the world is divided into twelve islands (*jazāʾir,* also peninsulas). These islands are identified with the different ethnic groups of the world: Arabs, Greeks (Rūm), Slavs, Nubians, Khazars, Indians, Sindis, East Africans (Zanj), Ethiopians, Chinese, Daylamites, and Berbers. According to al-Nuʿmān, these twelve "islands" encompass all the nations of the world. Each of the islands should have a regional head of the Ismaʿili mission, called a *naqīb* or a *ḥujjah,* who reported back to Cairo.[30] This Ismaʿili division of the world was referred to also by Ibn Ḥawqal, who mentions that Kirman belonged to the "island" of Khorasan in the missionary network.[31]

The promotion of the Ismaʿili *daʿwah* mission was a major motivation for collection of ethnographic and geographical knowledge, as is demonstrated by the work of Ibn Ḥawqal, almost certainly an Ismaʿili missionary himself, and the single most important influence on the author of the *Book of Curiosities.*[32] The focus of the *Book of Curiosities* on maritime spaces has much to do with the military and commercial orientation of the Fatimid Empire, in the eastern Mediterranean and in the Indian Ocean. But as a way of organizing knowledge about the world, it goes beyond the mere political interests of the dynasty. The hierarchy of seas, islands, bays, lakes, and rivers may have been influenced by unique Ismaʿili nomenclature, in which the world was divided into "islands", not provinces, and by themes of esoteric knowledge that affected the form as well as the substance of the treatise.

The Stand-Alone Map

The author of the *Book of Curiosities* broke new ground not only in the way he organized space, but also in the way he represented it. This was an author who gave much thought to the maps he made, and who had an unprecedented confidence in the power of maps to convey geographical information. Starting from the rectangular world map, and following through the maps of the great seas, lakes, and rivers, the geography of the *Book of Curiosities* is presented in the form of stand-alone maps. We have seen that the author released himself from the Ptolemaic model, with the aim of providing up-to-date, nonmathematical maps of seas, islands, and bays. But he also released himself from another convention of the Islamic geographical tradition, one which prioritized text over image. In the *Book of Curiosities,* maps are no longer there mainly to explicate the geographical data conveyed in prose. Instead, they are now the center of attention.

This was a radical departure from Islamic geographical literature as it had developed over the previous two centuries. In fact, not all geographical works contained maps. Some of the most important authors, like al-Masʿūdī, did not produce maps at all. Our first Islamic maps, by al-Khwārazmī, are primarily exercises in the methodology of cartography. The purpose of his diagram of the Encompassing Ocean, for example, is to introduce the cartographic nomenclature of coastlines, not to visualize any particular actual space. This is also the case with Suhrāb's diagram of how to make a world map based on coordinates, discussed earlier in this book.

The maps of the Balkhī School are of course much more informative. But the maps produced by al-Iṣṭakhrī and Ibn Ḥawqal were always attached to a long textual account and depended on it. Their maps show the contours of the land and the main routes, while the text is used to discuss more complex social and economic data.[33]

It is worth pointing out here that in medieval Europe of that period, maps were even more dependent on the text into which they were embedded. Medieval European maps first emerge as illustrations to Classical works of philosophy or history. These are the famous T-O diagrams, which show the three continents with Jerusalem prominently depicted in the middle of the map. Unlike Islamic maps, the theological element is paramount, and it is common for medieval European mapmakers to draw Paradise at the east, complete with Adam and Eve.[34] Most European maps in this period are world maps, and there are few examples of either itinerary maps or city plans. The primary role of these maps is often to visualize historical and religious narratives, not to convey topographical or social information.[35]

In the *Book of Curiosities*, on the other hand, maps are often independent of any textual description. They stand in for texts rather than illustrate them. Many chapters consist entirely of maps, without any additional information except the title. The rectangular world map and the maps of the three great seas are all stand-alone pieces, which required—so the author hoped—no further explanation beyond a title. Sometimes a map is introduced by a very brief note, providing geographical, historical or astrological context. This is the case with the map of Cyprus, introduced by a short passage on the Umayyad conquest of the island, or the five maps of the major rivers of the world, each introduced by a few lines on the course of the river. Other maps, such as that of Sicily, are preceded by historical accounts. But they are not dependent on the text that accompanies them. The map of Sicily gives far more topographical detail than can be found in the preceding narrative. The same is true for the maps of Mahdia and Tinnīs.

Instances in which the author freely switches from image to text are particularly instructive. As long as the labels were not too long, our author believed that a map could convey geographical information better than a narrative text. Once the map was in danger of becoming too cluttered, he had no compunction about ditching his image and resorting to pure narrative. The account of the Bays of the Aegean, with which we opened our discussion of navigation material in the *Book of Curiosities*, illustrates how text and image are interchangeable. It begins with a diagram of five finger-like bays in southwest Anatolia, with the labels on the diagram containing information on the size and direction of each bay. In the following pages, however, the remaining bays are described only in prose, without a visual

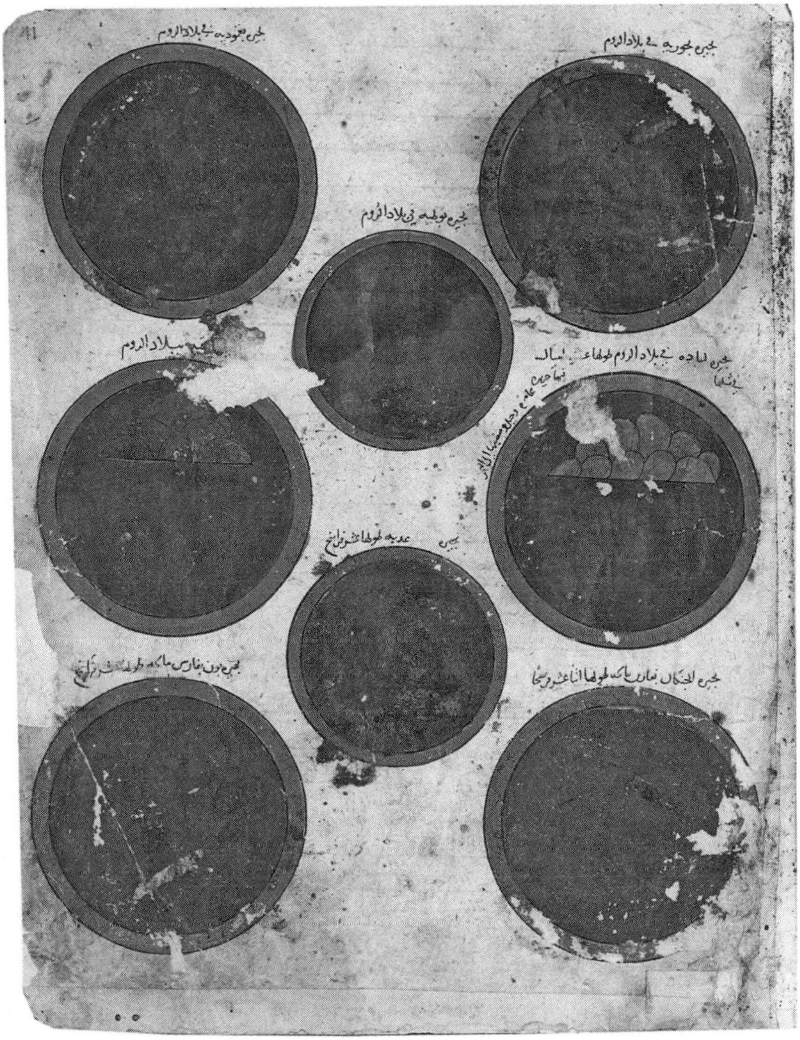

FIG. 10.3. Lakes of the world in the *Book of Curiosities*. Each lake is represented by a perfect circle. Blue represents sweet water, and green salt water. Short labels include name of the lake, its location and size. Oxford, Bodleian Library, MS Arab. c. 90, fol. 41a, copied ca. 1200.

diagram. Dropping the diagram format, with its limitations on space, allows the author to introduce more detail on each bay, such as key fortifications and islets along the coasts.[36]

Another example is the account of the lakes of the world, which similarly alternates between image and text. At the beginning of the chapter, lakes are shown visually as perfect circles of varying sizes (fig. 10.3). Each circle is independent of the others. The labels on the circles indicate the

FIG. 10.4. Lakes of the world (continued) in the *Book of Curiosities*. Here, each lake is described briefly, with the information corresponding to that found on the labels in the previous folios. The text is set in three columns. Note the word *buḥayrah* (بحيرة), "lake," at the beginning of most lines. Oxford, Bodleian Library, MS Arab. c. 90, fol. 41b, copied ca. 1200.

name, location, and size of the lake, and whether it is a freshwater or saltwater lake. This latter distinction is also shown visually by the use of colors, with blue for freshwater and green for salty water, in line with the conventions of other maps in the treatise. Yet on the following folio, the author decided to drop the visual medium and to list the remaining lakes without diagrams, arranging the text in three columns (fig. 10.4).[37]

The abstract design of the maps of the *Book of Curiosities* is our author's solution to the problem of accommodating big chunks of text onto visual representation. For many modern observers, the maps of the *Book of Curiosities* do not deserve to be called maps at all, because they are not an attempt to accurately represent geographical space. Perhaps this is so; it depends on what one calls a map. But for our author, the ovals, rectangles, and circles *were* maps. His discourse on cartographical methodology makes it clear that his aim is to visualize actual space, not to provide illustrative diagrams. But, precisely because he wanted maps to be the primary means of conveying geographical information, he had to simplify them in order to fit in long labels. He solved the problem of cluttering of information by straightening coastlines and rounding off islands.

Yet, these examples also show the limitations of his visual representations. On the one hand, our author prefers to resort to maps whenever possible. But his visual palette is limited to coastlines, lakes, rivers, mountains, and islands, to the green of the sea and the blue of the rivers. He expands this visual vocabulary when he depicts walls and palaces in the maps of the city islands, yet these are the exceptions that make these maps so remarkable. Visualizing more complex information is beyond his reach. He has no visual language to convey the presence of Byzantine galleys in the Bay of Miletos, or to indicate that the companions of the prophet Jonah lived around Lake Chad. These had to be described in narrative text, and as the text got longer and the information richer, the temptation to revert to prose was too powerful to resist.

Wonders and Curiosities

The last six chapters of the geographical section of the *Book of Curiosities* consist of an encyclopedic catalogue of natural wonders. Even the title of the treatise, the *Book of Curiosities* (literally, "the Curious Aspects of the Sciences," *gharāʾib al-funūn*), evokes a sense of the marvelous. These unillustrated chapters do not, however, form the intellectual or visual focus of the treatise. Despite its name, the *Book of Curiosities* is not primarily a collection of mirabilia.[38] But the marvels that are found here are both diverting and illuminating. They are not only fun to read, but also show an innovative approach of classifying and engaging with the exotic and the extraordinary. As such, the *Book of Curiosities* is also an important development in the history of the medieval Islamic literature on wonders, a literature which has been the subject of growing scholarly interest over the last decade.[39]

For medieval authors, anecdotes about marvels and wonders were intended to create a sense of bewilderment.[40] Rather than fantastic or supernatural, these were instances of concrete and even familiar creatures displaying extraordinary qualities. The most famous medieval collection of marvels, *The Wonders of Creation* by al-Qazwīnī (d. 1283), is an illustrated encyclopedia of the natural world, which includes a detailed discussion of the anatomy of the humble bee. At its best, the medieval Islamic marvel literature induced astonishment by looking at the most familiar.[41]

Islamic geographical literature contained material on wonders since its very beginning.[42] Ibn Khurradādhbih incorporated accounts of wonders into sections dealing with China and Byzantium. He narrated his wonders from the perspective of the Abbasid court, with the caliphs of Baghdad sending expeditions to the edges of the world in order to discover and possess exotic animals or monumental buildings.[43] Accounts of Muslim travelers into distant, non-Muslim lands also furnished much that was considered marvelous.[44] By the late ninth-century, the extraordinary and marvelous properties for each land or region come to the fore and become part of the geographical description. In the geography of Ibn al-Faqīh, the term "curiosity" signifies anything that is unusual or uncommon, including porcelain, Byzantine horses, the date palms of Basra, and the lighthouse of Alexandria.[45]

By the tenth century, some authors decided to devote entire treatises specifically to the collection of wonders. One early work of this kind is the *Marvels of India* by Buzurg ibn Shariyār, a captain who sailed in the Indian Ocean. The book consists of stories told by navigators about East Africa, India, and the islands of Southeast Asia. Some are an admixture of observations and mariners' tales, while others are definitely legendary and have their origins in the folklore of the people in question.[46] A second treatise, also composed in the middle of the tenth century, is the *Book of Marvels* by Ibrahīm Ibn Waṣīf Shāh, which has survived only in a reworking by al-Masʿūdī, known under the title "Relation of Past Times" (*Akhbār al-Zamān*).[47] The identity of Ibn Waṣīf Shāh remains open to debate.[48] The work consists of two parts: a collection of marvels and a longer discourse on the pre-Islamic, Pharaonic history of Egypt. The work as it has reached us reflects again an interest in the marvels of the Indian Ocean.[49]

In the *Book of Curiosities,* the section on the natural wonders of the world begins with a chapter on marvelous marine animals, followed by five chapters on deformed humans; magnificent rivers and springs; and marvelous plants, land animals, and birds. Like al-Masʿūdī's *Relation of Past Times,* it offers a catalogue of marvels compiled from earlier accounts. It is presented mostly in the form of short snippets, with little narrative or explanation. In terms of localities, most of the marvels are found in the remote, unexplored parts of the Indian Ocean and Africa. But unlike the narrative of wonders by Ibn Khurradādhbih, there is no imperial center. We do not find here a caliph, whether Abbasid or Fatimid, who sends embassies to explore the lands beyond the frontier. Nor is the scope limited to non-Muslim lands. Many of the marvels are in Iran, and some, such as the Mediterranean moray eel or the electric ray, could be seen very close to home.

The main criterion for inclusion seems to be unfamiliarity. The zoological sections in the *Book of Curiosities*, for example, do not discuss animals well known to the readers, and thus the material here has little in common with the *Book of Animals* by the litterateur al-Jāḥiẓ (d. 869), or with Aristotelian scientific observation and classification.[50] The author is only interested in those animals that fall outside the ordinary, whether they are real such as the giraffe and the hippopotamus, or legendary like the rook, a mythical beast with four legs below and four legs on top. The tone is similar to an account of a wonder in the extant fragment of the Fatimid chronicle of al-Musabbiḥī, in the entry for the Islamic year 415 (1024 AD): "a hippopotamus (*faras al-baḥr*) was seen in the Nile. Its color was that of an elephant, its legs resembled the legs of an ox, and it was humpbacked like a camel."[51]

Two novel features of the marvels material in the *Book of Curiosities* stand out. First, the scope is limited to the marvels of the natural world. Outlandish mores and costumes of human societies have been largely left out, or discussed as part of a separate chapter on the islands of the infidels. Neither does the author list man-made monuments and buildings, which were the first type of marvel to be discussed by Islamic geographers. In this, the *Book of Curiosities* is a precursor of the much later *Wonders of Creation* by al-Qazwīnī, also exclusively devoted to the natural world.

The second novelty is the classification of the marvels. In earlier geographical works, marvels are generally organized by their geographical location. The author of the *Book of Curiosities*, on the other hand, classifies the marvels of the natural world according to their inherent properties. It is not that they are detached from a physical space, for they are invariably located somewhere. But the location is of less importance than the mere incidence of the extraordinary. The author assumes that his audience would be interested in reading about strange fishes, or finding out about wondrous springs, without delving into long accounts of the geographical features of the region. In doing so he accords the genre of wonders, until now largely a subsidiary of the discipline of geography, a new autonomy. Here, too, the *Book of Curiosities* anticipates the more complex categorization of marvels by Abū Ḥāmid al-Gharnāṭī (d. 1169) and al-Qazwīnī.

The author of the *Book of Curiosities* occasionally attempts to corroborate the factuality of some of the wonders he relates. He cites the accounts of three informants, two of them with regard to African plants. First, he recounts seeing cushions filled with elastic kapok material in the house of a man from Ḥarrān called Abū al-Qāsim, who is said to have traveled extensively in East Africa. Second, a *naqīb*—an Ismaʿili missionary—called Aḥmad ibn al-Marzubān informed the author about the sweet taste of a large fruit produced by a Nubian palm tree, perhaps the doleib. Finally, a secretary called Shaykh Abū al-Ḥasan ibn Ṣabbāḥ confirmed to the author the marvelous account of the salamander, boasting that he has with him some salamander hair that fire cannot consume.

In these three instances, the role of the informant is to reinforce the remarkable character of the object described. This is different from the approach of Ibn Ḥawqal and al-Masʿūdī, who tended to be inherently sceptic. In an illuminating anecdote of tenth-century scientific empiricism, Ibn Ḥawqal relates a common belief that eating the flesh of a gilt-head bream brings about violent erotic nightmares. He then tests this claim by experimentation: He eats the fish with his friends and reports having a good night's sleep.[52] Al-Masʿūdī states in his *Relations of Times Past* that the accounts of wonders are presented for their curiosity value, but are not necessarily true.[53] Our author, on the other hand, does not share their scepticism. His wonders—whether the few he has collected from informants or the majority he has found in books—are all presented as matters of fact.

For the most part, our author takes familiar sources and reorganizes them. The Indian Ocean lore, mostly from the ninth-century *Account of China and India,* is broken up according to new categories. The list of Indian Ocean fishes comes in the chapter on marine creatures, while the account of the Wāq Wāq islands and their wondrous, sensual fruits comes in the chapter on wondrous plants. Indirect quotations from Aristotle's zoological works crop up in the chapters on marine creatures and on land animals, which include also paraphrases of the popular Alexander Romance, relating the marvelous journeys of Alexander the Great. The classification is not always neat: the entry on the octopus, a corrupt version of a Hellenistic source, is mentioned among the land animals (probably because the author had not realized that this is an octopus). Ethnographic material creeps up in an account of a fabulous Chinese bird. A long diversion on lunar mansions and their talismans is inserted into the chapter on the wonders of the seas, presumably because the talismans have monstrous forms.

The chapters on wonders contain some unique Hellenistic material which is not known to us from other sources. The chapter on quasi-human creatures is presented as a verbatim quotation of a *Sermon on the Races,* an ethnographic account attributed to ʿAlī ibn Abī Ṭālib, the Prophet's cousin and son-in-law and the first Shiʿī-Ismaʿili imam.[54] The attribution to ʿAlī is undoubtedly a means of crediting Hellenistic material with Islamic legitimacy, as attested for a comparable account of alchemy.[55] The chapter on deformed humans in the *Book of Curiosities* parallels a similar section in al-Masʿūdī's *Relation of Times Past.*[56] But there is also some unique material here on these monstrous races, including otherwise unattested names, and several semi-human hybrids, including one race of naked, horned people, born out of a union between humans and sea animals on the island of Thulé.[57] Some of the monstrous races were considered part of the nations of Gog and Magog. Their names and features are not found in parallel accounts of the Gog and Magog races in the Arabic Alexander Romance nor in the work of Ibn al-Faqīh.[58]

Much of the zoological material in the treatise is otherwise unknown to us. It may come from a lost translation of the work of Timotheus of Gaza, a Byzantine grammarian who composed a zoological treatise for the Emperor Anastasios (AD 491–516). Timotheus was not a scientist, but a collector of anecdotes about animals. His work is described in Byzantine sources as an account of the quadrupeds of Egypt and Libya and of portentous birds and snakes. Remke Kruk demonstrated that there are explicit references to Timotheus in a book of animals by the late eleventh-century author Sharaf al-Zamān Ṭāhir al-Marwazī.[59] In the *Book of Curiosities* three of the entries for land animals have exact parallels in the Greek fragments of Timotheus: the leopard, the octopus and the beaver. It is possible that other zoological material in the treatise comes from a translation of Timotheus, for several long entries for birds with magical powers of detecting poison and foretelling the future are reminiscent of Timotheus' section on "portentous birds."[60] The magical and medical properties also suggest parallels with the esoteric tradition associated with Hermes Trimegistus, which, as we have seen, is dominant in the section on the stars.[61]

The *Book of Curiosities* is thus an important milestone in the Islamic literature on wonders. While earlier geographical treatises subsumed sections on wonders under the geographical regions in which they occurred, the *Book of Curiosities* detached the marvels from the purely geographical material and classified them according to thematic categories. In this, the treatise anticipates the thematic organization by later authors such as al-Gharnāṭī and, most famously, al-Qazwīnī. But, unlike al-Qazwīnī, the author of the *Book of Curiosities* chose not to illustrate the chapters on wonders; the illustrations of the Wāq Wāq trees in the Bodleian manuscript are a later addition from the Mamluk or Ottoman periods (see fig. 2.5 above). The images that could count as wonders, such as the crocodile and the sand dunes at

FIG. 10.5. Page from a chapter on strangely formed breeds from Qazwīnī's *Wonders of Creation*. Unlike the eleventh-century *Book of Curiosities*, the quasi-human breeds are here invariably illustrated. *The Wonders of Creation and the Oddities of Existence*, copied in Wasit, Iraq, 1280. Munich, Bayerische Staasbibliothek, MSS cod. Arab. 464, fol. 208a.

the sources of the Nile, are normalized as part of the explictly geographical material. The clumsy crocodile in particular does not inspire much awe or wonder, at least in the way it is executed in the Bodleian copy.

It is instructive to compare the chapter on strangely deformed races in the Bodleian manuscript of the *Book of Curiosities* with the same text in al-Qazwīnī's autograph manuscript dated 1280. The account in the *Book of Curiosities* has a rich account of people who have faces on their chests; the one-legged *nisnās* who lives in Arabia; and the *J-ʿ-m-a*, a nation born from a union between humans and wild beasts, and which look like Turks. But these marvels are merely described and are not illustrated. Al-Qazwīnī, on the other hand, explicates the account of each hybrid race with a vivid sketch, which visually conveys a message about God's power behind all creation, grounded in Neoplatonic philosophy (fig. 10.5).[62] The absence of such images from the *Book of Curiosities* suggests either a lack of technical ability or that the iconography of Islamic art was yet to find a way to express messages about the divine. We should recall that the author of the *Book of Curiosities* was not averse to the power of images. But he preferred to produce maps and diagrams, not monsters.

CONCLUSION

Maps, Seas, and the Ismaʿili Mission

Who, then, was the author of the *Book of Curiosities?* His apparent access to naval military records suggests a direct connection to the Fatimid state. His personal acquaintance with an Ismaʿili missionary who had been to Nubia, and his unique information on itineraries in the world of the Indian Ocean, all suggest he was close to the Ismaʿili missionary network. The map of Palermo with its suburbs, the diagram of Tinnīs, and in particular the map of Mahdia, which is drawn from the perspective of someone looking at the city from a vantage point just outside of its walls, suggest that he had visited these port cities in person. In some ways, he is a successor to the geographer Ibn Ḥawqal and a predecessor of the poet Nāser-e Khosraw—both Ismaʿili missionaries, travelers and keen observers of human societies. Unlike them, however, his interest in trade is minimal, and he is more likely to have been a military man than a merchant. Nor was he a scholar of the caliber of his Egyptian contemporaries, the physician Ibn Riḍwān or Ibn Haytham, the founder of the science of optics. His grasp of mathematical concepts appears to have been quite poor, and he generally avoided technical discussions.

Rather than a scholar our author was, primarily, a mapmaker. It is the maps that make the *Book of Curiosities* such a distinct work of medieval scholarship and such an appealing manuscript for modern audiences. The author has unprecedented confidence in the ability of maps and diagrams to convey information. Unlike any other geographical treatise before this, the maps are stand-alone artifacts, unsupported by any accompanying text. This is true for some of the maps of the sky, but especially for the rectangular map of the world, the maps of the three great seas, and the maps of the rivers. Even when the maps are related to a text, such as those of the islands of Sicily and Cyprus, or the city of Mahdia, the information they contain goes well beyond that of the preceding prose sections. We do not have the original treatise, only a later copy, so we do not know how lavish it might have been when first penned. But the second part of the title literally translates as "that which is pleasant to the eyes" (*mulaḥ al-ʿuyūn*), indicating that this treatise was about the images as much as it was about the text.

Maps are at the center of this *Book of Curiosities*, and this anonymous mapmaker offers us his reflections on the craft of cartography. His chapter on mapmaking techniques introduces the maps of seas and islands, the most original maps in the treatise. He opens with a formulation of the purpose of his maps: "Although it is impossible for created beings to know the extent of God's creation, the knowledgeable and qualified among them are entrusted with witnessing or imparting a small part of it."[1] The maps that will follow will convey knowledge, albeit imperfect, of God's creation. He then continues to explain why his maps are intentionally "not accurate representations" of reality: the contours of coastlines change over time, the mapmaking instruments are not fine enough to reproduce reality on a small scale, and labels need to be legible. Here is a mapmaker explaining his choices and reflecting on the purposes and functionality of his maps. The results of his labor are unique medieval versions of "graphic representations that facilitate spatial understanding," to use the definition of "map" by the leading modern historian of cartography.[2] There is no parallel for this passage in any other medieval treatise known to us.

The *Book of Curiosities* is a profoundly Fatimid treatise. Like a *tirāz* armband, it wears its allegiance to the Fatimid caliphs on its sleeve. This is apparent from the opening dedication, from the blessings heaped on the Fatimid imams, and from the curses flung at the rebels who sought to overthrow them. The treatise also reflects some immediate political ambitions of the Fatimid state, especially in the Mediterranean. It depicts visually and in text the defenses of the strategic Fatimid holdings in Tinnīs, Mahdia, and Sicily. There are historical references to the early Islamic conquests of Cyprus, Crete, and Bari, with the inference that they may be ripe targets for Fatimid re-conquest. The mapping of anchorages, ports, and bays deep in Byzantine territory, some of them as far north as the Dardanelles, also reflect a military context. It is likely that much of the material here was actually drawn from the records of the Fatimid navy. And beyond the immediate political objectives, the maritime focus of the *Book of Curiosities* is also distinctly Fatimid. The unusual categories for organizing the geographical material, from seas to islands, and then to lakes and rivers, reflect the unique maritime orientation of the Fatimids, who, alone among the great medieval Muslim empires, preferred networks of ports, rivers, and islands over horses and land routes.

The treatise can be viewed as part of a westward shift in the geographical tradition and in the center of gravity in the Islamic world in general.

Most ninth-century works, such as those by Ibn Khurradādhbih and the *Account of China and India,* focused on Asia and the Indian Ocean. By the middle of tenth century, however, the gaze shifts to the Mediterranean, North and Sub-Saharan Africa. Al-Masʿūdī spent much of his later life in Egypt and Syria, and Ibn Ḥawqal provided an unprecedented account of the Maghreb. The eleventh-century *Book of Curiosities* focuses on the eastern Mediterranean. Later works of the Andalusian author Abū ʿUbayd al-Bakrī (d. 1094) and, of course, the Sicilian based al-Idrīsī (fl. 1154) have their focal point even farther to the west. The heavy reliance of the *Book of Curiosities* on the work of Ibn Ḥawqal is also suggestive, because the latter was, most likely, also a missionary. The focus on islands in the *Book of Curiosities* may have had special resonance against the backdrop of the Ismaʿili nomenclature of regional "islands."

In the preceding pages of this book we have devoted much more space to the maps of the Earth than we did to the maps of the sky, but this is a reflection of our modern interests. In the treatise itself, the Heavens and the Earth are perfectly balanced. The treatise begins with a diagram of the universe and proceeds down from the sphere of the fixed stars to those of the planets and comets, with the aim to explore the effect of the stars on earthly events. The astrological material is not in fact limited to the first half of the treatise. It spills over into the accounts of cities and rivers: The people of Tinnīs are full of happiness and joy because the city was founded when Pisces was in the ascendant; foreknowledge of the level of the Nile's annual flood comes about from observing Mars at the start of the year. For the author, the astrological mindset is paramount.

The Ismaʿili context of the *Book of Curiosities* invites comparisons with the influential *Epistles of the Brethren of Piety,* an encyclopedic corpus of science and Neoplatonic philosophy, composed in Iraq sometime before the middle of the tenth century. The Epistles are not cited in the *Book of Curiosities,* nor is there evidence for their circulation in Fatimid Egypt, despite their affinity with Ismaʿili teachings. Yet the Epistles seem to approach the subject matter of the sky and the Earth in a similar manner.[3] Following the Greek astronomer and geographer Ptolemy, geography is seen in the Epistles as an appendix to the study of the stars. But the Epistles also have a higher purpose: the reader is invoked to contemplate the design of the creator, "to ponder wonders (*ʿajāʾib*) of his creation and reflect on the curiosities (*gharāʾib*) of what he fashioned."[4] This desire to observe God's work explains some of the interest in marvels and wonders exhibited by the *Book of Curiosities,* as its title suggests.

Like the *Epistles of the Brethren of Piety,* the *Book of Curiosities* draws heavily and without compunction on the heritage of Greek science. Ptolemy, Hippocrates, and Galen provide our author with much of the material on the general structure of the Heavens and the Earth, and on the way the former influence the latter. Muslim scholars like al-Masʿūdī correct and add information, especially when one zooms in on the Earth's size and layout, but the general framework inherited from the Greeks is not questioned. And while God is omnipresent, the Qur'an is cited sparingly, only to invite reflection on creation or to buttress moral points about God's punishment meted out to the unbelievers. There is only one Tradition from the Prophet in the entire treatise, on the intrinsic purity of water. Such reliance on a Hellenistic heritage was not uncommon in eleventh-century Cairo. Mubashshir ibn Fātik, a wealthy and influential scholar, left us a remarkable collection of ethical sayings from the Greek sages, with special focus on the Late Antique and legendary Hermes.[5]

The author of the *Book of Curiosities* does not limit himself to Greek authorities, but is also acquainted with Persian, Indian, and Coptic knowledge. He cites an account of the birth of astrology in India, and Persian authorities on the ominous *bābānīyah* stars. He is also keen to show command of multiple languages. For example, the names of each planet are given in Persian, Classical Greek, "Indian," and Byzantine Greek; the names of each day of the week are given in Persian, Byzantine Greek, "Indian," Hebrew, and Coptic. The use of Coptic is of special significance. There is here a deep influence of Egyptian Coptic traditions, an influence which has been overlooked in modern scholarship on Islamic science. Coptic lore was already reworked by al-Masʿūdī, and the Coptic calendar seems to have been used very early when predicting the risings of the lunar mansions. There is even more Coptic material in the *Book of Curiosities*, including accounts of the moray eel and the foundation of Tinnīs. Most importantly, the explanations related by Copts regarding the flooding of the Nile—that the Nile floods are a result of the summer melting of the snow on equatorial mountains—are as close to the present understanding of the Nile system as was ever achieved by medieval Islamic scholarship.

The original copy of the *Book of Curiosities* probably ended up in the libraries of the Fatimid Palace or the state-sponsored House of Knowledge, established by the caliph al-Ḥākim in 1005. The anonymous patron to whom the treatise is dedicated may have been the caliph himself, or a wealthy patron such as Mubashshir ibn Fātik. In December 1068, however, the Fatimid libraries were looted by troops seeking pay. It is said that they took as many as eighteen thousand volumes on ancient sciences, as well as the luxurious silk world map made for the caliph al-Muʿizz a century earlier, and that the books from the House of Knowledge were taken by Lawāta Berber tribesmen, who used the book covers as sandals for their slaves.[6]

This is almost certainly an exaggeration, and it has been reasonably argued that the bulk of the Fatimid libraries were only sold off with the coming of Saladin a century later.[7]

We do not know what happened next to the original manuscript of the *Book of Curiosities*, but we do know that a lavish copy of it was produced sometime in the late twelfth or early thirteenth century, shortly after Saladin took over Egypt in 1171. That copy has now ended up in the Bodleian Library. The original manuscript, or perhaps a derivative version, then made its way to northern Syria, where several partial and unillustrated copies were produced from the sixteenth century onward. The rectangular map of the world, and the maps of seas, islands, and rivers—the maps that make this treatise so unique—are only found in the Bodleian manuscript. Even in the Bodleian copy, however, the labels are corrupt, often beyond recognition. This is evidence of the remarkably idiosyncratic nature of the *Book of Curiosities.* It was an exceptional product of an exceptional dynasty and sect, visualizing an unusual sea-based imperial worldview, and an encyclopedic intellectual outlook.

And so, as a result of its idiosyncratic nature, the *Book of Curiosities* had no discernible influence on the tradition of mapmaking in later centuries, except, perhaps, on al-Idrīsī's introduction of a western arm of the Nile. It is only the determination of modern scholars that brought this manuscript to public attention. Generations of medieval and early modern owners passed it on, perhaps with bewilderment but certainly without any real interest in its maps. After the Fatimids, medieval Islam followed different paths. The Crusades put an end to the Fatimids' Mediterranean dreams, and the Sunni armies of Saladin swept away their Ismaʿili–Shi'a teachings and scattered what was left of their missionary network.

Yet this treatise is not only one of the greatest achievements of medieval mapmaking. It is also a remarkable part of the story of Islamic civilization. Too often, the achievements of Islamic science are divorced from the culture that produced them and are only brought to light as a trophy in a sterile competition with the West, or as evidence of Islamic ingenuity. Islamic maps in particular get almost no attention in surveys of Islamic history, and even when they are shown they are rarely explained. Because we tend to view Islamic civilization through the prism of religion and faith, we find no use for these abstract diagrams that tell us nothing about God.

But what could be more foundational to any culture than the manner in which it conceived the sky and the Earth? We hope to have shown here that the discovery of the *Book of Curiosities* is also a timely rediscovery of those aspects of Islamic history which are too often neglected in academic and nonacademic visions of Islam. It is a rediscovery of the sea as an integral part of a civilization that supposedly originated in the desert, of an outward-looking scientific inquiry that was built on the foundation of the classical Greek legacy, and of the power of cosmographic and cartographic images in a culture that is too often reduced to texts. The *Book of Curiosities* opens an unexpected, unmediated window to the medieval Islamic view of the world.

APPENDIX

A Technical Discourse on Star Lore and Astrology

This appendix is for those who are particularly interested in the astrological significances of zodiacal signs or simply enjoy the mental gymnastics of aligning and realigning those signs and characteristics. Because of the importance of zodiacal signs and their relationship to the planets in forecasting events using horoscopes, the author of the *Book of Curiosities* devoted an extraordinary amount of time to that matter. Although there is no evidence he was a professional astrologer, he was very well versed in the role of zodiacal signs and planets in horoscopic astrology, and he assumed his readers were reasonably informed as well and equally interested in the subject.

The zodiacal signs are the twelve equal parts into which the zodiac is divided. From the viewpoint of an observer on Earth, the Sun appears to pass through one zodiacal sign in each month. The Sun, Moon, and the five planets visible to the naked eye (Mercury, Venus, Mars, Jupiter, Saturn) all appear to those on Earth to move along the pathway or belt of the zodiac (about a 18° wide band), never moving outside its boundary. The twelve divisions of the zodiac—that is, the zodiacal signs—are counted or numbered from the point of the vernal equinox. They are named after the twelve zodiacal constellations (Aries, Taurus, Gemini, Cancer, Leo, Virgo, Libra, Scorpio, Sagittarius, Capricornus, Aquarius, Pisces), and about two thousand years ago the signs lay in the constellations of the same name. However, the stars forming those classical constellations are no longer positioned within the zodiacal sign bearing its name, for the precession of the equinoxes has resulted over time in a displacement. At the present time, the vernal equinox (March 21, and thus the start of the zodiacal *sign* of Aries) occurs not in the *constellation* of Aries, but in the constellation of Pisces, and by the year 2597 it will occur in the constellation of Aquarius.

Folk astrology did not concern itself with this topic, for it relied upon the simple observance of a star or group of stars or the Moon in order to make a forecast, but the type of astrology that developed in Greece and continued to develop in Egypt relied heavily upon zodiacal signs as the major divinatory tool. Astrology was widely cultivated in the Greek-speaking communities of Egypt, building also upon earlier Babylonian and Indian practices. Horoscopes appear to have been an Egyptian invention of these Greek-speaking communities when Egypt was under Roman rule, and our author refers on occasion to horoscopes and assumes the reader understands what they are.[1]

What we call a horoscope is essentially what the Greeks called a *genethlios* and the Arabs *mīlād* (pl. *mawālīd*), both terms meaning a nativity. It is a diagram representing the configuration of the Heavens at the time of a given person's birth (or some other event). On a single grid, celestial bodies are related to each other and also to the Earth. The Sun, Moon, and visible planets are recorded by their positions within the zodiacal signs at a specified moment in time. This "map" of the skies is superimposed and frozen over the local horizon. The combined celestial/terrestrial diagram is then set within a frame of twelve astrological "houses" or "places" whose sequence begins at whatever point in the Heavens was on the eastern horizon of the given locality at the specified time. The astrological "places" or "houses" (Arabic *buyūt*; Greek *topoi*) were devoted to particular topics, such as illness or marriage or travel.

The zodiacal sign rising at the local eastern horizon was called a "rising sign" or "ascendant," and the Greek term for this rising sign was *horoscopos* ("observing the hour"), from which we get the word "horoscope." On the basis of such diagrams, claims were made to predict the fate and fortune of the person. Similar diagrams or horoscopes could be computed to determine the best moment to undertake an event or whether a proposed venture was auspicious or not.

As an example, let us take a horoscope produced in Egypt and contemporaneous with the time of our author (fig. appendix 1). It is a horoscope made in Cairo by a well-known physician and astrologer named Ibn Riḍwān (d. ca. 1061). In his "autobiography" quoted at length by the thirteenth-century historian of medicine, Ibn Abī Uṣaybiʿah, Ibn Riḍwān says:[2]

> It is appropriate for every person to take up the profession most suitable for him. The art of medicine is closest to philosophy with regard to obedience to God, the Mighty and Glorious. The astrological signs at my birth indicated that medicine should be my profession. Moreover, a life of merit is more pleasing to me than any other. I undertook the study of medicine when I was a boy of fifteen years, but it is best to relate to you the entire story:

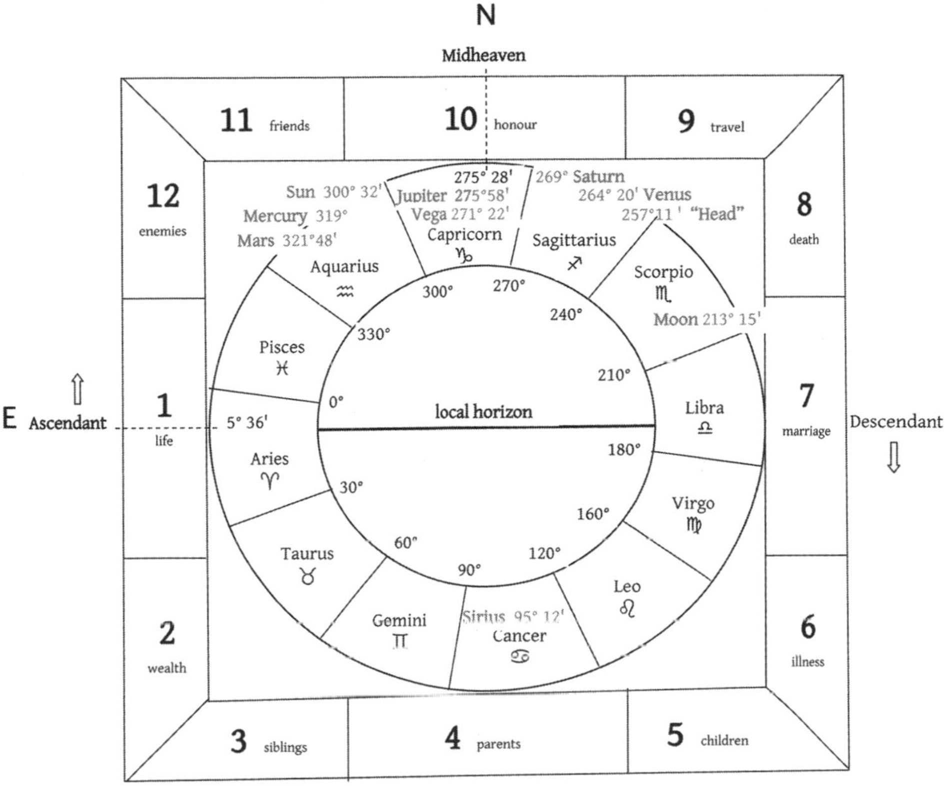

FIG. APPENDIX 1. A modern diagram illustrating the birth horoscope of Ibn Riḍwān, born in Cairo on January 15, 988, based upon detailed descriptions given in his "autobiography." Constructed by E. Savage-Smith.

> I was born in Egypt at a locality situated at 30° latitude and 55° longitude. The ascendant was, according to the tables of Yaḥyā ibn Abī Manṣūr,[3] Aries at 5° 36′ and the [mid-point] of the tenth house was Capricorn 5° 28′. The positions of the planets were the following: the Sun was in Aquarius at 5° 32′, the Moon in . . . [the full details follow].

The horoscope was constructed to show the configuration of the Heavens on the day of Ibn Riḍwān's birth, equivalent to January 15, 988.[4] Direct observation of what was on the local eastern horizon (the ascendant), as well as the positions of the planets, was of course desirable but in fact seldom possible—either because it was daytime, or there was cloud-cover, or because no one had the observational skills or equipment, or, as in the case of Ibn Riḍwān, it was constructed many years after the event.

Some historians have convincingly argued that it was an early but popular urban myth that observations were undertaken at the moment of birth.[5] In any case, the vast majority of horoscopes were calculated using tables of planetary (including solar and lunar) longitudes, after having determined the time and geographical latitude of the event. Numerous such tables (called *zīj* in Arabic) were calculated and updated by mathematical astronomers, and libraries today are full of manuscript copies of such tables testifying to the intense interest in the topic as well as the remarkable mathematical skills of medieval astronomers.[6]

Note that on this horoscope representing the nativity of Ibn Riḍwān (fig. appendix 1), in addition to the five planets visible to the naked eye, as well as the Sun and Moon, there is also indicated the "Head" of the Dragon.[7] The Dragon's Head is not an actual planet, but rather the point where the Moon crosses to the north of the ecliptic. It was often treated as a "planet" by astrologers, along with the point when it crosses to the south (the "Tail"). Note also that two stars are indicated: Vega (*α Lyrae*) in the sign of Capricorn and Sirius (*α Canis Majoris*) in the sign of Cancer. The indication of prominent stars on a horoscope is relatively unusual. It may reflect a particular interest among eleventh-century Cairene astrologers in the significance of prominent stars, particularly those known as the Thirty Bright Stars, to which the two on the horoscope belong—an interest borne out by the extraordinary amount of attention given the topic in the *Book of Curiosities.*

To a professional astrologer in the eleventh century, it was not the planets by themselves that indicated an outcome, but rather the relationship between the planet and the zodiacal sign in which it occurred and its relationship to other planets at the time of observation.[8]

The attributes of the zodiacal signs are given in the *Book of Curiosities* in excruciating detail. Zodiacal signs could be classified astrologically in many different ways and set into different groups, each arrangement having associations or attributes that could then be attributed to a particular birth, person, or event under discussion when that particular sign is in the ascendant—that is, rising at the local eastern horizon.

Each individual zodiacal sign is aligned with parts of the human body: for example, Aries with the face and head. The signs are also associated with cities, islands, or regions. Aries, for example, is associated with Babylon, Fars, Azerbaijan, Palestine, Cyprus, the coasts of Asia Minor, the lands of the Slavs, and Mosul. Many of the localities assigned a zodiacal sign in the *Book of Curiosities* appear to be derived from those given by Ptolemy in his *Tetrabiblos* (though Ptolemy is not named in this context). On the other hand, equally many localities (particularly more northern ones) specified by Ptolemy are omitted from the lists in the *Book of Curiosities*.[9] The zodiacal signs could also be classified in other ways. For example, grouped as brilliant/dark or female/male or fortunate/unfortunate, each with various favorable or unfavorable connotations.

Some of the categories assigned to zodiacal signs reflect astronomical phenomena. It was said, for example, that the zodiac consists of two unequal sections, the larger portion extending from Cancer to Sagittarius and the smaller from Capricorn to Gemini.[10] The statement reflects the fact that, in northern latitudes, the six signs (or 30°-segments of the ecliptic) from Capricorn through Gemini take less time to rise than those signs from Cancer to Sagittarius, due to the fact that the ecliptic is at an oblique angle to the equator, and the course of the Sun itself is an elliptical orbit.

Astrologers since antiquity were aware of the variation in rising times for the zodiacal signs and they calculated tables of oblique ascensions giving the amount of time required for each sign to rise over the local horizon at specific latitudes, because the length of their appearance depended also upon the geographical location of the observer.[11] Tables of oblique ascensions calculated for different latitudes were part of the standard manuals for medieval Islamic astronomers and mathematically inclined astrologers. Since horoscopes depended upon knowledge of the ascendant (the portion of the sign rising at the local eastern horizon), the ability to determine the passage of the rising sign was vitally important to a professional astrologer. Our author, however, makes no references to ascension tables nor to the complexities of compiling such tables or their use in calculating a horoscope, which shows again that the *Book of Curiosities* was not addressed to professional astrologers.

Shortly after discussing the two unequal divisions in terms of rising times of the zodiacal signs, our author states that the twelve signs are also divided into two groups in a different way: the six from Cancer through Sagittarius were "direct in rising" and the six from Capricorn through Gemini were "oblique, or crooked, in rising."[12] What he does not seem to realize is that this division reflects the same astronomical phenomenon as the previous one—namely, that the oblique ascension of those signs "oblique in rising" is shorter than those "direct in rising." He goes on to say that the signs oblique in rising "twist matters and corrupt them" and always defer to signs that are direct in rising, the latter being facilitating and reconciling signs.

Most of the groupings of zodiacal signs are rather simple, though somewhat arbitrary. For example, the *Book of Curiosities* divides them into four groups of three each, sequentially (such as Aries, Taurus, Gemini), each group associated with a season or a period of life. Or, the twelve signs can be divided into three groups of four, each according to the nature of their influence, with one group having a constant or "fixed" (*thābit*) influence, another dualistic in nature or "bi-corporeal" (*dhū jasadayn*), and the third changeable or "tropical" (*munqalib*).

As if that isn't enough, astrological meaning was assigned to subdivisions of each sign. Each zodiacal sign consists of a 30°-segment of the zodiac, and this could be divided into subsegments in three different ways: (1) into three equal parts, with each 10°-segment called a "face (*wajh*)"; (2) into five unequal parts called "terms" or "limits" (*ḥudūd*); and (3) into nine equal parts (*nūbahrāt* or *nawbahrāt*).[13]

Each "face" or 10°-segment of a zodiacal sign was said to be ruled by a particular planet, and these "rulers" were sometimes illustrated allegorically. See fig. appendix 2 for a late fourteenth-century manuscript in which the "faces" of Pisces are illustrated as being ruled by humanlike figures representing Saturn, Jupiter, and Mars.[14]

With the five unequal divisions of a sign, each "term" (or "limit") was alloted one of the five planets. The quality of a planet's influence in that sign was determined by the size—that is, the number of degrees—of the "term" allotted to it. According to al-Bīrūnī (a contemporary of our author, but working in the eastern provinces of Persia and modern Afghanistan), there were two different ways—the Egyptian and the Ptolemaic—of assigning the "terms" to planets, but he considered the Egyptian more accurate.[15]

As for the nine-part division, al-Bīrūnī reports that Indians regarded the ninth part of a sign (3° 20′) as very significant.[16] Our author, however, appears to have had little understanding of the concept and usually makes opaque, if not meaningless, statements such as "Aries has nine ninths." In general this manner of division is not as commonly employed as the divisions by "faces" and "terms."

Each zodiacal sign is also associated with a planet—that is, one of the "erratic" or "wandering" stars. Before the invention of the telescope, in addition to the Sun and Moon, only five planets were visible to the naked eye: Mercury, Venus, Mars, Jupiter, and Saturn. The five planets in Arabic were called *muṭaḥayyirah*, or "bedazzled," which al-Bīrūnī explained by saying that their motion "resembles confusion (*ṭaḥayyur*)." The planets appear to move around the Earth very near the ecliptic, deviating only slightly from the Sun's path. They generally seem to move around the zodiac in the same direction as the Sun and Moon, but (when viewed from Earth) the five planets occasionally appear to "wander" or stray, moving at times in the opposite direction, when it is called retrograde motion. The five planets plus the Sun and Moon were called "the moving stars" (*al-kawākib al-sayyārah*).[17]

FIG. APPENDIX 2. Manuscript illustration titled "Discourse on the First Face of the Sign of Pisces." In the upper roundel Jupiter is shown "domiciled" in Pisces. Beneath are the three "faces" or ten-degree segments of Pisces and their "lords" or dominant planets: Saturn (dark-skinned with a scythe), Jupiter (turbaned male sitting cross-legged) and Mars (kneeling man holding an animal-headed club in one hand and a human head in the other). From a copy of *Kitāb al-Bulhān* produced in Baghdad for the Mongol ruler of Baghdad, Sulṭān Aḥmad, who ruled 1382–1410. Oxford, Bodleian Library, MS Bodl Or. 133, fol. 22b.

When a planet is "domiciled" in a particular sign—that is, resides or is visible in that particular segment of the zodiac—then it has great power and dominates over all other planets that might be nearby. These are known as "planetary domiciles" or "houses" of the planets and are massively important in astrological divination. The five planets each have two such do-

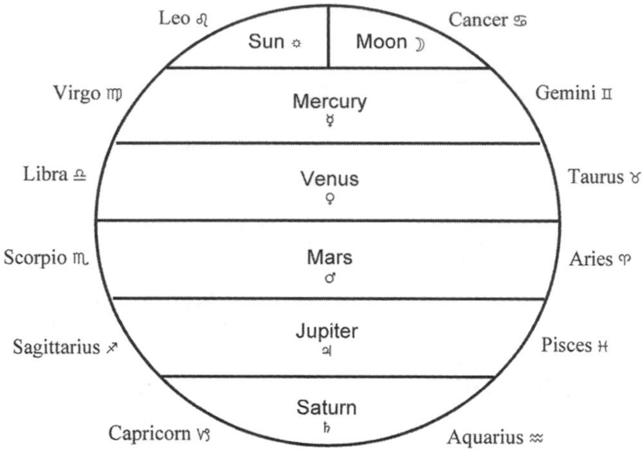

FIG. APPENDIX 3. Planetary domiciles: The twelve zodiacal signs with the planet whose domicile or house is in that sign.

miciles, while the Sun and Moon are associated with only a single sign each, as illustrated in fig. appendix 3.

So important is this topic of "domiciles" that our author presents this relationship of domiciled planets in three different ways. First, he explains it by saying:[18]

> God has created the planets out of the Sun and the Moon, but they became dazzled by the light and radiance of the Sun, and retreated at intervals from its powerful light. Finally, their arcs extended from the houses of the Two Luminaries—Leo and Cancer—to the point where each planet settled at the limit of its recurrent course. Thus the arc of Saturn is 210 degrees, for it has moved from the sign of Leo to the seventh sign, which is Aquarius, and settled there. Therefore, Aquarius is his house. If one counts the same angular distance in the opposite direction, starting from the sign of Cancer, it also reaches the seventh sign, which is Capricorn. Therefore Capricorn is also the House of Saturn and is associated with it. . . . In this way, each of the planets has attained its zodiacal signs according to their angular distance as they were dazzled by the light of the Sun.

This explanation is then illustrated with a circular diagram, labeled "Depiction of the arcs of the wandering planets and their associations with the twelve signs of the zodiac" (see fig. appendix 4).[19] A modern interpretation is given in fig. appendix 5.

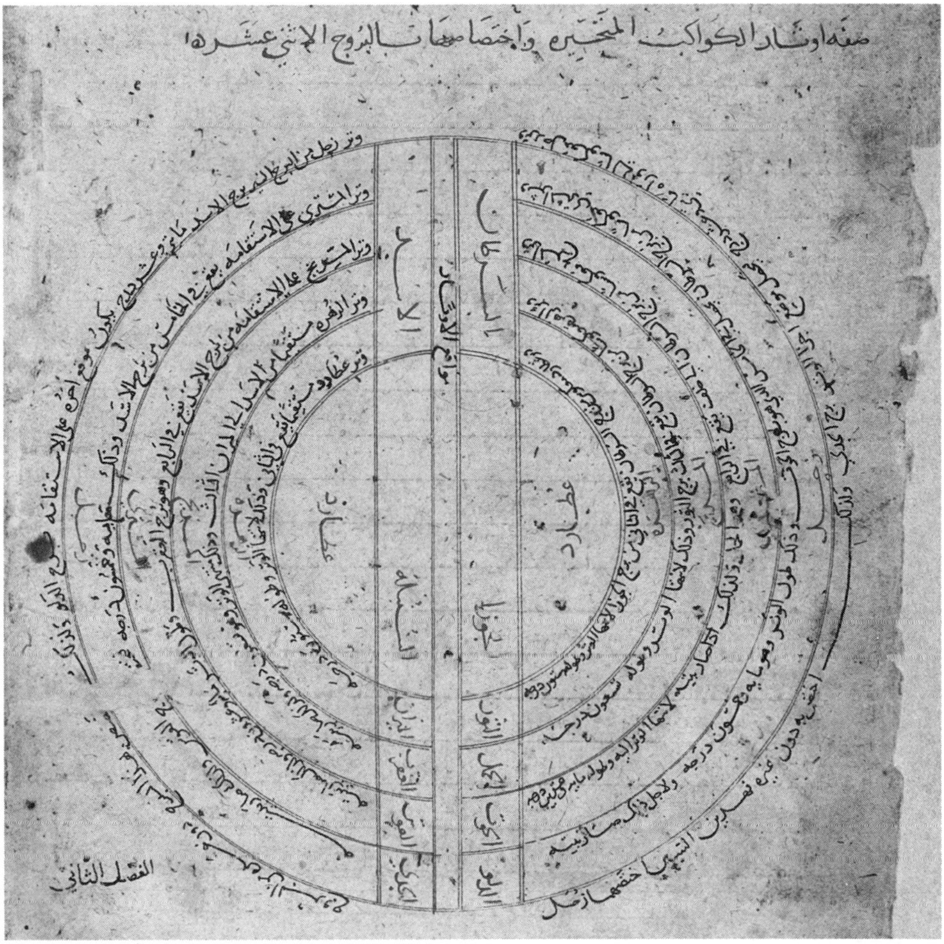

FIG. APPENDIX 4. Diagram labeled "Depiction of the Arcs of the Wandering Planets and Their Associations with the Twelve Signs of the Zodiac." Oxford, Bodleian Library, MS Arab. c. 90, fol. 5b.

By means of this diagram, the author represented the domiciles of the five planets in terms of the angular distance of their domiciles from the boundary between Leo and Cancer, the domiciles of the Sun and Moon, respectively.[20]

To explain the relationship of domiciles further, the author then provided the following allegory:[21]

> The philosophers have come up with a nice allegory for these planets. They have said that the Two Luminaries (the Sun and the Moon) are like kings. Every king must have a vizier to consult with, so the Sun in its

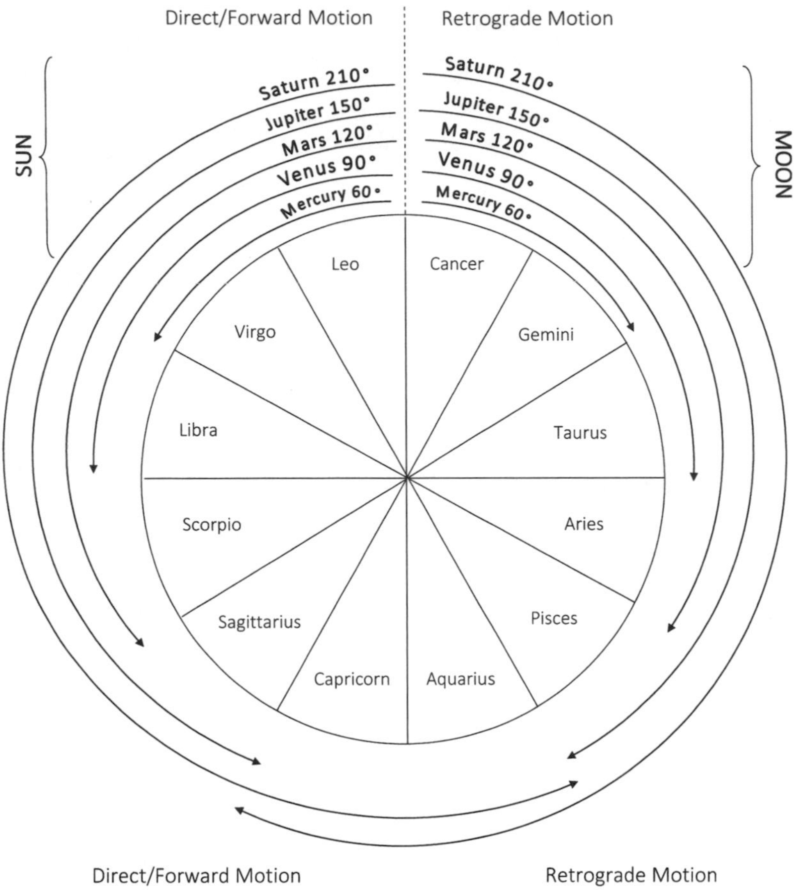

FIG. APPENDIX 5. An interpretation of the diagram labeled "Depiction of the Arcs of the Wandering Planets and Their Associations with the Twelve Signs of the Zodiac."

second house has taken Mercury as its vizier, while the Moon has also taken Mercury as its vizier in its second house in the opposite direction. That way, Mercury has attained two houses from both sides of the two Luminaries. A king must have a wife to rely on, so the Sun in its third house is associated with Venus, while the Moon in its third house in the opposite direction is also associated with Venus. Then they said: A king must have a swordsman to inspire awe, so the Sun in its fourth house has Mars while the Moon in its fourth house in the opposite direction has Aries, which [also] is the House of Mars. Then they said: A king must have a judge to pass judgments among his subjects, so the Sun in its fifth house is associated with Sagittarius, which is the House of Jupiter, while the Moon at five signs in the opposite direction is associated with Pisces, which is also the House of Jupiter. Then they said: A king must have someone to sow and cultivate the land, so the Sun in the sixth house is associated with Capricorn, which belongs to Saturn, while the Moon in its sixth house in the opposite direction is associated with Aquarius, which also belongs to Saturn. But God knows best.[22]

Yet more associations and attributes for each zodiacal sign and planet are then presented. In the ring of twelve zodiacal signs, those zodiacal signs opposite the signs of the "domiciles" of a planet are said to have the "detriment" (*wabāl*) of that planet, meaning that the specified planet is very weak when in those zodiacal signs (see fig. appendix 6). An example: Mars is domiciled in both Aries and Scorpio (see fig. appendix 3), while its "detriments" are in Libra and Taurus (see fig. appendix 6). The Indian astrologers, according to al-Bīrūnī, did not recognize this relationship, though our author clearly does.[23]

And there are still more ways to think about the sequence of zodiacal signs and their relationship to the planets. Each group of three zodiacal signs that are 120° apart from one another is called a triplicity (*muthallathah*): for example, Aries—Leo—Sagittarius, or, Cancer—Scorpio—Pisces (see fig. appendix 7). Each triplicity was assigned three ruling planets or "lords." One planet ruled by day, one by night, and one shared both day and night, according to the Baghdadi astrologer Abū Maʿshar (d. 886), and he is followed in this interpretation by al-Bīrūnī.[24] On the other hand, an astrologer working in Aleppo in the tenth century, al-Qabīṣī, states that

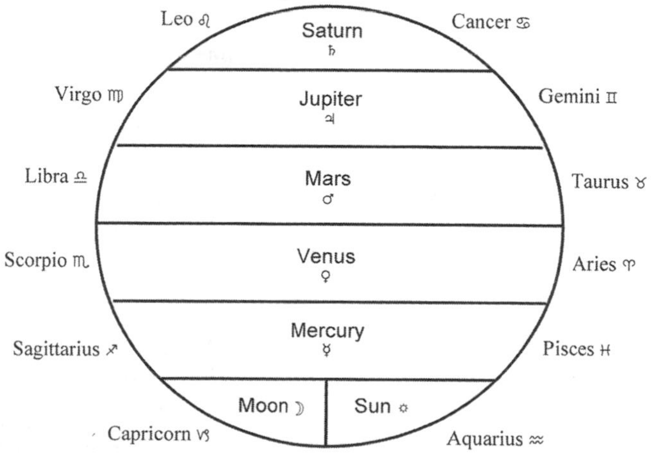

FIG. APPENDIX 6. Planetary detriments: The twelve zodiacal signs aligned with the planets whose "detriment (*wabāl*)" is in that sign.

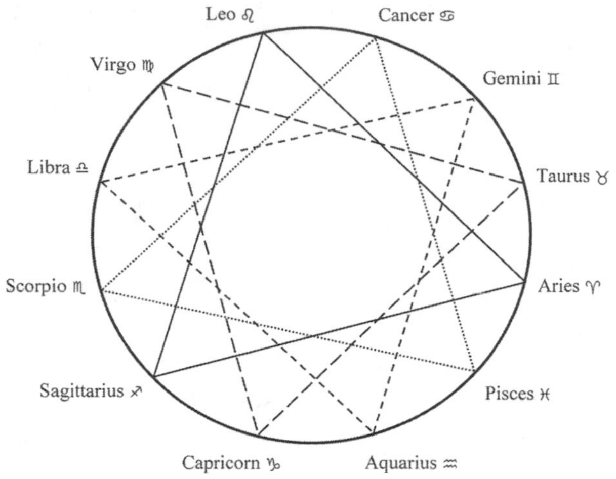

FIG. APPENDIX 7. Triplicities.

three planets rule by day and the same three by night, but in a different order.[25] This latter interpretation (which the author of the *Book of Curiosties* appears to be following) is said by al-Bīrūnī to be one used by "worthless (*ḥashwīyah*)" astrologers, whom he often criticizes.[26] It should be noted, however, that there is no evidence that the writings of al-Bīrūnī (d. *c* 1048), who worked in the far eastern provinces of Persia and what is now Afghanistan, were available in Egypt in the first half of the eleventh century.

The notion of triplicities is combined with that of domiciled planets in the method used to interpret one of the most important astrological concepts: the decans. The "decan" or "decanate" originates from Indian lore, and in Arabic it is commonly written as *darījān* (plural *darījānāt*), from the Indian *drekkāna,* which in turn derives from the Greek *dekanos.* Curiously, the *Book of Curiosities* repeatedly (and uniquely) uses the term *adaranjāt* instead of the common *darījān.* The word, however spelled, designates a third of a sign—that is, ten degrees. Each has a "lord," but it is different from the lord of the "faces" mentioned above. The lord for the first decan is the lord of the zodiacal sign itself—that is, the planet domiciled in it. The second decan has as its lord the domiciled planet of the fifth sign away (counting counterclockwise and including the starting sign as "one" in the sequence), while the third decan is assigned the lord (domiciled planet) of the ninth house in sequence (see fig. appendix 3 for the planetary domicles).[27]

Thus, in this Indian division by decans, the three 10°-segments of a given zodiacal sign are assigned as rulers those planets domiciled in the three signs forming the triplicity to which the one under consideration belongs. The triplicities are shown in fig. appendix 7, and the domiciled planets in fig.

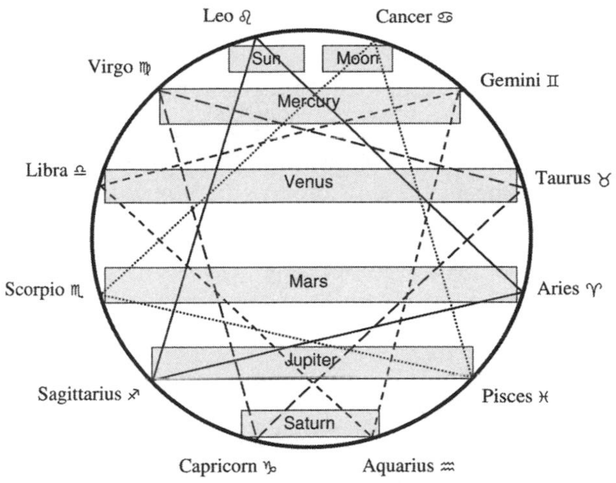

FIG. APPENDIX 8. The association of planets as "lords" with the three decans, or ten-degree segments, of a zodiacal house.

appendix 3, while fig. appendix 8 combines the two to give the alignments used in the interpretation of decans. For example, the three "lords" of the decans of Leo would be the Sun, Mars, and Jupiter.[28]

Yet another system of aligning zodiacal signs with groups of three planets is presented in the *Book of Curiosities*. This appears to be unique to our treatise, and it is called *adrijānāt*. This term is otherwise unknown, and its vocalization and meaning is uncertain. Its presentation always follows a listing of the "lords" of the decans, and the word *adrijānāt* also appears to be a corruption of a Hindu term. The alignments of *adrijānāt* with the twelve signs are given as follows:

Aries: Moon, Mercury, Venus
Taurus: Mercury, Moon, Venus
Gemini: Sun, Mars, Jupiter
Cancer: Saturn, Jupiter, Mercury
Leo: Venus, Sun, Mars
Libra: Mercury, Venus, Jupiter
Scorpio: Mars, Jupiter, Saturn
Sagittarius: Moon, Mercury, Venus
Capricorn: Sun, Mars, Jupiter
Aquarius: Saturn, Moon, Mercury
Pisces: Venus, Sun, Moon

The rationale behind this alignment has not been identified.

Yet different astrological indicators, but still very important, were the "exaltation" (*sharaf*) and "fall" (*hubūṭ*) of a planet, representing an older system of Babylonian origin.[29] Each planet was assigned one degree along the zodiac or ecliptic at which it has its greatest influence—called the "exaltation"—and a degree in the sign opposite, a degree of least influence (the "fall"). These peak moments of strength or weakness of a planet did not occur in the zodiacal sign that contained the planet's "domicile" or "detriment," but at other very specific points.[30] The assignment of exaltations and falls of planets to specific points of the ecliptic remained surprisingly consistent over time in the astrological literature.

PLANETS	EXALTATION	FALL
Sun	Aries 19°	Libra 19°
Moon	Taurus 3°	Scorpio 3°
Mercury	Virgo 15°	Pisces 15°
Venus	Pisces 27°	Virgo 27°
Mars	Capricorn 28°	Cancer 28°
Jupiter	Cancer 15°	Capricorn 15°
Saturn	Libra 21°	Aries 21°
Dragon's Head	Gemini 3°	Sagittarius 3°
Dragon's Tail	Sagittarius 3°	Gemini 3°

The two "pseudo-planets" (the Dragon's Head and Tail) were treated in the same manner as true planets and given similar attributes, as discussed above. Whenever a conjunction or opposition of the Sun and Moon occurs near these lunar nodes or "pseudo-planets," a solar or lunar eclipse occurs.[31] The exaltations and falls of planets were sometimes pictured symbolically in other treatises, along with the two "pseudo-planets," as illustrated in the late fourteenth-century manuscript shown in plate 25 and in fig. appendix 9.[32]

Like the zodiacal signs, planets could also be given qualities and attributes such as male/female, diurnal/nocturnal, single/dual, ill-omened and good-omened, as well as linked with "friendship" or "enmity." Moreover, planets were aligned with one or more bodily parts, as well as animals, birds, clothing, tastes, foods, activities, days of the week, and regions of the world.

According to Persian astrology, the years of a person's life are divided into set periods, called *fardārāt*, each governed by a specific planet. Knowing the planet dominating at the time of birth could enable the astrologer to determine the length of the life span of a newborn.[33] The years associated with the planets were of three types: great years, mean years, and least years. For example: Its great years might be fifty-seven years; its mean years thirty; and its least years eighteen. Astrologers used these to predict at

FIG. APPENDIX 9. Top right: exaltation of the Moon in Aries; top left: the fall of the Moon in Scorpio. Lower right: exaltation of the Dragon's Head in Pisces; lower left: fall of the Dragon's Tail in Sagittarius. From a copy of *Kitāb al-Bulhān* produced in Baghdad during the reign of Sulṭān Aḥmad (reg. 1382–1410). Oxford, Bodleian Library, MS Bodl Or. 133, fol. 27a.

the time of birth the probable length of life, although al-Bīrūnī cautioned against interpreting these numbers literally.

Each planet, according to the *Book of Curiosities*, also had an "area of influence." The Arabic expressed it as "its light behind it and in front of it," followed by a number that gave the degrees, both before and after, within which it had power and could affect another planet.

In the *Book of Curiosities* each discourse on an individual planet ends with its distance from the Earth and its size relative to other planets, distances being given in the unit of measure called *farsakh*.[34] The order in which they are discussed is the reverse order of their proximity to the Earth, in other words from Saturn down to the Moon. Discussions of planetary distances and sizes became a standard part of medieval astronomical treatises, often with recomputed values. For the most part, the values follow from results given by Ptolemy in the *Almagest* (for the Sun and Moon) and in his *Planetary Hypotheses* (for all the planets).[35]

Thus in the course of three out of the ten chapters on the Heavens, the author of the *Book of Curiosities* reminded the reader of the many interrelationships between zodiacal signs and planets that were fundamental for the interpretation of horoscopic astrology. Moreover, he also provided some allegories that might serve as an aide-memoire to what he considered the most important of the relationships, the domiciles. All of this was provided within the context of a highly illustrated but nontechnical discourse on the macrocosm and the microcosm.

Acknowledgments

One of the pleasures of completing a project or finishing a book is the opportunity it provides to thank the many people who have assisted at various times. Because this volume is the culmination of a decade-long project, the number of people who have generously assisted in one way or another is very long indeed. Those who were involved in the acquisition by the Bodleian Library of the remarkable *Book of Curiosities,* and those who have contributed to its detailed analysis, are singled out for acknowledgement in *An Eleventh-Century Egyptian Guide to the Universe: The "Book of Curiosities"* (Leiden: Brill, 2014). We remain grateful to every one of them.

As we prepared the present volume, in which we reflect upon the wider significance of the *Book of Curiosities,* several people have been particularly helpful. Lesley Forbes, now retired as keeper of Oriental Collections in the Bodleian Library, read through and amplified the account of the manuscript's "discovery" and acquisition. Catherine Delano-Smith, longtime editor of *Imago Mundi,* kept encouraging us to explain what the *Book of Curiosities* means for the history of cartography, and gave us the platform of the "Maps and Society" lectures at the Warburg Institute to present preliminary findings.

Many of our ideas for this book were presented at workshops associated with the Leverhulme Network "Cartography between Europe and the Islamic World 1100–1600," co-organized by Jerry Brotton and Alfred Hiatt, and we are grateful to both for their invaluable input. The students on the MA class on the history of cartography in medieval Islam and Europe, jointly taught by Alfred Hiatt and Yossi Rapoport in 2015–16, entreated us to discuss several draft chapters in their nearly final form.

This book explores a very wide range of topics in the history of cartography, history of navigation, history of science, Islamic history, and global history, and we have often called in the expertise of colleagues around the world. We are extremely grateful for those colleagues who have read and patiently commented on earlier drafts of some of the chapters: John Cooper, Jessica Goldberg, David Jacoby, Mark Horton, Stéphane Pradines, Robin Seignobos, and Leif Isaksen. Much of what we say about the representation of Sicily is indebted to the observations of Jeremy Johns, whose role in identifying the importance of the manuscript was discussed in chapter 1 of this book. We have also benefited immensely from the vast knowledge of Jean-Charles Ducène and Tony Campbell, as well as from the feedback of so many other colleagues who have heard us present aspects of this research over the past decade.

We have been fortunate in having two excellent anonymous referees for the press, who read through the manuscript for this book. In fact one of them was so intrepid and dedicated as to read through *three* versions of the book, providing wise and useful suggestions each time. We hope that the end result will merit all that effort.

Karen Merikangas Darling, executive editor at the University of Chicago Press, was very encouraging throughout the process of the z book's development, while Samuel Fanous, head of Bodleian Library Publishing, has steadfastly supported the effort. For supplying images of manuscripts and granting permission for their publication, we thank the Bodleian Library at the University of Oxford, the Maktabat al-Assad al-Waṭanīyah in Damascus, the Wellcome Library in London, the Nasser D. Khalili Collection of Islamic Art, the British Library, the Chester Beatty Library in Dublin, the Topkapı Sarayı Müzesi Kütüphanesi in Istanbul, the Bibliothèque nationale et universitaire de Strasbourg, the Bibliothèque nationale de France, the Universitätsbibliothek in Leiden, and the Bayerische Staasbibliothek in Munich. In addition, Mark Horton and the American Center of Oriental Research in Amman both generously supplied images of and permission to reproduce (respectively) the discovery of the Mtambwe hoard and the Madaba map, while *Imago Mundi* allowed us to reproduce a useful diagram. The diagrams and maps in figs. 2.1, 2.11, 5.4, 5.6, 7.1, 7.2, 7.3, 8.2, 8.6, 8.7, 9.1, and 9.2 were prepared for publication by Oxford Designers & Illustrators, Ltd.

Much earlier versions of chapters 3, 6, and 8 in this book were published as Yossef Rapoport and Emilie Savage-Smith, "The *Book of Curiosities* and a Unique Map of the World," in *Cartography in Antiquity and the Middle Ages: Fresh Perspectives, New Methods,* ed. R. J. A. Talbert and R. W. Unger (Leiden: Brill, 2008); Yossef Rapoport, "Reflections of Fatimid Power in the Maps of Island Cities in the *Book of Curiosities,*" in *Herrschaft verorten. Politische Kartographie des Mittelalters und der frühen Neuzeit,* ed. Martina Stercken and Ingrid Baumgärtner (Zürich: Chronos, 2012); and Yossef Rapoport, "The *Book of Curiosities*: A Medieval Islamic View of the East," in *The Journey of Maps and Images on the Silk Road*, ed. Philippe Forêt and Andreas Kaplony (Leiden: Brill, 2008). Six to eight years on, the interpretations of much of the material in these preliminary studies has broadened and deepened considerably.

As always, our families have shown infinite patience throughout a decade and a half of seemingly endless talk of skies and maps.

Yossef Rapoport, London
Emilie Savage-Smith, Oxford
September 2017

Notes

Introduction

1. The phrase "raised-up roof (*al-saqf al-marfūʿ*)" is found in Qurʾān 52:5.

2. Rapoport and Savage-Smith 2014.

3. Gautier-Dalché 2009; Pujades 2007.

4. Goldberg 2012; Horden and Purcell 2000.

5. Bass et al. 1982; Horton and Middleton 2000.

6. See, among others, Johns 2004; Edson and Savage-Smith 2004; Horden and Purcell 2006; Bloom 2007; Kaplony 2008; Koutelakis 2008; Rapoport and Savage-Smith 2008; Rapoport 2008; Savage-Smith 2009; Rapoport 2010; Savage-Smith 2011; Rapoport 2012; Savage-Smith 2014a.

7. For example, Barber 2005 and Brotton 2013.

8. Daftary 2007; Goldberg 2012; Seed 2014.

Chapter One

1. London, Christie's, *Islamic Art and Manuscripts*, October 10, 2000, lot 41.

2. Samuel Johnson, *A Dictionary of the English Language* (London, 1755), entry "map."

3. Harley and Woodward 1987, xvi. There is an important discussion of the meaning of "map" in the preface to the first volume of the monumental *History of Cartography* project undertaken in 1987 by J. B. Harley and David Woodward and still ongoing. Even Wikipedia, in its entry for "map" now opens with saying: "A map is a symbolic depiction emphasizing relationships between elements of some space, such as objects, regions, and themes" (https//en.wikipedia.org/wiki/Map; accessed September 8, 2017).

4. Legendary islands in the Indian Ocean were known for trees bearing human fruits (see *EI2* art. "Wāḳwāḳ"). The islands are indicated on the circular world map that immediately follows this illustration. For a rather disappointing introduction to the iconography of the *wāqwāq* tree, see Baer 1965, 66–68. There is also a second illustration on the lower half of the previous page, probably added by the same reader at the same time; it is an inhabited scroll in which some fruits are in the shape of animal heads, and it is labeled: "Concerning the marvellous watermelons of al-Hāwand"; see Rapoport and Savage-Smith 2014, 438.

5. See Seaver 2004.

6. The five-year grant from the Heritage Lottery Fund was later supplemented with funds from the Arts and Humanities Research Council (AHRC). For further details of the project, and its outcomes and impact, see Forbes 2014.

7. Johns and Savage-Smith 2003.

8. Reprinted in Pormann 2010, no. 52.

9. For example, Johns 2004, 428 and 433; Rapoport and Savage-Smith 2004; Rapoport 2008; the website with a preliminary edition and translation available since 2007 (www.bodley.ox.ac.uk); and Rapoport and Savage-Smith 2008.

10. Johns and Savage-Smith 2003, p. 16.

11. Rapoport and Savage-Smith 2014, 457.

12. Rapoport and Savage-Smith 2014, 2–4.

13. Conducted initially by Dr. Sandra Grantham, a consultant paper conservator, and continued later by Sabina Pugh in the conservation workshop of the Bodleian Library.

14. Carried out by Dr. Tracey Chaplin at the Christopher Ingold Laboratories, University College London. The results of the Raman spectroscopic analysis have been published in Chaplin et al. 2006.

15. www.bodley.ox.ac.uk/bookofcuriosities. Taken down in 2017.

16. Rapoport and Savage-Smith 2014.

17. Rapoport and Savage-Smith 2014, 4–12, 34. Several of these copies are attributed by cataloguers to an otherwise unknown Egyptian named Ibn al-ʿArabānī or Ibn al-ʿUryānī, who was likely a member of the fifteenth-century Ibn al-ʿUryānī scholarly family from Cairo.

18. All the illustrations preserved in the later copies are discussed and illustrated in Rapoport and Savage-Smith 2014, 16–28.

19. Oxford, Bodleian Library, MS Bodl. Or. 68, item 6; see Rapoport and Savage-Smith 2014, 5–8.

20. Gotha, Forschungsbibliothek, MS orient. A 2066, item 2; see Rapoport and Savage-Smith 2014, 9.

21. Edson and Savage-Smith 2004.

22. The Korean translation was published by eMorning Books and Danny Hong Agency (ISBN 89-90956-64-1), Korea, in 2006. For the German version, see Edson, Savage-Smith, and von den Brincken 2005.

23. Rapoport and Savage-Smith 2014, 441.

24. Rapoport and Savage-Smith 2014, 518.

25. Rapoport and Savage-Smith 2014, 443.

26. Rapoport and Savage-Smith 2014, 33–34.

27. Harley and Woodward 1992, 173–74 and plate 11. The copy estimated to have been made about 1300 is Paris, BnF, arabe 2221, with the world map not well preserved. Note that the Bodleian MS Pococke 375 is incorrectly said to be dated 1456, rather than 1553, while Bodleian MS Greaves 42 is assigned to the sixteenth century when it is more likely a fourteenth-century copy.

28. Rapoport and Savage-Smith 2014, 30–31. The map in the Bodleian manuscript occupies a full opening (a pair of conjugate leaves), with the back or verso of the left-hand side left blank and the back of the right-hand side also originally left blank but then filled in later by the reader in the fourteenth or even fifteenth century who drew the picture of the "wāqwāq" tree (see fig. 1.5).

29. That the back side of the bi-folio was originally left blank can be explained by the fact that the text for the chapter is quite short, beginning one-third way down folio 26a and continuing halfway down the 26b. A full opening was required for the map, leaving the back of the right-hand side of the map (fol. 27a) blank, to be filled in by a later reader with the drawing of the wāqwāq tree. The map for the following chapter also required a full opening

but its text was so short that it was written on the back side of its large map, leaving the back side (fol. 28b) of the left-hand portion of the circular world map also blank.

30. Daftary 2007; Goldberg 2012; Seed 2014.

Chapter Two

1. For examples of other mapping systems, see Harley and Woodward 1987, 54–93; Harley and Woodward 1992, 332–386; Harley and Woodward 1994, 511–578, 579–606, 701–740; and Woodward and Lewis 1998.

2. See Bloom 2001, 154. For the role of paper, book production, and ownership, see Hirschler 2011.

3. The particular aphorism occurs in some anatomical manuscripts and was repeated by the seventeenth-century Ottoman historian and bureaucrat, Ḥajji Khalīfah (Kātip Çelebī, d. 1657). For this aphorism and the role of "science" in medieval Islamic society in general, see Savage-Smith 2014.

4. Here, of course, anatomy does not mean dissection but rather the study of the structure of the body in order to demonstrate the design and wisdom of God

5. For the practice of the art of astrology, see *EI Three*, art. "Astrology" (C. Burnett) and Pingree 1990; de Callataÿ 2015, 137–56. For the role of the astrologer in Islamic Society, see Saliba 1992; Savage-Smith 2011.

6. Lemay 1997, 81, with slight changes in the punctuation and word order.

7. Neither term, it should be noted, is used by our author.

8. *EI Three*, art. "Astronomy," sect. 3 (F. Jamil Ragep); King 1993.

9. *EI2*, art. "Ismāʿīliyya" (W. Madelung); see also *Encyl. Iranica* art. "Cosmogony and Cosmology. vi. In Ismaʿilism" (W. Madelung); Netton 1989, 203–35.

10. Halm 1997, 86. For later criticisms of astrology by Muslim thinkers, see Michot 2000.

11. Sayılı 1960, 141–48.

12. Maqrīzī 1967–71 2:100; quoted in Halm 1997, 87. This event is confirmed by other sources as well; see Sayılı 1960, 134.

13. Oesterle 2013; Halm 1997, 74–76.

14. Maqrīzī 2002, 1:458–59; Halm 1997, 73–74.

15. In the Qurʾān only seven spheres are mentioned, while the Greek system requires eight (one each for the Sun, Moon, five planets, and the fixed stars) with an additional ninth added by some to account for the movement; see *Encyl. Qurʾān*, art. "Planets and Stars" (P. Kunitzsch); Heinen 1982; Janos 2011.

16. Ptolemy is usually said to have counted 1,025 stars, which included the three stars of Coma Berenices. For the confusion as to the total number depending upon how stars shared between constellations are counted, see Savage-Smith 1985, 114–16.

17. Rapoport and Savage-Smith 2014, 331 and 355. The sixth magnitude was sometimes divided into forty-nine plus five nebulous and nine "dark" stars.

18. For Hermes Trismegistus in the Arabic world, see van Bladel 2009; *EI Three*, art. "Hermes and Hermetica" (K. van Bladel).

19. Rapoport and Savage-Smith 2014, 355. See also Beeston 1963; *Enc. Qurʾān*, art. "Planets and Stars" (P. Kunitzsch).

20. Rapoport and Savage-Smith 2014, 326–31.

21. London, Wellcome Library MS Persian 474, fols. 18b–19a. For further details see Tourkin 2004 and Caiozzo 2005, and for other emblematic representations of zodiacal signs, see Caiozzo 2003, 61–109 and 199–212; Savage-Smith 1992, 63–65.

22. Rapoport and Savage-Smith 2014, 329–30. Ḥunayn ibn Isḥāq (d. *c* 260/873 or 264/877), following Ibn al-Biṭrīq (d. *c* 215/830), stated that the common people distinguished four winds: *ṣabā* as the east wind, *dabūr* the west wind, *shimāl* the north wind, and *janūb* the south wind (Lettinck 1999, 168); compare Qalqashandī 1913, 2:166–68; King 2004a, 812–13; *EI2* art. "*Maṭlaᶜ*" (D. King); and Varisco 1994, 111–17.

23. Our author repeats these dimensions a second time a bit later in the treatise; Rapoport and Savage-Smith 2014, 335. For data as given by al-Farghānī, see al-Farghānī 1998, 69 and 75.

24. Sezgin *GAS* VI, 149–51; A. I. Sabra, "al-Farghānī" in *Dictionary of Scientific Biograpy*, 4:541–45.

25. Rapoport and Savage-Smith 2014, 332.

26. Rapoport and Savage-Smith 2014, 332.

27. See Tripathi 1959, 230–35; *EI2*, art. "Ḳanawdj" (M. Longworth-[J. Burton-Page]).

28. See *EI2*, art. "Bilawhar wa-Yūdāsaf" (D. M. Lang).

29. Asoka's surviving edicts contain our first detailed information on the Indian calendars (Kulke and Ruthermund 1998, 62–67).

30. Rapoport and Savage-Smith 2014, 335–36.

31. The actual latitude of Kannauj is 27° 3' N, nearly 4° north of the Tropic of Cancer (at 23° 5'), and not "near the equator" as the text reads.

32. The "mean motions" of planets (*awsāṭ*, singular *wasaṭ*) were a standard feature of astronomical tables compiled into volumes called in Arabic a *zīj*; see King and Samsó 2001; *EI2*, art. "Zīdj" (D. A. King).

33. The term *kardajāt* (singular, *kardajah*) most often refers to trigonometric tables of sines occurring in astronomical tables based on Indian tables, where the argument is expressed in intervals of 3° 45' (the normal interval for Indian tables). The Arabic term *kardajah* comes (apparently through Pahlavī) from the Sanskrit *kramajyā*. It is, however, used in various ways by early Arabic writers, often ambiguously. In the present context, *kardajāt* refers to planetary functions.

34. Gold-water is gold-powder mixed with sizing or a gelatinous solution and used for ornamental writing (Lane 1863, 983).

35. Our Arabic text uses the term al-Budd for the Buddha. This is not the common Arabic name for Buddha, but rather the name given in *Kitāb Bilawhar wa-Yūdāsaf* to the prophet of the Indians. It can, however, also be identified with the Gautama Buddha (Gimaret 1971, 22). *EI2*, art. "*Bilawhar wa-Yūdāsaf* (D. M. Lang).

36. For the *Sūrya Siddhānta*, see al-Hāshimī 1981, 216–23.

37. The term *bābānīyah* does not actually occur in this table. It is, however, used by our author in one of the captions to his opening diagram (fig. 2.1 and plate 19) in which he is giving the total number of stars as 1022 *plus* the *bābābīyah* stars, nebulous (*saḥābiyah*) stars, hidden (*kafīyah*) or obscure stars, comets (*dhawat al-dhawāʾib*), and "rulers of events and changes" by which he probably meant the planets. The table giving the Thirty Bright Stars comprises chapter 4 of book 1; see Rapoport and Savage-Smith 2014, 356–59. This chapter is preserved only in MS Arab. c. 90, fols. 11a-11b and is missing from all later copies.

38. Ptolemy 1940, sect. I,9; Sezgin *GAS* VII, 43–44. The *Tetrabiblos* was translated into Arabic as *Kitāb al-Arbaᶜah* by Ibrāhīm ibn al-Ṣalt in the mid-ninth century and then revised by Ḥunayn ibn Isḥāq not long thereafter.

39. Kunitzsch 2001, 11–21.

40. van Bladel 2009, 237.

41. Rapoport and Savage-Smith 2014, 356.

42. An occasional Persian or Pahlavi (Middle Persian) name for a star occurs in some of the other fragments. It is evident that the tradition of Thirty Bright Stars came into Arabic through a Middle Persian mediation, further supported by the fact that the term *bābānīyah* is an Arabized form of the Middle Persian word *a-wiyābān-īg*, which literally rendered the Greek ἀπλανής, meaning "fixed star." Al-Bīrūni (d. *c* 1048) writes the word as *biyābānīyah* and says it means "desert stars" in Persian (based on the New Persian word *biyābān* meaning "desert"), adding the gloss "for finding the right way through deserts depends on them"; this is, however, a false etymology. See Bīrūnī 1934, 46 sect. 125; Kunitzsch 1981; Kunitzsch 2001, 16; *EI2*, art. "al-Nudjūm" (P. Kunitzsch).

43. In preparing this table, the fundamental studies of hermetic astrology by Paul Kunitzsch have been employed, and in particular his table comparing five of the fragments (Kunitzsch 2001, 26 Table 1).

44. Dublin, Chester Beatty MS Ar. 5399, fols. 206b–208b, copied in 1529 (935H); Kunitzsch 2001, 15–18, 56–80.

45. See Kunitzsch 2001, 18–21, 57–81.

46. Preserved in a unique manuscript in Oxford, Bodleian Library, MS Hunt. 546, fols. 29b-33b; see Kunitzsch 2001, 21–23, 84–99.

47. Kunitzsch 2001, 23–24.

48. Kūshyār 1997, 20–25; Bodleian Library MS Bodl. Or. 137, fol. 26b and, for Persian translation, al-Qummī 1997, 51.

49. Paris, BnF, hebr. MS. 1045, fols. 210a-211b; see Lelli 2001.

50. Forcada 1998; Varisco 1991; Varisco 1989; Varisco 2000; Forcada 2000.

51. *EI Three*, art. "*Anwāʾ*" (D. M. Varisco); Pellat 1955; Forcada 1998.

52. Abū Ḥanīfah al-Dīnawarī, d. *c* 895 (Sezgin *GAS* VII, 349), Ibn al-Aʿrābī, d. *c* 844 (Sezgin *GAS* VII, 345–46), Abū Yaḥyā ibn Kunāsah, d. *c* 822 (Sezgin *GAS* VII, 342).

53. Savage-Smith 2013 and sources cited there; Savage-Smith 1992; Dekker 2013, 286–307.

54. See Dekker 2013, 286–307, for al-Ṣūfī's (or copyists') technique.

55. Whether al-Ṣūfī's illustrated treatise was available to our author in Cairo between 1020 and 1050 is uncertain, although there is preserved today a sixteenth-century copy of al-Ṣūfī's treatise that has a colophon stating that one part had been based on a copy made in Cairo in 1011 (402H), which in turn is said to have been made from al-Ṣūfī's autograph copy. St. Petersburg, Institute of Oriental Studies, Russian Academy of Sciences, MS Arab. 185; see Caiozzo 2009, 118; Savage-Smith 2013, 147.

56. Chapter 5 of book 1 is devoted to this topic; Rapoport and Savage-Smith 2014, 361–73.

57. Samsó 2008; Forcada 1998, 305–308; Varisco 1995; Kunitzsch 1961, 9–20; *EI2* art. "al-Manāzil" (P. Kunitzsch); Ackermann 2004. For lunar mansions and early Greek and Coptic calendars, see Weinstock 1949.

58. Each lunar mansion was arbitrarily assigned twelve degrees fifty-one minutes of the ecliptic (360° divided by 28).

59. Chapter 9 of book 1 is devoted to this topic; Rapoport and Savage-Smith 2014, 391–409.

60. Ibn al-Ukhuwwah 1938, 177 (Arabic). See also *EI2* art. "Ibn al-Ukhuwwa" (Cl. Cahen) and King 2004, 637–39.

61. Rapoport and Savage-Smith 2014, 393–94.

62. Uncertain identification. Star groups called "the cattle" are described by *anwāʾ*-authors as being in various positions.

63. King 2004a; Ullmann 1972, 281–82; Sezgin, *GAS* VII, 38–41 and 80; Beck 2007, 101–11.

64. See King 2004a, 673; and King 1986, entry A21.

65. Kunitzsch and Ullmann 1992. We thank Professor Geert Jan van Gelder for taking time to search (unsuccessfully) for any similar proverb regarding the Pleiades.

66. They provide information on shadow lengths for the particular rising times as well as a table displaying the positions at thirteen points during the night of a lunar mansion with respect to the horizon and meridian.

67. King 2004a: 512–13; Schmidl 2006; Schmidl 2007, 221–22, 316–22.

68. King 1986, 57–58 entry C14. The opening of the poem from a copy in Istanbul (MS Hamidiye 1453, fol. 87b) is illustrated in King 2004a, 224 fig. 2.6a.

69. Oxford, Bodleian Library, MS Bodl. Or. 133, fols. 94b–130a; King 2004a, 299–302 and 510 and 641–42; Ullmann 1972, 344.

70. *ʿAyyūqāt*-stars are not a common feature of Arabic or Persian astronomical literature, and they do not form a part of the astrology of the great masters such as Abū Maʿshar or al-Bīrūnī. Nor are they part of the vocabulary used in three thirteenth-century treatises on folk astronomy written by al-Aṣbaḥī, Ibn Raḥīq and al-Fārisī composed in the Yemen and Mecca/Hijaz (Schmidl 2007).

71. Nash and Agius 2011, 178.

72. Rapoport and Savage-Smith 2014, 506.

73. The treatise has often incorrectly ascribed to the astronomer and mathematician Maslama ibn Aḥmad al-Majrīṭī (d. *c* 1007); see Fierro 1996; Boudet, Caiozzo, and Weill-Parot 2011; Samsó 2008, 121–22; Coulon 2014; de Callataÿ 2013.

74. Lippincott and Pingree 1987; see also Burnett 2004a; Juste 2007.

75. Or *Kitāb al-Istamāṭis*; Burnett 1987; Ullmann 1972, 374–75.

76. Oxford, Museum of the History of Science, inv. no. 37148; Ackermann 2004.

77. For similar images, see And 1998, 74. In India, this tradition of associating the lunar mansions with beasts is also evident in an illustrated astrological text of cloth made in India in the eighteenth century, but the animals are not particularly fabulous: the first three lunar mansions are represented as a horse, an elephant, and a sheep, respectively (see Khalili coll. TXT 225; Maddison and Savage-Smith 1997, 152–53; Savage-Smith 2011).

78. For knowledge of comets and meteors in medieval Islam, see *EI2* art. "nudjūm" (P. Kunitzsch); Kennedy 1957; Kennedy 1980; Rada 1999; and Cook 1999.

79. The word *dhawāʾib* more generally means locks of hair, or tufts, or anything that hangs down.

80. See Pasachoff 2000, 467–71; Ridpath 2011, 377–82.

81. See article "nudjūm" by P. Kunitzsch in *EI2*, 8:97–105, esp. sect. iii, c; Neuhäuser and Kunitzsch 2014.

82. Rapoport and Savage-Smith 2014, 374.

83. Rapoport and Savage-Smith 2014, 374.

84. Ptolemy does say that comets are sometimes given names such as "beams" (*dokidōn*), "trumpets" (*salpingōn*), and "wine-jars" (*pithōn*), but these names do not correspond to those given by our author; Ptolemy 1940, 192–93.

85. Rapoport and Savage-Smith 2014, 377–78.

86. The plumlike fruits of several varieties of myrobalan, a genus of tropical trees, came to be used extensively in compound remedies as well as in the dyeing and tanning industries; for medicinal uses, see Levey 1966, 342 no. 314. While the myrobalans are an important medieval Islamic medicament, they were unknown to the earlier Greco-Roman world.

87. *Risālah fī Dhawāt al-dhawāʾib wa-mā dhukira fīhā min al-ʿajāʾib.* Oxford, Bodleian

Library, MS Marsh 618, fols. 229b–231a [old 457–466] was employed for this study. See also King 1986, 34 no. B20 and pl. LXXX; Sezgin *GAS* VII, 328; Boudet 2016.

88. Another, *al-wardī*, is probably a variant of our *al-muwarrad* (no. 6)

89. *Dhikr mā jāʾ fī al-nayrūz wa-aḥkāmhu minhā fasarahū Baṭlamiyūs al-ḥakīm wa-wajadahū ʿan ʿilm Dāniyāl*; Hārūn 1951, 5:44–48.

90. The treatise begins: "Comets according to Aristotle are nine in number and according to Apuleius ten." This list of comet names was published in Tannery 1920, 356 and Pl. II, based on a manuscript in Paris (BnF grec 2424, fol. 189b). See also Bouché-Leclercq 1899, 357–69; CCAG 4:74. There is also a pseudo-Aristotelian treatise on comets said to have been translated by Ḥunayn ibn Isḥāq, *Kitāb al-Nayzak*; see Sezgin *GAS* VII, 215 also VI, 331–32.

91. See Thorndike 1950.

92. The facsimile edition of Ibn Hibintā's treatise presents the chapter on comets in two nearly identical versions taken from two different manuscript copies, both of which are closely related to the material given in the third portion of the sixth chapter of the *Book of Curiosities*; see Ibn Hibintā 1987, 1:356–65 and 2:134–44.

93. The date of the composition of his astrological treatise *al-Mughnī* has often been given as 829 (214H) on the basis of annotations found in a Munich copy of the treatise; see *EI2* art. "al-Kayd" (W. Hartner). However, Fuat Sezgin demonstrated that Ibn Hibintā mentioned the rebellion in 929 (317H) of Nāzūk against the caliph al-Muqtadir bi-llāh; see the editor's introduction to the first volume of Ibn Hibintā 1987. For the life and writings of Ibn Hibintā, see Sezgin, *GAS* VII, 162–64.

94. Kennedy 1957, 45; *EI2*, art. "Kayd" (W. Hartner); *EI2*, art. "al-nudjūm" (P. Kunitzsch).

95. *EI2* art. "Kayd" (W. Hartner).

96. Rapoport and Savage-Smith 2014, 379.

97. *EI2*, art. "ʿĀd" (F. Buhl); *EI2*, art. "Thamūd" (Irfan Shahīd).

98. *EI2*, art. "Madyan Shuayb" (C. E. Bosworth).

99. See, for example, Kennedy 1980, 164.

100. *Tuḥfat al-albāb fī bayān ḥukm al-adhnāb*; see King 1986, 105 no. D45 and pl. LXXXIV. The treatise is preserved in six recorded manuscripts, five in Cairo and one in Princeton (MS Mach 5065 [Yehuda 368]).

101. Rapoport and Savage-Smith 2014, 380. The Arabic phrase *al-kawākib al-khafīyah dhawāt al-ḥirāb al-marsūmah* might more literally be rendered as "obscure stars with impressed (or lightly-traced) lances."

102. Rapoport and Savage-Smith 2014, 384.

103. Rapoport and Savage-Smith 2014, 382–83.

104. Literally, "expanding (*yanmī*)."

105. *α Canis Minoris* (*al-shiʿrá al-shaʾmīyah*). According to all the later copies of the *Book of Curiosities*, however, the star in question is *al-shiʿrá al-yamāniyah,* that is Sirius, rather than Procyon.

106. See *EI2*, art. "al-Ḥākim bi-Amr Allāh" (M. Canard).

107. Rapoport and Savage-Smith 2014, 383–84.

108. A notable exception is Halley's Comet whose return every seventy-four to seventy-nine years, however, was not detected until Edmund Halley did so in the eighteenth century.

109. See Pasachoff 2000, 467–71.

110. For basic sources for astrological meteorology before the twelfth century, see Sezgin, *GAS* VII, 302–35. See also Jenks 1983; Bos and Burnett 2000; Burnett 2004a.

111. Rapoport and Savage-Smith 2014, 410–11. For traditional Arabic names of winds, see Lettinck 1999, 156–93; King 2004a, 812–13 and 815; *EI2* art. "Maṭlaʿ" (D. King)]; Varisco 1994, 111–17.

112. In the third century BC Timosthenes of Rhodes, admiral of the fleet of Ptolemy II Philadelphus, established the twelve-wind wind-rose, still with winds distributed on a circle representing the local horizon. Harley and Woodward 1987, 145–46, 153; Obrist 1997, 54–56 and fig. 14; see also de Callataÿ 2000.

113. Edson and Savage-Smith 2000; Bilić 2012.

114. Three of the copies are part of an anonymous astrological compendium, while the other nine are in anonymous scholia to Theon's commentary on Ptolemy's *Handy Tables*; Edson and Savage-Smith, 2000.

115. Rapoport and Savage-Smith 2014, 410.

116. This is likely a scribal error for the word *al-rajfah*, a common term for a particularly violent earthquake, also used in the Qur'an in reference to the fate of the Thamūd, a tribe destroyed by God with an earthquake for disobedience (Qur. 7:78, 7:91). See Lane 1863, 1042; *Enc. Qurʾān*, art. "Thamūd" (R. Firestone).

117. See Lettinck 1999, 209–24; Daiber 1975; *EI2* art. "Zalzala" (C. Melville); Hine 2002, 56–75.

118. For another divinatory collection that has an appendix making forecasts for occurrences in the Coptic month of Ṭūbah, see Fahd 1966, 488–95, and *EI2* art. "Nudjūm (Aḥkām al-)" (T. Fahd).

119. Rapoport and Savage-Smith 2014, 411.

120. Browne 1979, 54, compare 60.

121. This final section on the occurrence of winds on New Year's Day (6 Ṭūbeh) is close in form to another anonymous Coptic treatise also preserved today in fragments of the late ninth century; as only small fragments of the Coptic treatise are preserved, it is not possible to make a detailed comparison. The listing of weekday names in various languages, however, does not seem to be in the preserved part of the Coptic treatise. For these Coptic fragments, see Browne 1979, 45–63, and Till 1936.

122. Rapoport and Savage-Smith 2014, 412. The name is written as Dīqūs in the Bodleian copy; the text is also preserved in two later copies where it reads Diyāsqūrus and Dīsqūrus, respectively.

123. Sezgin, *GAS* VII, 310–11.

124. Rapoport and Savage-Smith 2014, 414.

125. The word *sh-n-s-r-w-a-r* is a reasonable attempt to transliterate the Hindi name for Saturday, *šanivar*. Al-Bīrūnī gives the name as *sanīchar wār* and states that it is also the name of the planet Saturn; al-Bīrūnī 1934, 165.

126. The word *sābāṭan* is a close transliteration of the Greek *sabbaton*, meaning "the sabbath."

127. *Bashmā* in the later copies.

128. We are very grateful to Professor Robert Simpson for his generous assistance in interpreting the Coptic terms in this treatise.

129. *EI2* art. "*malḥama*" (T. Fahd).

130. Fodor 1974.

131. Rapoport and Savage-Smith 2014, 473.

132. For a discussion of "exaltations" and planets as rulers of zodiacal signs, see the appendix.

133. Rapoport and Savage-Smith 2014, 467.

134. Rapoport and Savage-Smith 2014, 460–61.

135. See the appendix for a discussion of zodiacal signs considered "direct" or "oblique" in rising.

136. The term *sharaf* is here not used in its technical astrological sense of a planet's position of greatest influence, usually rendered as "exaltation." Rather, in this context *sharaf* is used in its general sense of high rank, eminence, honor, or glory. The sign of Leo, in fact, is not associated with the "exaltation" (*sharaf*) of any planet. For a discussion of "exaltations," see the appendix.

137. Rapoport and Savage-Smith 2014, 494.

138. *Fī masīrihi al-akbar.*

139. Rapoport and Savage-Smith 2014, 496, 498, 503.

140. See Stowasser 2014, 86–186; *EI Three,* art. "al-Bīrūnī" (Michio Yano).

141. Al-Bīrūnī 1934, para. 393; translation by the present author. There is no evidence that the writings of al-Bīrūni (d. *c* 1048) were available in Egypt in the first half of the eleventh century.

142. Another name for the domiciled planet. See the appendix for further details.

143. King 2004a, 643.

144. Ptolemy 1940; Sezgin *GAS* VII, 43–44. The *Tetrabiblos* was translated into Arabic by Ibrāhīm ibn al-Ṣalt in the mid-ninth century and then revised by Ḥunayn ibn Isḥāq not long thereafter.

145. According to historian David King: "Virtually no Egyptian astronomical works survive from the period between Ibn Yūnus and the middle of the thirteenth century" (King 2004a, 299). While Ibn Yūnus was an observational and mathematical astronomer, he is credited with one astrological treatise, *Bulūj al-umnīyah fī-mā yataʿallaq bi-ṭulūʿ al-shiʿrá al-yamānīyah* (On the Moon in the twelve zodiacal signs at the time of Sirius rising); Sezgin *GAS* vii,193.

146. Sabra 1998.

147. Ibn Abī Uṣaybiʿah 1882, 2:91; *Dictionary of Scientific Biography* art. "Ibn al-Haytham" (A. I. Sabra); *New Dictionary of Scientific Biography* art. "Ibn al-Haytham" (A. I. Sabra).

148. Langermann 2017; the attribution of this treatise to Ibn al-Haytham has been questioned; Rashed 2007a; Rashed 2007b; Sabra 1998.

149. Seymore 2001; Dols and Gamal 1984, 54–66; Sezgin *GAS* VII, 44–45.

150. Seymore 2001.

151. The Arabic is preserved in Oxford, Bodleian Library, MS Marsh 663, fols. 339–46. See also Sezgin *GAS* vii, 175–76. The Latin version was titled *Liber Capitulorum Almansoris.*

152. *Kitāb al-Qaḍāyá al-ṣāʾibah fī maʿānī aḥkām al-nujūm* and *Kitāb al-Amthilah lil-duwal al-muqbilah*; Sezgin vii, 185–86; Köhler 1994, 184.

153. Al-ʿAzīz, al-Ḥākim, and al-Ẓāhir; see IAU 1882, 2:90; Köhler 1994, 184; Sezgin *GAS* vii, 287.

154. Rapoport and Savage-Smith 2014, 589.

155. Köhler 1994, 184.

156. IAU 1882, 2:98–99; Ibn al-Qifṭī 1903, 269; *EI2* art. "al-Mubashshir b. Fātik" (F. Rosenthal).

157. Rosenthal 1961, 136.

158. Van Bladel 2009, 94–95, 184–96, 200.

159. Rapoport and Savage-Smith 2014, 332. For the identification of Idrīs with Hermes, see van Bladel 2009; Michot 2000a, 170–74; *EI2* art. "Idrīs" (G. Vajda).

160. Translations of Qurʾānic passages are those of ʿAbdullah Yūsuf ʿAlī (ʿAlī 1975).

161. *EI Three*, art. "Brethren of Purity (Ikhwān al-Ṣafāʾ)" (G. de Callataÿ); de Callataÿ 2005, 36–41; de Callataÿ 2016.

162. Nos. 3, 16, and 36; see Ikhwān al-Ṣafāʾ 2015a; Ikhwān al-Ṣafāʾ 2013; Ikhwān al-Ṣafāʾ 2015b.

163. Ikhwān al-Ṣafāʾ 2011.

164. Pingree 1980; Boudet, Caiozzo, and Weill-Parot 2011.

165. Fierro 1996; de Callataÿ 2013; Coulon 2014.

166. The leading scholar of the Brethren of Purity epistles, Godefroid de Callataÿ, has said, "It is puzzling that the *Epistles* appear to have remained completely unknown to the Fāṭimids"; *EI Three*, art. "Brethren of Purity (Ikhwān al-Ṣafāʾ)" (G. de Callataÿ).

167. Rapoport and Savage-Smith 2014, 325.

Chapter Three

1. Westrem 2011; Kline 2001.

2. Russell 1991, 51–57; Edson and Savage-Smith 2004, 67.

3. For overviews of the Balkhī School maps, see Tibbetts 1992b; Savage-Smith 2003; Pinto 2016.

4. For one famous nonmathematical world map of Late Antiquity, see Kominko 2013.

5. Al-Khwārazmī 1926.

6. See al-Khwārazmī 1926, 139, line 4, reading "a city that has no name on the map"; and 77, line 9, "other (rivers ?) which are not named on the map." See Nallino 1939–48; Tibbetts 1992a, 100.

7. Strasbourg, Bibliothèque Nationale et Universitaire, MS 4247, fols. 11b, 21a, 30b–31a, 47a. The four maps are reproduced in the edition of the text by Hans von Mžik: see al-Khwārazmī 1926. The maps are also reproduced in Tibbetts 1992a, 105–6 and plates 4–5.

8. For this island al-Khwārazmī generally used the name *Jazīrat al-Jawhar* (Island of the Jewel), as can be seen on his map of the island; in the tables he also used *Jazīrat al-Yāqūt* (Island of Sapphires). For an illustration, see al-Khwārazmī 1926, Tafel 1; and Tibbetts 1992a, 105, fig. 4.8. This island probably originated in an episode of the Alexander Romance. See Tibbetts 1979, 185–86; Tibbetts 1992a, 105–6; Ducène 2008, 294.

9. Al-Khwārazmī provides coordinates for various coastlines of the island, for several localities in it, and for the surrounding mountain range. See al-Khwārazmī 1926, 7–8, 40–42, and 83.

10. The parachute-like depiction of the Mountain of the Moon is later incorporated into the world maps accompanying copies of al-Idrīsī's *Nuzhat al-Mushtāq fī ikhtirāq al-āfaq* composed about 1154 for Roger II, the Norman king of Sicily. It is also to found on the circular world map in the *Book of Curiosities* (see plate 3).

11. For example, the account of the fifth clime mentions a total of twenty-nine mountains in this clime, including the Mountains of Ḥārith and Ḥuwayrith (Great Ararat and Lesser Ararat); a mountain between Mosul and Shahrazur; and the mountain of Ṭabaristān, between Nīsābūr (Nīshāpūr) and Jurjān. See Rapoport and Savage-Smith 2014, 433. This list of mountains reproduces a sequence of twenty-eight mountains in al-Khwārazmī's table of mountains of the fifth clime, which also includes their coordinates (al-Khwārazmī 1926, 54–56).

12. Rapoport and Savage-Smith 2014, 414.

13. Kennedy and Kennedy 1987, 377.

14. The *Geography* of Marinus, as known to us through the work of Ptolemy, had the *oikoumenē* begin even farther south of the equator, at 24° S.

15. Al-Farghānī 1998, 22 [of the Arabic text]; Tibbets 1992a, 102 and note 59.

16. Harley and Woodward 1987, 1:182; Berggren and Jones 2000, 20–22, 65; see also Fontaine 2000. I would like to thank Leif Isaksen for pointing out this issue to me.

17. Pedersen 2011.

18. Suhrāb 1930. See also Tibbetts 1992a, 104–5; King 1999, 33n61; and Kennedy 1987, 113–19.

19. Most of al-Bīrūnī's discussion concerns maps of the stars, but he explains that precisely the same methods apply to maps of the Earth; his various proposed methods of projection, however, had no apparent impact upon subsequent celestial or terrestrial mapping in the Islamic world. See Savage-Smith 1992, 34–38; Tibbetts 1992c, 140–42; Berggren 1982. His preserved world map is not plotted (Tibbetts 1992c, p. 142, fig. 6.4).

20. For an analysis by Raman spectroscopy of the pigments used in the manuscripts, see Chaplin et al. 2006.

21. Rapoport & Savage-Smith 2014, 442.

22. Al-Khwārazmī 1926, Tafel II; reproduced in Tibbetts 1992a, 106, fig. 4.9. For discussion, see Kahlaoui 2008, 113–15.

23. Tibbetts 1992b, 120.

24. Gautier-Dalché 2009, 58–60.

25. Gautier-Dalché 2009, 58.

26. Arabic: *muṣawwarah min ghayr kitāb*. Tibbetts 1992a, 96: "without a text." The sense of *kitāb* as label, or legend, corresponds with its use by al-Khwārazmī 1926, 101. But "*min ghayr kitāb*" could also mean: "in more than one book."

27. Al-Masʿūdī 1894, 33, and a summary of the same information in p. 44; a slightly different translation by Tibbetts 1992a, 96.

28. The map made for al-Maʾmūn may have been richer, but it had very little in common with the world map in the *Book of Curiosities* or with the map used by al-Khwārazmī. In the *Book of Curiosities'* world map there are no indications of stars and spheres, and its focus is on showing only the inhabited climes, not the entire world. The same was true of the map used by al-Khwārazmī. That al-Maʾmūn's world map represented a different cartographic tradition is confirmed by another account, coming from the twelfth-century geographer al-Zuhrī. Al-Zuhrī claims to have seen a copy of the world map made for al-Maʾmūn with depictions of the wonders of the world, famous and marvellous things, and historical monuments. Nothing here seems to correspond with the world map of the *Book of Curiosities*, and it seems highly unlikely that it was one of the sources used by our author. See al-Zuhrī 1968, 306; Tibbetts 1992a, 95.

29. Tibbetts 1992a, 104–5; Sezgin 2000–2007, 1:73–140. While Sezgin argued that al-Masʿūdī describes in this passage the map made for the caliph al-Maʾmūn, nothing here suggests a link with that royal map, which had a very different structure and where place-names would have been written in Arabic, not Greek.

30. Al-Masʿūdī 1965, 1:101–2 (nos. 191–93); translations in al-Masʿūdī 1962, 1:76–77; Sezgin 2000–2007, 1:80; Kahlaoui 2008, 114.

31. The vast majority of names on the world map can be deciphered with reference to Ibn Ḥawqal's maps, including labels that were horribly misspelled by the late twelfth-century copyist of the Bodleian manuscript. For example, a large label along one of the rivers of Iraq clearly reads *sawād filasṭīn* (سواد فلسطين, "the marshes of Palestine"), and is very much

out of place. But comparison with Ibn Ḥawqal's regional map of Iraq shows that the label in this location should read *bawādī wāsiṭ* (بوادي واسط, "the deserts of Wasit"). If you try your hand at writing these pairs of words in cursive style without diactrical dots, you'll find this a very plausible miscopying.

32. Ibn Khurradādhbih 1889, 99.

33. Several reconstructions have been published. See Sezgin *GAS* XII, no. 1b (using a modern projection, and shown with north at the top); Daunicht 1968; Tibbetts 1979, fig. 1b.

34. This large peninsula is described by Ptolemy, as can be seen in the late medieval reconstructions of Ptolemy's maps, as well as in later printed Ptolemaic maps. See Milan, Biblioteca Ambrosiana, MS gr. 997, D 527 inf. (early fourteenth century), where the large peninsula can be seen in the world map on fols. 94v-95r, and on the separate map of Asia and the Indian Ocean, fols. 99v–100r. See reproductions in Sezgin *GAS* XII, 63 no. 31, and 66 no. 31c.

35. The periphery of the circular island is decorated in red and gold. The red pigment has been analyzed as cinnabar (or vermilion), but the analysis of the reflective gold decoration was inconclusive. It is probably metallic gold, but the actual presence of metallic gold cannot be detected by Raman spectroscopy. Similar reflective golden flakes were used on one of the comets in book 1 (fol. 14a). See Chaplin et al. 2006, 866 and 870.

36. Tibbetts 1992a, 102.

37. Isaksen 2012.

38. Simon 2013, 36.

39. Arabic: *ishhārān li-maʿālim rasūl Allāh.*

40. Al-Maqrīzī 2002, 2:305 (describing the looting of the Fatimid palaces in 461/1068).

41. Kaplony 2008.

42. Kahlaoui 2008.

43. Tibbetts 1992c, 138.

44. Manuscripts include BnF MS arabe 2214, MS Ayasofia 2934, and MS Topkapı 3347. The dating is based on Ayasofia 2934, copied ca. 600/1200. MS arabe 2214 was copied in 847 H (1445).

45. The oval shape and the presence of elements from both Ptolemy and al-Khwārazmī led Gerald Tibbetts to speculate that this might reflect a Ptolemaic projection (Tibbetts 1992c, 138). While all copies of this world map are very sparsely labeled, the earlier of the two Istanbul copies has a slightly more complex design, with boundaries for the first three climes and one south of the equator. In that Istanbul version of this oval map there are also zodiac names along the borders, an element that suggests either an origin in an astrological treatise or a later inclusion (Kamal 1926, 3.3:805–9). Another version of this world map has the same features of the inhabited world but drawn within a semicircle having its base formed by the equator. This variant, found in the opening folios of the same Paris copy of the twelfth-century Ibn Ḥawqal III set, also has clime lines drawn (Tibbetts 1992c, 149, fig. 6.10 [MS 2214, fol. 3]). Gerald Tibbetts also noted the close similarity of this semicircular world map in Ibn Ḥawqal III to a semicircular world map found in *Menāreth qudshēh*, a Syriac treatise by the thirteenth-century scholar Bar Hebraeus (Tibbetts 1992c, 148, fig. 6.9 [Staasbibliothek Berlin Orientabteilung, MS Sachau 81, fol. 37b]).

46. For illustrations of maps from Ibn Faḍlallāh al-ʿUmarī's *Masālik al-abṣār fī mamālik al-amṣār*, see King 1999, 35, fig. 1.7.5, and 93, fig. 2.8.3; Tibbetts 1992c, 153, fig. 6.14; Sezgin *GAS* XII, 23, 1a (rotated with north at top).

47. Fuat Sezgin, who has rightly highlighted the importance of this circular world map to the history of cartography, has linked this map to the world map made for al-Maʾmūn

(Sezgin 2000). He also found that the coordinates provided by Ibn Faḍlallāh match the work of al-Khwārazmī, not Ptolemy's (Sezgin 2000, vol. 1, 17–19 and passim, 301–3). Because there are some late place-names on this map, such as Delhi, the noted historian of Islamic cartography Gerald Tibbetts wrongly suggested that Ibn Faḍlallāh's map was influenced by late medieval European models (Tibbetts 1992c, 153). Jean-Charles Ducène has recently argued that the *Geography* cited by Ibn Faḍlallāh is not Ptolemy's, but an Arabic work, which was also used by the Andalusi geographer Ibn Saʿīd in the thirteenth century (Ducène 2011c, 21).

Chapter Four

1. See chapter 1 above for a discussion as to whether this map of the al-Idrīsī type was part of the original treatise or whether it might have been added by the copyist of the Bodleian manuscript.

2. The entry by J. H. Kramers in the Encyclopaedia of Islam is a useful starting point despite its considerable age, but the emphasis is on geographical texts and not on visual representations, and the discussion is brief. See *EI2*, art. "al-Nīl" (J. H. Kramers). The article was originally published in 1936. An insightful short article by Levtzion reexamines the accounts of the sources of the Nile in Arabic geographical texts, especially with regard to West Africa (Levtzion 2000). Only over the last decade did maps of the Nile receive more sustained attention. Jean-Charles Ducène and John Cooper have written detailed studies of Ibn Ḥawqal's and al-Idrīsī's maps of the Nile Delta, demonstrating that the maps make an attempt to faithfully represent the complex canal and branch system as it existed in the tenth century (Ducène 2004a and 2004b; Cooper 2012 and 2014). A recent issue of Cartes et Géomatique was devoted to the medieval European and Islamic cartography of Africa, including a contribution by Ducène on the depiction of Africa in Islamic world maps (Ducène 2011c). Most recently, Robin Seignobos has produced comprehensive studies of the depiction in European and Islamic maps of the island of Meroe in the Upper Nile and of the imagined western origins of the Nile in North Africa. These studies take into account the maps of the Nile in the *Book of Curiosities* (Seignobos 2011 and 2017).

3. *EI2*, art. "al-Nīl" (J. H. Kramers) and Seignobos 2011, Figure 3: "Le cours du Nil selon Ptolémée." Ptolemy knows only of two lakes, not lying on the same latitude. He also does not mention a number of rivers coming from the Mountains of the Moon, just one.

4. Ducène 2011c, 21; Seignobos 2011, 7, and the sources cited there. EI2, art. "al-Nīl" (J. H. Kramers) needs correction, as the entry states that "al-Khwārazmī took over from Ptolemy a *western* tributary of the Nile" (emphasis mine).

5. Gautier-Dalché 2009, 58.

6. Seignobos 2011, 7.

7. See al-Khwārazmī 1926, 106–7; Levtzion 2000, 71; Hopkins and Levtzion 1981, 9.

8. Dzhafri 1985, 88; Sezgin, *GAS* XII, 4; Ducène 2011c, 31 and Figure 1, based on Mžik 1916.

9. Al-Masʿūdī 1965, 1:112 [no. 215].

10. See also, for example, the world map in Leiden MS Cod Or 3101, copied 1173; reproduced in Ducène 2011c, 31 (fig. 2).

11. Al-Iṣṭakhrī 1870, 50.

12. The comparison between the maps of al-Iṣṭakhrī and Ibn Ḥawqal is illustrated in Tibbetts 1992b, 121, fig. 5.14.

13. Ducène 2004a; Cooper 2012.

14. Ibn Ḥawqal 1873, 89–94. The map also depicts the river island that lies between Giza and Fustat, also discussed in al-Iṣṭakhrī 1870, 48–49.

15. See, for example, the world map in Ahmet III MS 3346, copied 1086, reproduced in Tibbetts 1992b, 123, fig. 5.16.

16. Ibn Khurradādhbih 1889, 176; Ibn al-Faqīh 1885, 64–65 (cited in Levtzion and Hopkins 1981, 27 and Levtzion 2000, 72). Cf. confused account in Ibn Rustah 1892, 90. Both Ibn Khurradādhbih and Ibn Rustah suggest that the Mountain of the Moon is in the Yemen (*al-b.b.n* in de Goeje's edition of Ibn Rustah).

17. Arabic: *wa-lam yaʿzi aṣla-hu ilā mawḍiʿ*. Ibn Ḥawqal 1873, 98.

18. The text of the side panel has the third lake feed into eight outlets, or rivers, rather into the main branch of the Nile: "[The course of the Nile] consists of ten streams, of which five are to the east and five to the west. It then empties into two marshes and from the two marshes into one large marsh at the equator. It then descends to its eight outlets [*sic*]" (Rapoport and Savage-Smith 2014, 87, 494 [fol. 42a]). The last sentence is almost certainly a copyist error. It should have said that the eight rivers feed into the third lake, as is depicted on this map and on al-Khwārazmī's map.

19. Rapoport and Savage-Smith 2014, 87, 494.

20. Seignobos 2017.

21. Ibn al-Faqīh 1885, 64; translated in Levtzion and Hopkins 1981, 27; Levtzion 2000, 72; and Seignobos 2017, 385–86. See also Norris 1972, 76.

22. Norris 1972, 72.

23. Seignobos 2017, 386.

24. Zadeh 2011.

25. Ibn al-Faqīh 1885, 65.

26. Art. "ʿUḳba b. Nāfiʿ," *EI2*.

27. Norris 1972, 33, 151.

28. This point has been suggested to me by R. Seignobos.

29. Troupeau 1954; Maqbul 1995, 117.

30. Al-Maqrīzī 1998, 1:355.

31. See the reproductions in Tibbetts 1992b, 121, fig. 5.14.

32. Ducène 2009.

33. Ducène 2009, 277.

34. For the coordinate values of the Fayyum in medieval Islamic tables, see Kennedy and Kennedy 1987, 119. The values of the Fayyum in the *Book of Curiosities* are significantly different from those of later tables.

35. Al-Masʿūdī 1894, 57.

36. Tibbetts 1992c, 138.

37. Also reproduced in Tibbetts 1992c, 139, fig. 6.2. The maps of the Nile in the other copies of the Ibn Ḥawqal III set, Ayasofia 2934 and Topkapı 3347, are reproduced in Kamal 1926, 3.3:805, 810, 812.

38. The text of these marginal comments is published by Kramers in Ibn Ḥawqal 1938, 1:147–48; the map is reproduced in the following page.

39. On the map of the Nile, the label is at the top left-hand side, and refers to a river that comes out from the Mountain of the Moon.

40. See the world map in BnF arabe 2214, fols. 52v–53, reproduced in Tibbetts 1992c, 140, fig. 6.3.

41. Rapoport and Savage-Smith 2014, 488–89.

42. Al-Masʿūdī 1965, 1:112 [no. 215].

43. Al-Masʿūdī 1965, 2:79 [no. 796].

44. Ibn Ḥawqal's map of the Nile is reproduced in Ibn Ḥawqal 1938, 1: 139 and Ibn Hawqal 1964, map 6. Label no. 7 reads: *madīnat al-nūbah*, "the city of the Nubians."

45. Seignobos 2011, 7.

46. This was suggested to me by John Cooper. On the Red Sea-Nile canal, see Cooper 2009.

47. Levtzion 2000; Seignobos 2017, 391–92; Ahmad 1992, 156.

48. Also reproduced in Sezgin *GAS* XII, plate 8b. For more detailed images of al-Idrīsī's sectional maps from the Istanbul Aya Sofya 3502 manuscript, see Ahmad 1992, 165, figs. 7.10, 7.11, and 7.12. Note that Miller's reconstruction of the sectional maps in al-Idrīsī's second work, the *Uns al-muhaj wa-rawḍ al-furaj* [reproduced in Sezgin *GAS* XII, plate 8c], shows two eastern tributaries, one coming from the equator, the other much farther north, in a source in the first clime. This is so because the sectional map of Upper Egypt in the Hekimoglu Ali Pasa MS. 688 copy of *Uns al-muhaj* [also reproduced in Ahmad 1992, 167, fig. 7.18] shows the Nile making a sharp turn near Ṭūḥ. As a result of the way Miller reconstructed the world map from sectional maps, the impression is that the Nile section from Qinā through Luxor is an eastern tributary.

49. The entire "atlas" (interactive, with composite map of 70 parts) is available at http://classes.bnf.fr/idrisi/explo/index.htm as part of on-line publication of exhibition *al-Idrisi la Méditerranée au xiie siècle* at the Bibliothéque nationale de France. The maps in questions are in Section 4, climes 1–3, of the composite map that opens the site.

50. Al-Idrīsī 1970, 33; Levtzion 2000, 73–74; Levtzion and Hopkins 1981, 115.

51. For the relevant sectional map from the Istanbul Aya Sofya 3502, see Ahmad 1992, 165, fig. 7.10.

52. Ahmad 1992, 160–62. On the relationship of the circular world map with the rest of the treatise, see the introduction above, and Rapoport and Savage-Smith 2014, 30–31.

53. See reproductions in Ahmad 1992, 161 (figs. 7.1–7.4), 171 (7.21); Sezgin *GAS* XII, plates 5, 6, 9.

54. See Seignobos 2011, nt 35. On the copies of al-Idrīsī's maps, see Ducène 2004b, 64 (following the study by Rubinacci).

55. Rapoport and Savage-Smith 2014, 30–31.

56. The label is very faint against the brown background.

57. Another feature of the circular world map in the *Book of Curiosities* is an eastern tributary, fed by an eastern lake on the equator; as noted above, this eastern tributary is absent from copies made in the fifteenth century or later.

58. On Gog and Magog in the Islamic tradition, see von Donzel and Schmidt 2010.

Chapter Five

1. Rapoport and Savage-Smith 2014, 483.

2. The incomplete "[..]*baʿīn*" could be read either as 40 or 70.

3. Al-Bakrī 1992, 754–62 (list of anchorages from Morocco to Syria along the North African coast); al-Idrīsī 1970, 623–25 (list of anchorages and the distances between them along the coasts of Sicily).

4. See Tibbetts 1992b, 118–20, and fig. 5.12 (p. 120).

5. Compare with the maps of the Mediterranean by al-Iṣṭakhrī and Ibn Ḥawqal in Tibbetts 1992b, 120; Pinna 1996. See also recent overviews of Islamic mapping of the Mediterranean in Savage-Smith 2014b; Kahlaoui 2008.

6. For maps by al-Iṣṭakhrī and Ibn Ḥawqal that show the Black Sea as narrow straits or canal, see Tibbetts 1992b, 120–23. On the Burjān, see *EI2*, art. "Bulghār."

7. Strobilos is on the northwestern tip of the Ceramic Gulf, ten kilometers southwest of modern Bodrum; see Foss 1988; Malamut 2004, 23–24.

8. Agios Georgios is mentioned as a stop on the way to Constantinople by Saewulf in 1102, and then by fourteenth-century portolans. Pryor identified it with the Byzantine town of Ganos (mod. Gaziköy). See Pryor 1994, 55–56. Al-Idrīsī mentions Shanṭ Jirjī in the Dardanelles, south of Gallipoli (al-Idrīsī 1970, 800).

9. Al-Anṭākī 1990, 233, 240–42. See also al-Masʿūdī 1894, 141, and Qudāmah 1981, 188. Compare with Goitein 1967, 1:307–8, where the evidence appears inconclusive. On the term *usṭūl*, see also Lev 1984, 247; Picard 1997, 117.

10. See Picard 2009.

11. On the arsenals and other naval structures of the Muslim Mediterranean, see Jansen et al. 2000, 159, 168–69. D. Bramoullé has argued that the Fatimids maintained arsenals only in Cairo and Fustat, to retain their central control over naval operations, although he notes that the Ayyubid-era authors al-Makhzūmī and Ibn Mammātī report there were also arsenals in Alexandria, Tinnīs, and Damietta (Bramoullé 2009a, 267).

12. Al-Idrīsī uses the Arabic names, in Magrebi dialect. These wind names would later appear later in Latin texts: *shaluq* (scirocco) for SE wind, *al-libag* (libeccio) for the SW wind (Gautier-Dalché 1995, 58). A different twelve-point Greek wind system, associated with astronomical observations, is described earlier in the *Book of Curiosities*, in a chapter dealing with winds and earthquakes; see chapter 2 above.

13. Arabic: *wa-dukhūluhu fī al-buryās al-layyinah*.

14. The Arabic text here poses a problem of interpretation. In many cases, the copyist of the manuscript wrote "*tasīru min* X," which means "you go" or "you enter with X wind." According to this reading, the mapmaker indicates winds favorable for entering a harbor. But in some labels the manuscript reads "*yasturu min* X," meaning "[the harbor] protects from [X wind]." In the edition of the *Book of Curiosities* we have preferred the second reading and consistently amended "*tasīru min*" to "*yasturu min*." This emendation is chiefly in light of the text of al-Idrīsī, where the form "*yasturu min*" is used to indicate protection from winds offered by North African ports. See al-Idrīsī 1970, 252 lines 11 (Wahrān), 272 line 14 (Waqūr), 280 line 10 (Gabes).

15. See Pryor 1988, 88–89.

16. On other maps of the island, see Stylianou and Stylianou 1980.

17. Damascus, Maktabat al-Assad al-Waṭāniyah, MS 16501. fol. 105b. For an illustration see Rapoport and Savage-Smith 2014, 27. The Damascus copy contains sentences omitted by the copyist of the Bodleian copy, so is not derived from it (Rapoport and Savage-Smith 2014, 12).

18. Nordenskiöld 1897, 12, nos. 297 and 309.

19. Nordenskiöld 1897, 10–15.

20. Ibn Ḥawqal 1938, 205.

21. The standard Greek term for westerly wind was *Zephuros* (Kemp 1976, 941). See also discussion of variants in Gautier-Dalché 1995, 70ff.

22. Ibn Ḥawqal 1938, 179, 205.

23. Pryor and Jeffreys 2006, 373ff.

24. Goitein 1967, 211–213, 318–320; Gertwagen 1996, 73–92.

25. Pryor 2003, 87–88; Pryor and Jeffreys 2006, 105, 333–55.

26. Bramoullé 2009a, 263.

27. Qudāma 1981, 48; paraphrased in Fahmy 1950, 141–42; Christides 1982, 100.

28. Pryor 2003, 93. On the author of this manual, Pryor and Jeffreys 2006, 181–83; McGeer 1995, 79–86. On intelligence gathering by naval forces, see also Pryor and Jeffreys 2006, 393–94, and the sources cited there.

29. Al-Bakrī 1992, 754–62; Lewicki 1977, 451.

30. Al-Bakrī 1992, 754.

31. Al-Bakrī 1992, 16ff.; Lewicki 1977, 439–69. A subsequent section lists seventeen stopping points from Alexandria to Antioch in Syria. These are mostly listed by name only, and no distances are given (al-Bakrī 1992, 758). A final maritime itinerary leads from Atlantic islands to Azīlá (modern Asilah), on the Atlantic coast of Morocco. Here distances are mentioned only sporadically, and in days of sailing; if the ports are near each other, there is no mention of the distance between them. Again, some of the stopping points are noted for their protection from the elements, allowing ships to moor over the winter, and prominent landscape features (al-Bakrī 1992, 758–61).

32. Al-Idrīsī 1970, 623–625 (Sicily); 271–275 (Oran-Bone).

33. Al-Idrīsī 1970, 272 (Waqūr)

34. Kahlaoui 2008, 194–196; al-Idrīsī 1970, 252 (Wahrān),

35. Arrian 2003. See, for example, "From the Artane to the River Psilis it is 150; small boats can be moored near the rock that emerges not far from the mouth of the river" (p. 69); "From the Achaious to Herakleia Point is 150 stades. From there to the point which is a shelter from the Thraksian and north winds, 180" (p. 77).

36. Nordenskiöld 1897, 11–12. Example: "From Paphos to Numenium—an island with a spring; the crossing short; when you have approached the islet, hug the land on the right. 55 stadia."

37. Rapoport and Savage-Smith 2014, 469.

38. Savage-Smith 2003.

39. Rapoport and Savage-Smith 2014, 484.

40. The Arabic *shalandiya* is derived from the Greek *chelandion*, a boat used by the Byzantines for military and commercial purposes in the Mediterranean, and adopted by the Fatimids and the Almohads; see Agius 2001.

41. Rapoport and Savage-Smith 2014, 486.

42. On Slav migrations to the Peloponnesus in the early medieval period, see Avraméa 1997, 67–86.

43. Only two examples are cited: A small bay to the north of Ḥamdīs (?), a site to the east of modern Piraeus, which allows ships to moor protected from all winds; and a harbor that protects from all winds is found in the bay of Methone in the Peloponnesus (Rapoport and Savage-Smith 2014, 485–86).

44. A single exception is the port of Tracheia, opposite Rhodes, which is discussed both in the account of the Aegean Bays and on the map of the Mediterranean. This was because the port gave its name to two nearby gulfs. On the location of Tracheia, see Hild 2000, 109.

45. The names of two bays along the western coasts of Anatolia may refer to the major micro-asiatic islands of Lesbos and Strongyle (*Miṭilṭālās* = Mitylini, *Isṭarnkīlih* = Strongyle). However, the size and direction of the bays leaves no doubt that they are located on the Anatolian mainland and not on the islands. The first is twenty miles long and seven miles wide, and is entered from the west proceeding east. The second is forty miles long, and twenty miles wide, and it also entered from the west to the east. See Rapoport and Savage-Smith 2014, 484.

46. On Slavs in the Fatimid navy, see Bramoullé 2007, 14, 24ff.

47. See a translation of an excerpt from the earliest preserved portolan guide, dating to the late thirteenth century, in Gautier-Dalché 2002, 62.

48. Pryor 1998, 97ff.; Jacoby 1997; Malamut 2004, 21–29.

49. See examples in Pryor and Jeffreys 2006, 333ff.; Huxley 1976; Pryor 2003, 92–94; Lewicki 1977, 449ff.

50. Pryor 1994, 53–55.

51. Rapoport and Savage-Smith 2014, 442.

52. Here we depart from the views of Tibbetts (1992d), and we concur with Pujades 2007.

53. Al-Muqaddasī 1877, 10–11, Muqaddasī 1994, 9 (translation by Collins). See also Agius 2007, 192.

54. Arabic: *rabbānīn wa-ashātimah wa-riyāḍiyīn wa-wukalāʾ wa-tujjār* (Muqaddasī 1877, 10). Note the similarity with the informants supposedly used by the author of the *Book of Curiosities.*

55. Al-Muqaddasī 1877, 10.

56. Al-Muqaddasī 1877, 11.

57. Sezgin 2000–2007, 2:3–59. See the resounding refutation by Pujades 2007, 508–10.

58. Gautier-Dalché 2002, 65ff.; Gautier-Dalché 1995, 39–67; Pujades 2007, 516.

59. Gautier-Dalché 2005.

60. Gautier-Dalché 1995.

61. The interpretation of the Latin text here is disputed. Pujades claims that the *Liber* is used to prepare a map, while Gautier-Dalché argues that the map has been consulted during the preparation of the text; see Gautier-Dalché 2011.

62. See the insightful remarks by Pujades 2007, 507, with regard to the possibility of Roman maritime itineraries being projected graphically: "their linear nature and detachment from surrounding space would render them useless when it came to representing routes over the waters."

63. Pujades 2007, 418–20.

64. Pujades 2007, 456–57, 521. See also the comments by in Campbell 1987 [2010]. One still is faced with the technical question of whether the precision of the portolan charts could have been an accidental by-product of recording distances and compass bearings on a chart, without any recourse to coordinates or mathematical form of projection. See the models suggested in Lanman 1987 and Gaspar 2008, recently challenged by Nicolai 2015. Some of the arguments by Nicolai, such as the inability of medieval mariners to measure distances effectively, are refuted by the evidence of the *Book of Curiosities.*

Chapter Six

1. Antrim 2012, 108, 125.

2. See the accounts in al-Maqrīzī 2002, 2:305; al-Maqrīzī 1967–71, 2:285, 292–93.

3. Bloom 2007, 46–47; Lev 2012. The map was carried with the Fatimid court to Egypt, and was put on display in the Fatimid palaces in Cairo. It was lost when the Fatimid treasuries were looted by rebelling Turkish troops in AD 1068–69. For earlier discussions, see King 1999, 35. The earlier attribution of the map to Ibn Yūnus (d. 1009) is incorrect (Tibbetts 1992a, 96).

4. Another example of a medieval Islamic map carrying a decidedly political message was found on the decorations of the walls of the Cairo Citadel dating from the Mamluk period, as discussed in Rabbat 1995, 166. According to the chronicler Ibn ʿAbd al-Ẓāhir, the

sultan al-Qalāwūn (r. 1279–90) ordered some of the walls of the Cairo Citadel to be depicted with the likeness (*ṣifāt*) of each of his castles and citadels, surrounded by mountains, valleys, rivers, and seas. Rabbat points out the link between the representations of the forts, the location of the images on the walls of the citadel, and the power of the person of the sultan. Like the Fatimid world map, these decorations are no longer extant.

5. Rapoport and Savage-Smith 2014, 457.

6. On this chapter and the map of Sicily see also Johns 2004; Metcalfe 2009.

7. The correct name is Abū al-Ḥusayn Aḥmad ibn Ḥasan ibn Abī al-Ḥusayn, the second Kalbid emir (r. 954–69).

8. Called the *Bāb Rūṭah* by Ibn Ḥawqal 1938, 121. See also Idrıs 1962, 2:419.

9. Compare this list of gates with Ibn Ḥawqal 1938, 121–22.

10. Rapoport and Savage-Smith 2014, 457. Compare the more expansive account in Ibn Ḥawqal 1938, 119_{4-7}: "Next to [Palermo] is a city called al-Khāliṣah. It has a wall of stone that is not like the wall of Palermo. The ruler and his followers live there. It has two baths, but no markets or *funduq*s ("hotels for merchants"). It has a small, modest congregational mosque. The ruler's army is there, and the naval arsenal and the administration. It has four gates to the north, south and west, but to the east is the sea and a wall without a gate." There have been several attempts to reconstruct the line of the walls of this palatial complex (Zorić 1998).

11. Compare Ibn Ḥawqal 1938, 119: "Most of [Palermo's] markets lie between the mosque of Ibn Siqlāb and the New Quarter, including the markets of: the oil-vendors and their corporation; the flour-merchants; the money-changers; the apothecaries; the blacksmiths; and the polishers; the markets of the wheat-vendors, the embroiderers, the fish merchants, and the grain-sellers; a group of butchers; the vegetable-sellers; the fruit-vendors; the sellers of aromatic plants; the jar-makers; the bakers; the rope-makers; the corporation of perfumers; the butchers; the shoe-makers; the tanners; the carpenters; and the potters. The wood-merchants are established outside the town. In Palermo [itself] is a group of butchers, jar-makers, and shoe-makers."

12. Rapoport and Savage-Smith 2014, 457. Compare Ibn Ḥawqal 1938, 119.

13. One quarter, that of the Prayer-ground of Abū Ḥajar (*ḥāra tusammmā muṣallā Abī Ḥajar*), is not framed by a yellow box. However, this is probably a copyist mistake, since yellow borders do frame a nearby rural locality (Abū Sālim).

14. Nef 2013, 47; Bagnera 2013, 78.

15. Bagnera 2013, 82.

16. Nef 2013, 53.

17. Bagnera 2013, 78–79.

18. Nef 2013, 53. It may have been the site of al-Khāliṣah too (Bagnera 2013, 69). Elena Pezzini argues, on the basis of the *Book of Curiosities* map, that Palermo and al-Khāliṣah were surrounded by a joint perimeter wall (Pezzini 2013, 197).

19. Ibn Hawqal 1938, 119.

20. See the map of Arabia in the Bologna manuscript of al-Iṣṭakhrī, copied 589/1183 (Bologna, Bibioteca Universtaria, Cod. 3521, fol. 5v); reproduced in Tibbetts 1992b, 118, fig. 5.9.

21. Daftary 2007.

22. On the foundation of Mahdia, see Halm 1996, 214; Brett 2001, 142; Lézine 1965; Golvin 1979. See also a recent translation of a tenth-century account in al-Nuʿmān 2006, 228–29.

23. Compare the early accounts cited by the traveler al-Tijānī, who visited Mahdia in 1309 (Soudan 1990; Golvin 1979, 80).

24. On Abū Yazīd and his rebellion, see Brett 2001, 167–75; Halm 1996, 298–309; Halm 1984.

25. Rapoport and Savage-Smith 2014, 467.

26. Rapoport and Savage-Smith 2014, 467.

27. Rapoport and Savage-Smith 2014, 468.

28. On the topography of medieval Mahdiya, see Lézine 1965; Golvin 1979; Djelloul 1993.

29. Al-Bakrī 1965, 30; Soudan 1990, 142. See also Bloom 2007, 31.

30. Marçais 1954, 91; Idris 1962, 2:450; Djelloul 1993, PL I.

31. Lézine 1965, 45.

32. Al-Bakrī 1965, 30. Al-Tijānī, writing in the early fourteenth century, also notes the opulent gilded decoration on the exterior of al-Mahdī's palace (Soudan 1990, 143). See the interpretations by Idris 1962, 2:449; Golvin 1979, 87; Djelloul 1993, 91, and PL I; Bloom 2007, 30.

33. Golvin 1979, 87–88.

34. This distinction is repeated in the parallel, more abstract plan in the later Damascus copy. The depiction of the two palaces in the Bodleian copy also attempts to create a sense of depth, unlike the flat facades common in Iraqi manuscripts of the time. For comparison with the depiction of architecture in other medieval illustrated Arabic manuscripts, see Rabbat 1995, 161ff.; Barrucand 1994.

35. Ibn Ḥawqal 1938, 71.

36. Al-Bakrī 1965, 29; cited in Golvin 1979, 83.

37. Al-Idrīsī 1970, 282.

38. Lézine 1965, 24–38. This is also the view in Djelloul 1993, 83, and Bloom 2007, 23.

39. Marçais 1954, 90; Golvin 1979, 76–86.

40. Djelloul 1993, 77–78, and fig. 3. The map does not support Djelloul's hypothesis of a second, exterior port serving galleys and located next to the arsenal.

41. Antrim 2012, 41.

42. Al-Idrīsī 1970, 283. Most modern scholars interpreted this passage as referring to a moat flanking the western walls of Mahdia (Idris 1962, 2:449; Marçais 1954, 89; Golvin 1979, 86; Djelloul 1993, 81).

43. Bagnera 2013, 70–71.

44. Cooper 2012, 68–70.

45. Lev 1999; Ibn Bassām 1967.

46. Rapoport and Savage-Smith 2014, 470.

47. Gascoigne 2007.

48. See map in Gascoigne 2007, 162.

49. Gascoigne 2007, 170.

50. I owe this interpretation of the archaeological survey to John C. Cooper.

51. Tibbetts 1992b, 115; Antrim 2012, 41.

52. Rapoport and Savage-Smith 2008. Walls around Constantinople are also depicted in a copy of Ibn Ḥawqal's map of the Maghreb, which is closely related to the rectangular world map in the *Book of Curiosities* (see fig. 3.5 and plate 2 above; Istanbul, Topkapı Sarayı Müzesi Kütüphanesi, Ahmet III MS 3348, fols. 20a, 20b, and 21a). One other early example is found in the map of the Sea of Azov by al-Khwārazmī, where two named cities are depicted as shields, probably standing for the city walls, with three towers on top (Bibliothèque Nationale et Universitaire de Strasbourg, MS 4247, fol. 47a; reproduced in Harley and Woodward 1992, plate 5). In the maps of a 684/1285 copy of al-Iṣṭakhrī's treatise., almost all towns are

represented by brick walls topped by towers of various shapes and colors. A major city, such as Córdoba, is depicted as larger structure (Istanbul, Topkapı Sarayı Müzesi Kütüphanesi, Ayasofia MS 3348, copied 684/1285; see reproductions of the Maghreb and the Mediterranean maps of this manuscript in Pinna 1996, 64, plate 20; and 111, plate 48).

53. Johns and Savage-Smith 2003. Two copies of this map (Paris, Bibliothèque Nationale de France, MS Arabe 2221, fol. 204, and Oxford, Bodleian Library, MS Pococke 375, fols. 187[b]–188[a]) are reproduced in Pinna 1996, 2:52–53 and 62–63.

54. British Library MS Or. 3623, fol. 49b, copied 729/1329. The map is reproduced in al-Qazwīnī 1960, 176.

55. British Library MS Or. 3623, fol. 119v. Reproduced and discussed in Tibbetts 1992c, 152–54, fig. 6.16; and in al-Qazwīnī 1960, 434.

56. The maps and plans of Ibn al-Mujāwir have yet to receive attention by historians of cartography. These maps are not mentioned in any of the contributions to Harley and Woodward 1992. They survive in the earliest extant copy made in 1595 (Ayasofia 3080).

57. MS Ayasofia 3080, f. 53a. The map is reproduced in Rex Smith 2008, 148, fig. 5; and Ibn al-Mujāwir 1951, 129, tabula V. See Vallet 2010, 747, Carte 5, who reproduces a sketch of the map, and also places the localities and inscriptions on a modern plan of the port.

58. Bowersock 2006.

59. Bowersock 2006, 77. A representation of Verona from the ninth or tenth century, which survives in an eighteenth-century copy, similarly shows the civic institutions cramped between the walls. This appears to be the only extant medieval European city map produced before the end of the thirteenth century (Miller 2003, 84–85).

60. See also Rabbat 1995, 168.

61. Contadini 2007.

62. Rapoport and Savage-Smith 2014, 476, 479.

63. For discussion of the term "*thaghr*" in geographical literature, see Brauer 1995, 14.

64. In an earlier study I emphasized the defensive elements of the urban representation as against the more aggressive tone of the text. I now believe that the maps reinforce the Fatimid aspirations for Mediterranean dominance (Rapoport 2012).

Chapter Seven

1. Horden and Purcell 2000; Savage-Smith 2014b.

2. Muslim geographers extended the term Galicia to large areas of Christian Spain (*EI2*, art. "D̲jillīḳiyya").

3. *EI2*, art. "Īṭaliya" (U. Rizzitano).

4. For maps by al-Iṣṭakhrī and Ibn Ḥawqal that show the Black Sea as narrow straits or canal, see Tibbetts 1992b, 120–23.

5. *EI2*, art. "Bulghār" (I. Hrbek).

6. The ruins are today known as Aspat or Chifut Kalesi, "the Jew's Castle" (Foss 1988). Mentioned as *Istrūbilī* by al-Idrīsī (Idrīsī 1970, 648).

7. *EI2*, art., "Algiers."

8. Al-Idrīsī 1970, 268; Idris 1962, 2:496.

9. Rapoport and Savage-Smith 2014, 112, 479.

10. Goldberg 2012, 202.

11. Goldberg 2012, 219, map 8.1; and 255, Map 9.5

12. Goldberg 2012, 255, Map 9.5.

13. Udovitch 1987; Goldberg 2012, 203, 243.

14. Picard 2009.

15. Including the well-known account of the port of Acre by al-Muqaddasī, whose grandfather was reposnsible for the renovation of the ramparts, the archway at the entrance and the chain. See Bramoullé 2009a, 267, and the sources cited there.

16. Picard 2009, 222. The oriental model was only introduced to the ports of the western Mediterranean from the thirteenth century onward.

17. Nāser-e Khosraw 1986, 16; see also Bramoullé 2009a, 269.

18. Koutelakis 2008.

19. Koutelakis suggests also the alternative of Ployaegos, meaning "many goats," in the southwestern Cyclades (Koutelakis 2008).

20. The argument is presented most comprehensively in Jacoby 2000a.

21. Reinert 1998, 130ff. See also Jacoby 2000a, 35, 65; and Jacoby 2000b, 96n73, 97n75.

22. Nāser-e Khosraw 1986, 15–17; Bianquis 1986, 535–9.

23. Cheynet 2003, 71–108; Ahrweiler 1966, 163–171.

24. Bass et al. 2004.

25. Jacoby 1998, 83–95.

26. Goldberg 2012, 107, and passim.

27. The question of the meaning of the term *Rūm* in Islamic geographical literature has been recently addressed in Durak 2010. According to Durak, the meaning of *bilād al-Rūm* was extended to include all of Christian Europe, mirroring the shift in Islamic self-image from one centered on the Abbasid dynasty to a polycentric world of Islam, *bilād al-Islām*.

28. Goldberg 2012, 306–7, 333.

29. Al-Masʿūdī 1965, 1:151 (no. 305).

30. Lev 2012, 134.

31. Lev 2012, 135–136; Bramoullé 2009a, 258.

32. For accounts of Byzantine-Fatimid warfare, see Lev 1984; Lev 1995; Farag 1990; Bianquis 1986, 208, 236–37, 478ff.; Pryor and Jeffreys 2006, 75ff.

33. Al-Anṭākī 1990, 233–35, 240–42; Maqrīzī 2002, 3:619–21. The incident has received much attention in secondary literature, as it triggered riots against the Amalfitan merchants accused of setting fire to the fleet. See Lev 1984; Jacoby 2000a, 103; al-Tadmurī 1978, 1:295–6.

34. Bramoullé 2009a, 259; Lev 2012, 135–36.

35. On Muslim raids in the eleventh century, see Felix 1981, 202ff.; Cheynet 2003, 97–98, 102–3; Lev 1984, 251; Pryor and Jeffreys 2006, 87; Ahrweiler 1966, 130–34; Jacoby 1998. The Geniza letters dealing with the capture of Jewish merchants from Attaleia and Strobilos are edited and translated in Starr 1939, 186 (no. 128), 190 (no. 132), 191 (no. 133).

36. Al-Muqaddasī 1877, 15 lines 5–8; al-Muqaddasī 1994, 16–17.

37. On the fate of Cretan navy and its commanders after the fall of Crete, see Christides 1984, 184–85. This possibility was suggested to me by Yaacov Lev.

38. Ramaḍān 2009.

39. Ibn Ḥawqal 1938, 197–98. This interpretation is significantly different from Kramers' translation of this passage in Ibn Ḥawqal 1964, 192–93.

40. Koutrakou 1995, 132.

41. Jacoby 2000b, 84.

42. Horden and Purcell 2000, 155.

43. Horden and Purcell 2000, 159, 172 (in relation to the Geniza); see also Goldberg 2012.

44. See articles by Andrea Baubin, Dimitrios G. Letsios, and Christos G. Makrypoulias in al-Hijji and Christides 2002. On maritime law see Khalilieh 2006; Udovitch 1993.

Chapter Eight

1. Qutbuddin 2011, 206–15, 235.

2. Wink 1990, 1: 213.

3. Qutbuddin 2011, 218.

4. See especially Sauvaget 1948, 3–5.

5. *EI2*, art. "al-Ṣīn." Khānfū has altenrativaly been identified with Fuzhou rather than Quanzhou (Leslie 1982, 8).

6. Al-Masʿūdī reports that an Arab merchant visited the Chinese emperor at a capital city called *Anṣū* in the 870s; this must be Chang'an (Xi'an). *EI2*, art., "al-Ṣīn"; al-Masʿūdī 1965, 1:307–12, nos. 336–41. See also Leslie 1982, 6.

7. An active volcano near the island of Zābaj (or Jāba) in the Indian Ocean is frequently mentioned in the early Arab geographic literature (Sauvaget 1948, 10 [no. 20]; Ibn Khurradādhbih 1889, 66; Ibn al-Faqīh 1885, 13; Tibbets 1979, 104–10). Suarez suggests that the volcano described by the Arab geographers may be the monumental volcano of Krakatau, or one of the smaller volcanic Indonesian islands known as the "fire mountains" (Suarez 1999, 52).

8. Ibn Khurradādhbih 1889, 69.

9. Daunicht 1968, 3: 268, 361

10. Suarez 1999, 48.

11. Hansen 2012, 164–65; Park 2012, 56–90.

12. Sen 2003, 176.

13. Hansen 2012, 165.

14. *EI2*, art. "al-Ṣīn"; al-Masʿūdī 1965, 1:308, no. 336.

15. Sen 2003, 167, 181.

16. Silverstein 2007a.

17. Forêt and Kaplony 2008.

18. The most detailed discussion is in *Ḥudūd* 1970, 225.

19. de la Vaissière 2002, 318–323; Sen 2003, 169.

20. The label is wrongly written as a-*k-b-y-t jibāl*, undoubtedly a mistake for *al-Tubbat jibāl*. It is typical of the pervasive copying mistakes in the manuscript.

21. The label is wrongly written Futūḥ instead of *Qannawj*. The Sanskrit name of the city, Kanaakubdja or Kanyakubdja, was rendered by Arab geographers as *Qannawj* or *Qinnawj*. *EI2*, art. "Kanawdj or Kannawdj."

22. *EI2*, art. "Bahāwalpur." We have previously suggested identification with Dāvalpur (or Diplapur), two hundred kilometers east-northeast of Multān (Rapoport and Savage-Smith 2014, 501).

23. Al-Masʿūdī gives the variants *barūzah* and *baʾūzah*, and claims that the name is associated with every king of Kannauj (al-Masʿūdī 1965, no. 412). Another suggestion is that the name refers to the title *barāha*, one of the titular names of the Pratīhāra ruler Mihira Bhoja (r. 836–ca. 888) (Majumdar 1955, x).

24. The Hindu name Thāneswar (meaning "a place of the god") was later corrupted to Thānesar, and in the medieval Islamic sources the name usually appears as *Tānīsar* or *Tānīshar* (*EI2*, art. "Thānesar).

25. For Mahāʾūn in Muslim sources, see Jackson 1999, 131, 134, 143; Habib 1982, 4a (Mahoban). It is also possible to read the label as Mahodaya or Mahodayā ("full of high prosperity"), one of the names associated with the city of Kannauj (Mishra 1977, 38, 70; Majumdar 1955, 29).

26. MacLean 1989, 59–63; *EI2*, art. "Arūr."

27. Ibn Ḥawqal 1938, 2: 327.

28. *Ḥudūd* 1970, 89–90; 123.

29. Al-Bīrūnī 1958, 161, 164. For al-Bīrūnī's itineraries, see also Sircar 1971, 241; Schwartzberg 1992b, plate IV.3 (2).

30. Wink 1990, 1: 213.

31. Wink 1990, 1: 186–87, citing al-Iṣṭakhrī 1870, 173–75.

32. Al-Masʿūdī gives the name as a variant of Khmer, which makes more sense as a source of the material (al-Masʿūdī 1965, 1: 199).

33. *Ḥudūd* 1970: 89–90; 123.

34. Al-Balādhurī 1958, 617–18.

35. *EI2*, art. "Multān" (Y. Friedmann); Wink 1990, 1:187; Majumdar 1955, 125, 402.

36. *EI2*, art. "Multān" (Y. Friedmann); Wink 1990, 1: 214–15; Flood 2009, 155; Daftary 2007, 166 (where the date is given as 347/958).

37. Flood 2009, 42; Stern 1955, 14–16.

38. Flood 2009, 30; Stern 1955.

39. Wink 1990, 1:216; al-Muqaddasī 1877, 485.

40. Wink 1990, 1:216; Flood 2009, 50 (circulation of Egyptian artifacts in Sind, including an Abbasid Dinar minted in Egypt between 842 and 847).

41. Lev 2012, 145

42. Daftary 2007, 167.

43. Wink 1990, 1:216, 218.

44. Flood 2009, 19.

45. Elverskog 2010, 40.

46. Al-Bīrūnī 1958, 157.

47. Wink 1990, 1:288; illustrated in Schwartzberg 1992, plate IV.3 (2).

48. *EI2*, art. "Allāhābād."

49. Maqbul 1960, 99.

50. Al-Masʿūdī describes the kingdom of Lakshmībūr as landlocked kingdom near the seafaring Dahram (Dahrampāla) in Bengal (al-Masʿūdī 1965, 1: 205). See also Sauvaget 1948, 14, 54; Tibbetts 1979: 90.

51. The Arabic sources call this kingdom Dharma (دهرم). See al-Masʿūdī 1965, 1: 205; Sauvaget 1948, 3, 14, 35–36; Wink 1990, 1:255–6.

52. See Dale 2009; Bielenstein 2005, 72–77; Wink 1990, 1: 260, 273–74.

53. Sen 2003, 171–74.

54. Hansen 2012, 187.

55. Sen 2003, 173, Map 5 (Kathmandu—Zongga—Tingri—Shigatse—Gyantse—Lhasa—Nagqu—Zhutuokalun—Chang"an).

56. Elverskog 2010, 92; *Ḥudūd* 1970, 93–94.

57. *Ḥudūd* 1970, 93, 258. Al-Masʿūdī mentions a merchant who, around 875 AD, sailed to India, and then proceeded partly by water and partly by land to Khanfū (al-Masʿūdī 1965, 1: 303).

58. Al-Bīrūnī 1958, 159–60; Schwartzberg 1992, plate IV.3 (2).

59. Akasoy et al. 2011, 13.

60. Akasoy et al. 2011, 40.

61. Compare similar accounts in al-Bakrī 1992, 270 [411]; al-Damīrī 1906, 2:265–66. This account is not cited by al-Jāḥiẓ.

62. Wink 1990, 1: 284–85.

63. On Greek geographical knowledge of India, see Gole 1983, 28–34; Madan 1997; Suarez 1999, 60–77.

64. *EI2*, art. "Mihrān."

65. Al-Masʿūdī 1965, 1: 113.

66. Al-Masʿūdī 1965, 1: 202.

67. Gole 1983, 31.

68. British Library, I.O. 4725 and I.O. 4380; discussed and reproduced in Schwartzberg 1992, 435 and plate 32. The scrolls are twenty meters by twenty-five centimeters and twelve meters by twenty centimeters, respectively.

69. Sen 2003, 181.

Chapter Nine

1. *EI2*, art. "Berberā" (I.M. Lewis).

2. *EI2*, art., "Guardafui."

3. Chittick 1976; Red Sea Pilot 1967, 480.

4. Tibbetts 1971, 423. If the original reading is retained (*Raʾs al-Fīl*), this could also be Capo Elefante, located about forty miles west of of Cape Guardafui. The cape is so-called because of its shape (Red Sea Pilot 1967, 481), but the name is not recorded in medieval Arabic texts.

5. Tibbetts 1971, 423.

6. Injār is mentioned in a late medieval Arabic navigational text (Tibbetts 1971, 422).

7. I am indebted to John Cooper for explaining this point to me.

8. Tibbetts 1971, 423. Modern Maiṭ, also called Maydh on modern maps, is a small village about 4.5 miles east-northeastward of Ras Jilao (Red Sea Pilot 1967, 472).

9. Appears on modern maps as Xiis. Situated on the shore of a small bay about fourteen miles east-northeastward of Ras Shulah, and provides a good anchorage point (Red Sea Pilot 1967, 472; Chittick 1976).

10. Tibbetts 1971, 426.

11. Trimingham 1975, 278; Pradines 2004, 292. Alternatively, *M.l.n.d.a* / *M.l.n.d.s* may be identified with Manda Island, the main port in the Lamu archipelago during this period.

12. Only the final, unconnected "*yn*" are legible.

13. Rapoport and Savage-Smith 2014, 443.

14. Idrīsī 1970, 61; *EI2*, art., "Zandjibār."

15. See Crowther et al. 2014, and the references cited there.

16. *EI2*, art., "Mtambwe Mkuu."

17. Tibbetts 1971, 436; Pradines 2004, 293.

18. *EI2*, art., "Mafia."

19. Horton and Middleton 2000, 56; Pradines 2004, 46, 307–8; Pradines and Blanchard 2005; *EI2*, art., "Kilwa."

20. Pradines 2004, 308.

21. Rapoport and Savage-Smith 2014, 489.

22. Casson 1989, 59 (I owe this reference to Mark Horton).

23. *EI2*, art. "Sofāla."

24. Rapoport and Savage-Smith 2014, 482. Compare Tibbetts 1979, 50, 80; Sauvaget 1948, 3 (no. 4); al-Masʿūdī 1965, 1:179–80 (nos. 366–68); Idrīsī 1970, 69; al-Masʿūdī 1938, 37.

25. Rapoport and Savage-Smith 2014, 519. The illustration of the Wāq Wāq trees in the Bodleian copy is a late medieval addition.

26. *EI2*, art. "Wāḳwāḳ" (G. R. Tibbetts and Shawkat Toorawa). The map of the Indian Ocean has no mention of the islands of al-Qumr, the name given by al-Idrīsī and later geographers to either Madagascar or the Comoros. The island of al-Qumr does appear on the Circular world map in the Bodleian manuscript, but this map may not be part of the original treatise. On the identification of al-Qumr as either the Comoros or Madagascar, see *EI2*, art. "Ḳumr." The Comoros yielded rock crystal remains associated with long distance trade (Pradines 2013). On crystal from East African, see now also Horton et al. 2017.

27. Margariti 2007, 27; al-Muqaddasī 1877, 85.

28. Nāser-e Khosraw 1986, 43.

29. Vallet 2010, 558.

30. Vallet 2010, 402–4; Margariti 2007, 49 (water shipments from Zaylaᶜ to Aden).

31. Goitein and Friedman 2008, 21.

32. Ibn Ḥawqal 1938, 48; cited in Bramoullé 2012, 128.

33. Kawatoko 1993a; Kawatoko 1993b; Power 2008; Power 2009.

34. Margariti 2009; Vallet 2012, 359–60.

35. Nāser-e Khosraw 1986; Cooper 2011.

36. Bramoullé 2012; Nāser-e Khosraw 1986, 43, 65.

37. Al-Masᶜūdī, however, included the Black Sea / the Sea of Azov and the Encompassing Sea in his list of the great seas of the world (al-Masᶜūdī 1894, 50–70). As we have seen, al-Masᶜūdī is major source of inspiration for the author of the *Book of Curiosities.*

38. Rapoport and Savage-Smith 2014, 481.

39. Boussac et al. 2012; Horton and Middleton 2000, 33; Casson 1989.

40. See however the different intepretation of this work in Arnaud 2012, who argues for a scholarly work with predominantly geographical and cartographical concerns.

41. *EI2*, art. "Somali, the name of a people of the Horn of Africa, and Somalia, Somaliland."

42. Horton and Middleton 2000, 33–36.

43. Marcotte 2012, 13; Horton and Middleton 2000, 36; Smith and Wright 1988.

44. Horton and Middleton 2000, 37.

45. Horton and Middleton 2000, 31–33, 75–78.

46. Pradines 2009.

47. Horton and Middleton 2000, 78–80; Horton 1987.

48. On the analysis of the hoard, *EI2*, art., "Mtambwe Mkuu." The locally minted silver coins contained the names of local rulers with rhyming couplets on their reverse (Horton 2007, 75; Horton 1987, 81). Horton further argues that no Persian Gulf coinage has been found in East African sites for the tenth and eleventh centuries (Horton and Middleton 2000, 78); this is disputed in Pradines 2004, 266.

49. Horton 1987, 76; Horton and Middleton 2000, 80.

50. Nāser-e Khosraw 1986, 53.

51. Contadini 1998, 16–18.

52. Pradines 2013.

53. Contadini 1998, 18.

54. Nāser-e Khosraw 1986, 53. Horton also suggests that the high quality Fatimid dinars were made from gold brought from the regions between the Zambezi and Limpopo basins (Horton 1987, 76).

55. Vallet 2010, 557.

56. Pradines 2009, 57, 62. But see Horton and Middleton 2000, 78; Chittick 1974, 2:305.

57. Bramoullé 2012, 132; Vallet 2010, 485; al-Qalaqashandī 1913, 3:524.

58. In the Fatimid period the term refers to convoys traveling to the Indian Ocean from Egyptian ports, not exclusively Muslim (Margariti 2007, 152; Bramoullé 2012, 135).

59. A comparison made in Vallet 2012, 363. Vallet notes that the *Periplus* is far more interested in ports and specific commercial goods.

60. There is also an intriguing numismatic correlation between the miniscule coinage found in Shanga, on the East African coasts, and a series of Fatimid silver coins from Sind, carrying the name of the Fatimid caliphs al-Muʿizz and al-ʿAzīz, acquired by the British Museum in 1980. The two sets of coins from Shanga and Sind share a unqiue and consistent die-alignment, where the obverse and reverse are precisely at 3, 6, 9, or 12 o'clock—a feature that is absent from the rest of the Islamic world at this time (Lowick 1983; Horton and Middleton 2000, 50; Horton 2007, 75).

61. The expansion of the network of the Geniza merchants toward the Indian Ocean only occured in the 1080s, after the consolidation of Ṣulayḥid Ismaʿili rule in the Yemen (Goldberg 2012, 305, 346).

62. Naysābūrī 2011, 55 (no. 59).

63. Naysābūrī 2011, 56 (no. 62).

64. Tibbetts 1979, 77–81.

65. See the insightful remarks in Picard 2011, 43. Picard argues that the Indian Ocean came to be identified with Islamic civilization.

66. Horden and Purcell 2006, 736.

67. More on Indian Ocean islands on this map and more generally in medieval Islam, see Margariti 2012.

Chapter Ten

1. Rapoport and Savage-Smith 2014, 325. This is a reference to the term *al-saqf al-marfūʿ* found in Qur'an 52:5.

2. Silverstein 2010.

3. Tolmacheva 2006; Silverstein 2010; Antrim 2012.

4. Heck 2002, 94–95. For earlier typologies, see Maqbul 1995; Miquel 1967; Hopkins 1990. Heck also adds a fourth approach, a literary one, *adab*—making geography a means of transmitting moral education and cultural refinement.

5. *Kitāb Ṣūrat al-Arḍ*; Tibbetts 1992a, 98.

6. See the insightful remarks by Antrim 2012, 106.

7. Silverstein 2010, 283–84; Antrim 2012, 108; Durak 2010.

8. Rapoport and Savage-Smith 2014, 414.

9. Rapoport and Savage-Smith 2014, 417.

10. Translation and analysis in Langermann 1985; Mercier 1992, 178–81; and King and Samsó 2000.

11. Agapius 1912. A much truncated version is also preserved in the abridgement to Ibn al-Faqīh's (fl. ca. 290/903) *Kitāb al-Buldān* (Ibn al-Faqīh 1885). See Romm 2010 for this type of literature. Zayde Antrim argues that the account of the climes in the *Book of Curiosities* appears to favor the third clime, that of Fatimid North Africa and Egypt, over the fourth clime of Abbasid Iraq, which is usually seen as the most temperate (Antrim 2012, 95–96). This is not entirely accurate, as the inhabitants of the fourth clime are credited with the best constitution, disposition and learning. In any case, the account of the climes is evidently not original to the treatise.

12. See Miquel 1967, 3:288, 294.

13. Silverstein 2010, 274–79; Heck 2002, 107–9. Al-Muqaddasī tries to reconcile geographical knowledge with the statements in Q 55:19, 25:53, and others, concerning the two seas (Silverstein 2010, 279). Nor does our author dawdle with the alternative Persian division of the Earth into *kishvars*, which previous authors often present as an alternative to the division into climes.

14. Rapoport and Savage-Smith 2014, 442.

15. Picard 2011.

16. Antrim 2012, 108.

17. Tibbetts 1992b, 120; Antrim 2012, 117.

18. Al-Masʿūdī 1894, 50–70.

19. *EI2*, art. "Ibn al-Raqīq" (M. Talbi).

20. Bramoullé 2009a; Lev 2012. See also the account of the Fatimid interest in naval affairs in al-Qalqashandī 1913, 3: 523–24.

21. Bramoullé 2009a, 258–59.

22. Bramoullé 2009a, 260.

23. Bramoullé 2009a, 267; See al-Qalqashandī 1913, 3: 523–4.

24. Bramoullé 2009b, 306; Ibn Shaddād 1962, 93.

25. Bramoullé 2009a, 264.

26. Silverstein 2007, 121–25.

27. Al-Munajjid 1958.

28. Miquel 1967, 1:309–12; Maqbul 1995, 110.

29. Troupeau 1954; Maqbul 1995, 117.

30. Daftary 2007, 118, 217–18, and 594n182; al-Nuʿmān 1969–72, 2:74.

31. Daftary 2007, 218; Ibn Ḥawqal 1938, 310_7.

32. For an assessment of Ibn Ḥawqal's relations with the Fatimids, see Garcin 1983.

33. Antrim 2012, 114–15.

34. Scafi 2006.

35. Edson and Savage-Smith 2004; Savage-Smith 2003.

36. In the Damascus copy of the treatise, the fingerlike diagram is missing together with its labels, an omission demonstrating that the shift from a visual format to a textual format was part of the original treatise.

37. Again, the Damascus copy confirms that this transition between image and text was part of the original treatise. It takes up the text of the chapter that corresponds to folio 41b of the Bodleian manuscript, but does not reproduce any of the labels that accompany the diagrams in folios 40a–41a.

38. As Syrinx von Hees argues, the tendency to lump together all works that happen to have the term *ʿajāʾib* in their title makes no historical sense, and artificially creates a genre of *mirabilia* that was not recognized as such by medieval Muslim writers (von Hees 2005).

39. See von Hees 2005; Berlekamp 2011, esp. 22–26; Zadeh 2011.

40. Berlekamp 2011, 22; Von Hees 2005, 113.

41. Berlekamp 2011.

42. Al-Khwārazmī included an account on monumental buildings, both Islamic and pre-Islamic (Miquel 1967, 1:12).

43. Zadeh 2011; Heck 2002, 123.

44. Mirabilia accounts include the aforementioned *Akhbār al-Ṣīn wa'l-Hind*, the continuation of this work by Abū Zayd al-Sīrāfī in the early fourth/tenth century, and Ibn Faḍlān's account of his embassy to the Volga Bulghars in 309/921. See Montgomery 2006; Tolmacheva 2006.

45. Ibn al-Faqīh 1885, 251–55. For attempts at definition of the term in Ibn al-Faqīh's work, see Miquel 1967, 162–79; for general definitions of the term is medieval Arabic literature, see Fahd 1978, 117–35. The Balkhī School geographers also included a section on the special products and unique natural phenomena of each region (Miquel 1967, 1:296, 326).

46. Translation in Freeman-Grenville 1981. See also the discussion in Silverstein 2007a; Shafiq 2013.

47. The work has been published in French by Carra de Vaux as Ibn Waṣīf Shāh 1898; and in Arabic as al-Masʿūdī 1938.

48. Ursula Sezgin has suggested that he be identified with a Baghdadi ophthalmologist called Ibn Waṣīf al-Ṣābiʾ, who was active in Baghdad in the middle of the 4th/10th century. See Sezgin (U.) 2001; *EI2* art. "al-Waṣīfī".

49. Al-Masʿūdī 1965, 1:127 (no. 249): "we have decided not to report in this book the marvels found in waters (*ʿajāʾib al-miyāh*)." See also statements about relegating the *mirabilia* material to his *Akhbār al-Zamān* in al-Masʿūdī 1965 1:138 (no. 274); 1:173 (no. 355).

50. Similarly, the zoological sections in the works of Ibn Qutaybah and al-Tawḥīdī discuss domestic or proverbial species (Ibn Qutaybah 1949, Kopf 1956). For a non-Aristotelian zoology by a Muslim scientist, see Kruk 2001c. .

51. Al-Musabbiḥī 1978, 1:57.

52. Ibn Ḥawqal 1938, 156–57

53. Arabic: *alā ṭarīq al-taʿajjub lā min ṭarīq al-taṣdīq* (al-Masʿūdī 1938, 17).

54. This account and its attribution to ʿAlī was known to al-Maqrīzī, who cites it in connection with the Beja tribes in the Eastern desert of Upper Egypt (al-Maqrīzī 2002, 1:535).

55. *Khuṭbat al-bayān*, attributed to ʿAlī ibn Abī Ṭālib (d. 40/660), is an introduction to alchemy, based on Greek texts, found in the work of Jābir ibn Ḥayyān (Jābir 1928, 22; Sezgin, *GAS* IV:22). See also ʿAlī's account of the *jinn* communities that existed before the creation of Adam, included in al-Masʿūdī's *Akhbār al-Zamān* (al-Masʿūdī 1938, 11).

56. Al-Masʿūdī 1938, 16–17; translated by Miquel 1967, 3:358–59.

57. Greek geographers viewed Thule as unreachable at the far north, and did not provide any information in its inhabitants (Romm 1992, 157–58).

58. The passages in the Alexander Romance which deal with his building of the wall to retain Gog and Magog mention the names of twelve, or sometimes twenty-two, nations which were enclosed by the wall. Lists of these names are reproduced in A. R. Anderson's study (Anderson 1932, 31–36). They do not appear to correspond with the names mentioned here. The Arabic tradition attributed to the enclosed peoples the Biblical names of the sons of Yaphet: Nawil, Taris/Tiras, Minsak (Meshech), Kumara (Gomer). See Anderson 1932, 97; Ibn al-Faqīh 1885, 298–99; al-Masʿūdī 1938, 68–69. On Gog and Magog in the Islamic tradition see now also von Donzel and Schmidt 2010; Doufikar-Aerts 2010; Zadeh 2011.

59. Kruk 2001a; Kruk 1999; Timotheus 1949.

60. They do not, however, match the marvelous Hellenistic birds listed by al-Marwazī. See Kruk 1999, 104–6.

61. We have not been able to identify a Hermetic source for the marvels in the *Book of Curiosities*. They are not in the Hermetic Kyranides, and have almost nothing in common with the influential *Physiologus*. See Toral-Niehoff 2004; van Bladel 2009; Muradyan 2005; Physiologus 2009.

62. See Berlekamp 2011.

Conclusion

1. Rapoport and Savage-Smith 2014, 442.

2. Harley and Woodward 1987, xvi.

3. For recent assessments of the date of composition of the Epistles, and their structure and audience, see *EI Three*, art. "Ikhwān al-Ṣafāʾ" (de Callatäy); de Callatäy 2005.

4. Ikhwān al-Ṣafāʾ 1928, 1:110–131.

5. *EI2*, art. "Mubashshir ibn Fātik" (F. Rosenthal); Ibn Fātik 1958.

6. Al-Maqrīzī 1967–71, 2:294–95; Halm 1997, 77–78.

7. Bora 2014; Walker 1997 [2008], 28–30.

Appendix

1. For a general discussion of the topic, see Beck 2007; also Barton 1994.

2. Ibn Abī Uṣaybiʿah 1882, 2:99.

3. An astronmer (d. 216/832) who carried out astronomical observations as part of the program instigated by the Abbasid caliph al-Maʾmūn. He was a major contibutor to the tables known as the "Tested Maʾmūnic Tables." See *EI2* art. "Marṣad" (J. Samsó); *Dictionary of Scientific Biography* art. "Yaḥyā ibn Abī Manṣūr" (J. Vernet).

4. See Seymore 2001, 244–49 and 130–39 for a discussion of the data in this horoscope. For Latin versions of this same horoscope, see North 1986, 84–87. The values for some of the planetary positions differ in different manuscripts and between those given by Ibn Riḍwān in his commentary on the *Tetrabiblos* of Ptolemy and in his "autobiography" reproduced in the *ʿUyūn al-anbāʾ fī ṭabaqāt al-aṭibbāʾ* of Ibn Abī Uṣaybiʿah. The latter is used here as a source for the values.

5. Beck 2007, 26.

6. King and Samsó 2001; *EI2*, art. "Zīdj" (D. A. King).

7. The horoscope of Ibn Riḍwān included *tasyīr* or prorogation calculations (not shown in fig. Appendix 1), a technique used by Ptolemy for predicting length of life by calculating an arc separating two celestial bodies at a given time and viewed from a particular locality on Earth. It also included a "lot of fortune," another complicated technique but one never mentioned by the author of the *Book of Curiosities*. See Seymore 2001, 168, 246–49.

8. See Beck 2007, 20–26.

9. There are also similarities with geographical localities assigned by al-Bīrūnī in his book on astrology written in the eastern provinces in Persia and what is now Afghanistan. Ptolemy aligns Aries, for example, with Britain, Gaul, Germania, Bastarnia [Moldova], Northern Syria, Palestine, Idumaea [Negev], and Judea, while Bīrūnī gives Babylon, Fars, Palestine, Azerbaijan, and Alān for the zodiacal sign of Aries. See *Tetrabiblos* ii.3 (Ptolemy 1940, p. 157) and Bīrūnī 1934, 365–66.

10. Rapoport and Savage-Smith 2014, 334.

11. For oblique ascension (*al-maṭāliʿ al-baladīyah*), see King 2004a, 37–38; and *EI2* art. "al-Maṭāliʿ" (D. A. King).

12. *Mustaqīmah fī al-ṭulūʿ* and *muʿwajjah fī al-ṭulūʿ*. See Abū Maʿshar 1994, 27; al-Bīrūnī 1934, 229 sect. 378; al-Qabīṣī 2004, 20–21 sect 1[8]; Kūshyār ibn Labbān 1997, I,12[5].

13. Called in Sanskrit *navāṃśas* (*nowbahra*); see *Encyl. Iranica*, art. "Astrology and Astronomy in Iran," sect. "iii. Astrology in Islamic Times" (D. Pingree).

14. See Carboni 1988 for a discussion of the *Kitāb al-Bulhān* and its remarkable images.

15. Al-Bīrūnī 1934, 265 sect. 453. See also al-Qabīṣī 2004, 26–27 sec 1[19]; Ellwell-Sutton 1977, 62.

16. Al-Bīrūnī 1934, 266–67 sect. 455.

17. *EI2* art. "al-Nudjūm, sect. "Planets" (P. Kunitsch).

18. Rapoport and Savage-Smith 2014, 337.

19. Rapoport and Savage-Smith 2014, 300–302 and 337–39.

20. For the counterclockwise direction the term *ʿalá al-istiqāmah* (going in forward motion) is used, while the angular distance in the opposite, clockwise, direction is indicated by the term *maʿkūsan* (reversed). In this context, the directional terms do not refer to the phenomena of apparent retrograde motion that the five planets display in their orbits when viewed from Earth, but rather to the direction of the planetary domicile when calculated from signs of Leo and Cancer.

21. Rapoport and Savage-Smith 2014, 337.

22. A rather similar assignment of roles to the seven planets—but not within the context of explaining their domiciles—is found at the end of the third epistle of the Brethren of Purity, where it is said that the Sun is like the king, the Moon like a vizier or heir to the throne, Mercury a secretary, Venus a servant, Mars chief of the army, Jupiter a judge, and Saturn the head of the treasury. See de Callataÿ 2005, 38; Ikhwan al-Safaʾ 2015a, 79.

23. Al-Bīrūnī 1934, 257 sect. 442.

24. Compare Abū Maʿshar 1995, 2:328; Abū Maʿshar 1994 sect.1.86; al-Bīrūnī 1934, 259 sect. 445.

25. Compare al-Qabīṣī 2004, 2.

26. Al-Bīrūnī 1934, 259 sect. 445.

27. Abū Maʿshar 1995, 2:329–30; al-Qabīsī 2004, 131; al-Bīrūnī 1934, 363 para 451. The alignments for the *adaranjāt* presented by our anonymous author correspond to those for the Indian *darījānāt* listed by al-Bīrūnī, with the exception of three errors that occur for the signs Aries, Taurus, and Leo.

28. For a comparison of the role of decans in Islamic versus European astrological texts, and their allegorical depictions, see Berlekamp 2011, 128–30.

29. Rochberg-Halton 1988, 53–57.

30. Rapoport and Savage-Smith 2014, 339–49; see also al-Qabīṣī 2004, 25; Kūshyār ibn Labbān 1997, 35; al-Bīrūnī 1934, 258, sect. 443; Ellwell-Sutton 1977, 90.

31. Hartner 1938.

32. The author of the *Book of Curiosities*, however, did not include the two "pseudo-planets" of the Dragon's Head and Tail in his chapter on planets.

33. See *Encycl. Iranica* art. "Astrology and Astronomy in Iran" (D. Pingree); Abū Maʿshar 1994, 81–82; Abū Maʿshar 1995, 3:551; al-Bīrūnī 1934, 239 and 255; al-Qabīṣī 2004, 65; Kūshyār ibn Labbān 1997, 214.

34. *Farsakh* is the Arabic form of the Persian unit of measure *farsāng* or *parsāng*. It usually equaled three Arabic miles (*mīl*), when a mile was about 4,000 cubits (*dhirāʿ*); see Mercier 1992.

35. See Swerdlow 1968, Goldstein 1967, Goldstein and Swerdlow 1970.

References

Abū Maʿshar. 1994. *The Abbreviation of "The Introduction to Astrology": Together with the Medieval Latin Translation of Adlard of Bath*. Edited and translated by Charles Burnett, Keiji Yamamoto, and Michio Yano. Islamic Philosophy, Theology and Science: Texts and Studies 15. Leiden: Brill.

———. 1995. *Kitāb al-madkhal al-kabīr ilá ʿilm aḥkām al-nujūm: Liber introductorii maioris ad scientiam judiciorum astrorum*. Edited and translated by R. Lemay. 9 vols. Naples: Istituto Universitario Orientale.

———. 2000. *Abū Maʿshar on Historical Astrology: The Book of Religions and Dynasties (On the Great Conjunctions)*. Edited and translated by Keiji Yamamoto and Charles Burnett. 2 vols. Islamic Philosophy, Theology and Science: Texts and Studies 33–34. Leiden: Brill.

Ackermann, Silke. 2004. "The Path of the Moon Engraved: Lunar Mansions on European and Islamic Scientific Instruments." *Micrologus* 12: 135–64 and figs. 1–14.

Agapius (Mahboub) de Menbidj. 1912. *Kitāb al-ʿunvān*. Edited by A. Vasiliev. In Patrologia Orientalis (Paris, 1910–12), vol. 5: 605–21 (Arabic only).

Agius, Dionisius A. 2001. "The Arab Šalandī." In *Egypt and Syria in the Fatimid, Ayyubid, and Mamluk eras. III: Proceedings of the 6th, 7th and 8th International Colloquium Organized at the Katholieke Universiteit Leuven in May 1997, 1998, and 1999*, edited by U. Vermeulen and J. van Steenbergen, 49–60. Leuven, Belgium: Uitgeverij Peeters.

———. 2007. *Classic Ships of Islam: From Mesopotamia to the Indian Ocean*. Leiden: Brill.

Ahmad, S. Maqbul. 1992. "Cartography of al-Sharīf al-Idrīsī," in Harley & Woodward 1992, 156–74.

Ahrweiler, H. 1966. *Byzance et la mer: La marine de guerre, la politique et les institutions maritimes de Byzance aux VIIe-XVe siècles*. Bibliothèque byzantine. Études, no. 5 Paris: Presses universitaires de France.

Akasoy et al. 2011. *Islam and Tibet: Interactions along the Musk Routes*. Edited by Anna Akasoy, Charles Burnett, and Ronit Yoeli-Tlalim. Farnham, UK: Ashgate.

ʿAlī, ʿAbdullah Yūsuf . 1975. *The Meaning of the Holy Qurʾān*. London: Nadim & Co.

Amari, Michele. 1933. *Storia dei Musulmani di Sicilia*. Edited by Carlo Alfonso Nallino. 2nd rev. ed., 3 vols. Catania: Romeo Prampolini, 1933–39.

And, Metin. 1998. *Minyatürlerle Osmanli-ı Mitologyası*. Istanbul: Akbank.

Anderson, Andrew Runni. 1932. *Alexander's Gate, Gog and Magog, and the Inclosed Nations*. Monographs of the Medieval Academy of America 5. Cambridge, MA: Medieval Academy of America.

al-Anṭākī, Yaḥyā ibn Saʿīd (d. 1065 or 1066). 1990. *Tārīkh al-Anṭākī. Edited by ʿUmar ʿAbd al-Salām al-Tadmurī.* Tripoli, Lebanon: Jarrūs Press.

Antrim, Zayde. 2012. *Routes and Realms: The Power of Place in the Early Islamic World.* Oxford: Oxford University Press.

The Arabian Nights II: Sindbad and Other Popular Stories. 1995. Translated by Husain Haddawy. New York: Norton.

Arnaud, P. 2012. "Le Periplus Maris Erythraei: une œuvre de compilation aux préoccupations géographiques." In Boussac et al. 2012, 27–61.

Arrian. 2003. *Periplus Ponti Euxini.* Edited with introduction, translation, and commentary by Aidan Liddle. London: Bristol Classical.

Avraméa, A. 1997. *Le Péloponnèse du IVe au VIIIe siècle: Changements et persistances.* Paris: Publications de la Sorbonne.

Baer, Eva. 1965. *Sphinxes and Harpies in Medieval Islamic Art: An Iconographical Study.* The Israel Oriental Society, The Hebrew University of Jerusalem, Oriental Notes and Studies 9. Jerusalem: Central Press.

Bagnera, Alessandra. 2013. "From a Small Town to a Capital: The Urban Evolution of Islamic Palermo (9th–Mid-11th Century)." In *A Companion to Medieval Palermo: The History of a Mediterranean city from 600 to 1500*, edited by A. Nef, 61–88. Leiden: Brill.

al-Bakrī, Abū ʿUbayd. 1965. *Description de l'Afrique Septentrionale.* Traduction par Mac Guckin de Slane. Édition revue et corrigée. Paris.

———. 1992. *Al-Masālik wa-al-mamālik.* Edited by A. P. Van Leeuwen and A. Ferre. 2 vols. Tunis: al-Dār al-ʿArabiya lil-Kitāb.

al-Balādhurī, Aḥmad ibn Yaḥyá. 1866. *[Futūḥ al-buldān] Liber expugnationis regionum.* Edited by Michael Jan de Goeje. Leiden: Brill.

———. 1916. *The Origins of the Islamic State, Being a Translation from the Arabic Accompanied with Annotations, Geographic and Historic Notes of the Kitâb futûh al-buldân of al-Imâm abu-1 'Abbâs Ahmad ibn-Jâbir al-Balâdhuri.* Translated by P. K. Hitti and F. C. Murgotten. New York: Columbia University.

Barber, Peter, ed. 2005. *The Map Book.* London: Weidenfeld & Nicolson.

Barrucand, Marianne. 1994. "Architecture et espaces architecturés dans les illustrations des Maqâmât d'al-Harîrî du XIIIe siècle." In *The Art of the Saljuks in Iran and Anatolia*, edited by Robert Hillenbrand, 79–88. Costa Meza, CA: Mazda.

Barton, Tamsyn. 1994. *Ancient Astrology.* London: Routledge.

Bass, George, et al. 2004. *Serçe Limani: An Eleventh-Century Shipwreck.* The Nautical Archaeology Series1, no. 4, The Ship and Its Anchorage, Crew and Passengers. College Station: Texas A&M University Press.

Beck, Roger. 2007. *A Brief History of Ancient Astrology.* Oxford: Blackwell.

Beeston, A. F. L. 1963. *Baiḍāwī's Commentary on Sūrah 12 of the Qurʾān: An Annotated Translation of the Theoretical Chapters.* Princeton, NJ: Princeton University Press.

Berggren, J. Lennart. 1982. "Al-Biruni on Plane Maps of the Sphere." *Journal of the History of Arabic Science* 6: 47–112.

Berggren, J. Lennart, and Alexander Jones. 2000. *Ptolemy's Geography: An Annotated Translation of the Theoretical Chapters.* Princeton, NJ: Princeton University Press.

Berlekamp, Persis. 2011. *Wonder, Image, and Cosmos in Medieval Islam.* New Haven, CT: Yale University Press.

Bianquis, T. 1986–89. *Damas et la Syrie sous la domination fatimide: 359–468/969–1076; Essai d'interprétation des chroniques arabes médiévales.* Damas: Institut français de Damas.

Bielenstein, H. 2005. *Diplomacy and Trade in the Chinese World, 589–1276*. Leiden: Brill.

Bilić, Tomislav. 2012. "Crates of Mallos and Pytheas of Massalia: Examples of Homeric Exegesis in Terms of Mathematical Geography." *Transactions of the American Philological Association* 142: 295–328.

al-Bīrūnī. 1878. *[Kitāb al-Āthār al-bāqiyah] Chronologie orientalischer Völker von Albērūnī.* Edited by Eduard Sachau. Leipzig: Gedruckt auf Kosten der Deutshen Morgenländischen Gesellschaft; reprinted Leipzig: Harrassowitz, 1923.

———. 1879. *The Chronology of Ancient Nations: An English Version of the Arabic Text of the "Athār-ul-bākiya" of Albīrūnī, or "Vestiges of the Past," Collected and Reduced to Writing by the Author in A. H. 390–1, A. D. 1000*. Edited and translated by Eduard Sachau. London: W. H. Allen; reprinted Frankfurt: Minerva, 1969.

———. 1888. *[Kitāb Ta'rīkh al-Hind] Alberuni's India: An Account of the Religion, Philosophy, Literature, Geographhy, Chronology, Astronomy, Customs, Laws and Astrology of India about A.D. 1030*. Edited by Eduard Sachau. 2 vols. London: Trübner; reprinted Delhi: S. Chand, 1964.

———. 1934. *[Kitāb al-Tafhīm li-awā'il ṣināʿat al-tanjīm] The Book of Instruction in the Elements of the Art of Astrology*. Edited and translated by Robert Ramsay Wright. London: Luzac.

———. 1974. *Kitāb al-Tafhīm li-avā'il ṣināʿat al-tanjīm*. Edited by Jalāl al-Dīn Humā'ī. Tehran.

———. 1976. *[Ifrād al-maqāl fī amr al-ẓilāl] The Exhaustive Treatise on Shadows by Abū al-Rayḥān Muḥammad b. Aḥmad al-Bīrūnī.* Translated and with commentary by Edward Stewart Kennedy. 2 vols. Aleppo: Institute for the History of Arabic Science.

Blok, Josine H. 1994. *The Early Amazons: Modern and Ancient Perspectives on a Persistent Myth*. Leiden: Brill.

Bloom, Jonathan M. 2000. "Walled Cities in Islamic North Africa and Egypt with Particular Reference to the Fatimids (909–1171)." In *City Walls: The Urban Enceinte in Global Perspective*, edited by James D. Tracy. Cambridge: Cambridge University Press.

———. 2001. *Paper before Print: The History and Impact of Paper in the Islamic World*. New Haven, CT: Yale University Press.

———. 2007. *Arts of the City Victorious. Islamic Art and Architecture in Fatimid North Africa and Egypt*. New Haven, CT: Yale University Press.

Blue, Lucy, John P. Cooper, Ross Thomas, and Julian Whitewright eds. 2009. *Connected Hinterlands: Proceedings of Red Sea Project IV, Held at the University of Southampton, September 2008*. Society for Arabian Studies Monographs 8, British Archaeological Reports S2052. Oxford: Archaeopress.

Bon, Antoine. 1951. *Le Péloponnèse Byzantine jusqu'en 1204*. Paris: Presses universitaires de France.

Bora, Fozia. 2014. "Did Ṣalāḥ al-Dīn Destroy the Fatimids' Books? An Historiographical Enquiry." *Journal of the Royal Asiatic Society* (July): 1–19.

Bos, Gerrit, and Charles Burnett. 2000. *Scientific Weather Forecasting in the Middle Ages: The Writings of al-Kindī; Studies, Editions, and Translations of the Arabic, Hebrew and Latin Texts*. London: Kegan Paul.

Bouché-Leclercq, Auguste. 1899. *L'astrologie grecque*. Paris: E. Leroux; reprinted Bussels: Culture et civilization, 1963.

Boudet, Jean-Patrice. 2016. "Les comètes dans le Centiloquium et le De cometis du pseudo-Ptolémée." In *Micrologus: Nature, Sciences and Medieval Societies* 24, "The Impact of Arabic Sciences in Europe and Asia": 195–226.

Boudet, Jean-Patrice, Anna Caiozzo, and Nicolas Weill-Parot, eds. 2011. *Images et magie: Picatrix entre Orient et Occident.* Paris: Honoré Champion.

Boussac, M.-Fr., et al. 2012. *Autour de périple de la mer Erythrée.* Textes édités par M.-Fr. Boussac, J.-Fr. Salles, et J.-B. Yon. Lyon: Maison de l'Orient et de la Méditerranée.

Bowersock, Glen Warren. 2006. *Mosaics as History. The Near East from Late Antiquity to Islam.* Cambridge, MA: Harvard University Press.

Bramoullé, David. 2007. "Recruiting Crews in the Fatimid Navy (909–1171)." *Medieval Encounters: Jewish, Christian and Muslim Culture in Confluence and Dialogue*, vol. 13, no. 1: 4–31.

———. 2009a. "Activités navales et infrastructures maritimes: Eléments du pouvoir fatimide en Méditerranée orientale (969–1171)." In *Les ports et la navigation en Méditerranée au moyen âge: Actes du colloque de Lattes*, 12, 13, 14 novembre 2004, musée archéologique Henri Prades. Sous la direction de Ghislaine Fabre, Daniel Le Blévec, Denis Menjot, 257–75. Paris: Manuscrit.

———. 2009b. "Les populations littorales du Bilād al-Šām fatimide et la guerre, IVe/Xe-VIe/XIIe siècle." *Annales islamologiques* 43: 303–34.

———. 2012. "The Fatimids and the Red Sea (969–1171)." In *Navigated Spaces, Connected Places: Proceedings of Red Sea Project V; Held at the University of Exeter, 16–19 September 2010*, 127–36. Edited by Dionisius A. Agius et al. Oxford: Archaeopress.

Brauer, R. W. 1995. *Boundaries and Frontiers in Medieval Muslim Geography.* Transactions of the American Philosophical Society 85, pt. 6. Philadelphia: American Philosophical Society, 1995.

Brett, Michael. 2001. *The Rise of the Fatimids: The World of the Mediterranean and the Middle East in the Fourth Century of the Hijra, Tenth Century CE.* Leiden: Brill.

Brotton, Jerry. 2013. *A History of the World in Twelve Maps.* London: Penguin.

Browne, Gerald M. 1979. *Michigan Coptic Texts.* Papyrological Castroctaviana 7. Barcelona: Papyrological Castroctaviana.

Burnett, Charles. 1987. "Arabic, Greek, and Latin Works on Astrological Magic Attributed to Aristotle." In *Pseudo-Aristotle in the Middle Ages: The Theology and Other Texts*, edited by J. Kraye, W. F. Ryan, and C. B. Schmitt, 84–96. London: Warburg Institute; reprinted in Magic and Divination in the Middle Ages. Ashgate: Variorum, article 3.

———. 2004a. "Lunar Astrology: The Varieties of Texts using Lunar Mansions, with emphasis on Jafar Indus." *Micrologus* 12: 43–133.

———. 2004b. "Weather Forecasting in the Arabic World." In *Magic and Divination in Early Islam*, edited by E. Savage-Smith, 201–10. Aldershot, UK: Ashgate.

Caiozzo, Anna. 2000. "Les talismans des planètes dans les cosmographies en person d'époque médievale." *Der Islam* 77: 221–62.

———. 2003. *Images du ciel d'Orient au Moyen Âge: Une histoire du zodiaque et de ses représentations dans les manuscrits du Proche-Orient musulman.* Paris: Presses de l'Université de Paris-Sorbonne.

———. 2005. "The Horoscope of Iskandar Sultān as a Cosmological Vision in the Islamic World." In *Horoscopes and Public Spheres: Essays on the History of Astrology*, edited by Günther Oestmann, H. Darrel Rutkin, and Kocku von Stuckrad, 115–44. Berlind: Walter de Gruyter.

———. 2009. "Iconography of the Constellations." In *Images of Islamic Science: Illustrated Manuscripts from the Iranian World*, edited by Ziva Vesel, Serge Tourkin, and Yves Porter, 106–33. Tehran: Institut Français de Recherche en Iran.

Campbell, Tony. 1987 [2010]. "Portolan Charts from the Late Thirteenth Century to 1500:

Additions, Corrections, Updates to Volume 1, *The History of Cartography*," 371–463. University of Chicago Press. http://www.maphistory.info/portolanchapter.html. (Accessed December 20, 2014.)

Carboni, Stefano. 1988. *Il Kitāb al-bulhān di Oxford. Quaderni del Dipartimento di Studi Eurasiatici.* Università degli Studi di Venezia 6. Turin: Editrice Tirrenia.

———. 1997. Following the Stars: Images of the Zodiac in Islamic Art. New York: Metropolitan Museum of Art.

Casson, L. 1989. *Periplus Maris Erythraei.* Princeton, NJ: Princeton University Press.

Catalogus Codicum Astrologorum Graecorum (CCAG) (various editors). 1898–1953. 12 vols. in 20 parts. Brussels: Lamertin.

Chaplin, Tracey D., Robin J. H. Clark, Alison McKay, and Sabina Pugh. 2006. "Raman Spectroscopic Analysis of Selected Astronomical and Cartographic Folios from the Early 13th-Century Islamic Book of Curiosities of the Sciences and Marvels for the Eyes." *Journal of Raman Spectroscopy* 37: 865–77.

Cheynet, J-C. 2003. "Basil II and Asia Minor." In *Byzantium in the Year 1000*, edited by P. Magdalino, 71–108. Leiden: Brill.

Chittick, Neville. 1974. *Kilwa: An Islamic trading city on the East African Coast.* With a foreword by Mortimer Wheeler. 2 vols. Nairobi: British Institute in Eastern Africa.

———. 1976. "An Archaeological Reconnaissance in the Horn: The British Somali Expedition, 1975." *Azania* 11: 117–33.

Christides, V. 1982. "Two Parallel Naval Guides of the Tenth Century: Qudama's Document and Leo VI's Naumachica; A Study on Byzantine and Moslem Naval Preparedness." *Graeco-Arabica* 1: 51–103.

———. 1984. *The Conquest of Crete by the Arabs (ca. 824): A Turning Point in the Struggle between Byzantium and Islam.* Athens: Akadēmia Athēnōn.

Christie's. 1996. *Islamic Art and Indian Miniatures and Rugs and Carpets* (sale catalogue). London, October 15–17.

Comes, Mercè. 1996. "The Accession and Recession Theory in al-Andalus and the North of Africa." In *From Baghdad to Barcelona: Studies in the Islamic Exact Sciences in Honour of Prof. Juan Vernet*, edited by Josep Casulleras and Julio Samsó, 2:349–64. Barcelona: Instituto "Millás Vallicrosa" de Historia de la Ciencia Arabe.

———. 2001. "Ibn al-Hāʾim's Trepidation Model." Suhayl 2: 291–408.

Contadini, Anna. 1998. *Fatimid Art at the Victoria and Albert Museum.* London: V&A Publications.

———. 2007. *Arab Painting: Text and Image in Illustrated Arabic Manuscripts.* Edited by Anna Contadini. Leiden: Brill.

———. 2011. *A World of Beasts: A Thirteenth-Century Illustrated Arabic Book on Animals (the Kitāb Na't al-Ḥayawān) in the Ibn Bakhtīshū' Tradition.* Leiden: Brill.

Cook, D. 1999. "A Survey of Muslim Material on Comets and Meteors." *J. Hist. Astronomy* 30: 131–60.

Cooper, John. P. 2009. "Egypt's Nile-Red Sea Canals: Chronology, Location, Seasonality and Function." In Blue et al. 2009, 195–209.

———. 2011. "No Easy Option: Nile versus Red Sea in Ancient and Medieval North-South Navigation." In *Maritime Technology in the Ancient Economy: Ship-Design and Navigation*, edited by W. V. Harris and K. Iara. *Journal of Roman Archaeology Supplementary Series, 189–210*. Portsmouth, RI: Journal of Roman Archaeology.

———. 2012. "'Fear God; Fear the Bogaze': The Nile Mouths and the Navigational Landscape of the Medieval Nile Delta, Egypt." *Al-Masāq* 24: 53–73.

———. 2014. *The Medieval Nile: Route, Navigation, and Landscape in Islamic Egypt.* New York: American University in Cairo Press.

Coulon, Jean-Charles. 2014. "Autour de Ġayat al-ḥakīm (Le but de sage): Compte rendu critique des actes du colloque Images et magie." *Arabica* 61: 89–115.

Daftary, Farhad. 2005. *Ismailis in Medieval Muslim Societies.* London: I. B. Tauris.

———. 2007. *The Ismaʿilis: Their History and Doctrines.* 2nd ed. Cambridge: Cambridge University Press.

Daiber, Hans. 1975. *Ein Kompendium der aristotelischen Meteorologie in der Fassung der Ḥunain ibn Isḥâq.* Amsterdam: North-Holland.

Dale, Stephen F. Dale. 2009. "Silk Road, Cotton Road, Or . . . Indo-Chinese Trade in Pre-European Times." *Modern Asian Studies* 43, no. 1: 79–88.

al-Damīrī, Muḥammad ibn Mūsā. 1906. *Hayat Al-Hayawan: A Zoological Lexicon.* Translated by Ātmārāma Sadāsiva Jayakar. 2 vols. London: Luzac.

———. 1994. *Ḥayāt al-ḥayawān al-kubrá: Wa-yalīhi 'Ajāʾib al-makhlūqāt wa-gharāʾib al-mawjūdāt li-Zakarīyā Muḥammad ibn Maḥmūd al-Qazwīnī.* Qum: Manshūrāt al-Raḍī; Tehran: Manshūrāt-i Nāṣir Khusraw, 1364–1415 [1985 or 1986–1994].

Daniel, Norman. 1975. *The Arabs and Mediaeval Europe.* London: Longman; Beirut: Librairie du Liban.

Daunicht, Hubert. 1968. *Der Osten nach der Endkarte al-Ḫuwārizmīs: Beitrage zur Historischen Geographie und Geschichte Asiens.* Bd. I: Rekonstruktion der Karte, Interpretation der Karte: Südasien. Bonner Orientalistische Studien, Neue Serie 19. 4 vols. Bonn: Selbstverlag des Orientalischen Seminars der Universität.

de Callataÿ, Godefroid. 2000. "Οικουμένη ʿυποθράνιος: Réflexions sur l'origine et le sens de la géographie astrologie." *Geographia Antiqua* 8–9: 25–70.

———. 2005. *Ikhwān al-Safaʾ: A Brotherhood of Idealists on the Fringe of Orthodox Islam.* Oxford: Oneworld.

———. 2013. "Magis en al-Andalus: Rasāʾil Ijwān al-Ṣafāʾ, Rutbat al-Ḥākim y Ġāyat al-Ḥākim (Picatrix)." *al-Qantara* 34: 297–344.

———. 2015. "On Cycles and Revolutions" and "Introduction to Epistle 36." In *Epistles of the Brethren of Purity: Sciences of the Soul and Intellect,* Part I: *An Arabic Critical Edition and English Translation of Epistles* 32–36. Edited and translated by Paul E. Walker, Ismail K. Poonawala, David Simonowitz, and Godefroid de Callataÿ, 137–233. Oxford: Oxford University Press.

———. 2016. "Who Were the Readers of the Rasāʾil Ikhwān al-Ṣafāʾ?" In *Micrologus: Nature, Sciences and Medieval Societies* 24, "The Impact of Arabic Sciences in Europe and Asia": 269–302.

de la Vaissière, Ètienne. 2002. *Histoire des marchands Sogdiens.* Paris: Collège de France, Institut des hautes études chinoises.

Dekker, Elly. 2013. *Illustrating the Phaenomena: Celestial Cartography in Antiquity and the Middle Ages.* Oxford: Oxford University Press.

Der Neue Pauly: Enzyklopädie der Antike. 1996–2003. Edited by Hubert Cancik, Helmuth Schneider, and Manfred Landfester. 19 vols. Stuttgart: J. B. Metzler. Available online as Brill's New Pauly, ed. Hubert Cancik, Helmuth Schneider, and Manfred Landfester with an English tranlation by Christine F. Salazar and Francis G. Gentry at http://referenceworks.brillonline.com.

Dhahabi, Muḥammad ibn Aḥmad. 1963. *Mizān al-Iʿtidāl fī naqd al-rijal.* Edited by ʿAlī Muḥammad al-Bajawī. Cairo: ʿĪsá al-Bābī al-Ḥalabī.

Dictionary of Scientific Biography (DSB). 1970–80. Edited by C. Gillispie. 18 vols. New

York: Charles Scribner's Sons. Available online through Gale Virtual Reference Library.

Djelloul, Neji. 1993. "Histoire topographique de Mahdia et de ses environs au Moyen-Age." *Cahiers de Tunisie* 162–63: 71–108.

Dols, Michael W. Dols, and Adil S. Gamal. 1984. *Medieval Islamic Medicine: Ibn Riḍwān's Treatise "On the Prevention of Bodily Ills in Egypt."* Berkeley: University of California Press.

Doufikar-Aerts, Faustina. 2000. "'Epistola Alexandri ad Aristotelem' Arabica." In *La diffusione dell'eredita classica nell'eta tardoantica e medievale: Filologia, storia, dottrina,* edited by Carmela Baffioni Edizioni, 35–51. Alessandria, Italy: Edizioni dell'Orso.

———. 2010. *Alexander Magnus Arabicus. A Survey of the Alexander Tradition through Seven Centuries; From Pseudo-Callisthenes to Ṣūrī.* Leuven, Belgium: Peeters.

Dozy, R. P. A. 1881. *Supplément aux dictionnaires arabes.* 2 vols, Leiden: Brill.

Drawnel, Henryk Drawnel, ed. 2011. *The Aramaic Astronomical Book from Qumran: Text, Translation and Commentary.* Oxford: Oxford University Press.

Drory, Rina. 2000. "The Maqama." In *The Literatures of al-Andalus,* edited by M. Rosa Menocal, R. P. Scheindlin, and M. Sells, 190–210. Cambridge: Cambridge University Press.

Dubler, C. E. 1954. "ʿAdjāʾib." *EI2.*

Ducène, Jean-Charles. 2004a. "Le delta du Nil dans les cartes d'Ibn Ḥawqal." *Journal of Near Eastern Studies* 63, no. 4: 241–56.

———. 2004b. "Le delta du Nil dans les cartes du Nuzhat al-muštāq d'al-Idrīsī." *Zeitschrift der Deutschen Morgenländischen Gesellschaft* 154, no. 1: 57–67.

———. 2008. "Les extrémités du monde chez les géographes arabes." *Interprétation: Mythes, croyances et images au risque de la réalité;* Roland Tefnin (1945–2006) in memoriam, edited by Michèle Broze, Christian Cannuyer, and Florence Doyen. Brussels: Société Belge d"études Orientales.

———. 2009. "Les coordonnées géographiques de la carte manuscrite d'al-Idrîsî (Paris, Bnf ar 2221)." *Der Islam* 86: 271–85.

———. 2011a. "Du nouveau sur les Amazones dans les sources arabes et persanes médiévales." *Rocznik Orientalistyczny* 64, no. 2.

———. 2011b. "The Nomenclature and Dimensions of the Indian Seas in Medieval Near Eastern Sources." Paper presented: "Mapping the Indian Ocean: Transfers of Knowledge, East and West, from Antiquity to the Sixteenth Century Conference." Bibiliothèque nationale de France, October 28. http://median.hypotheses.org.

———. 2011c. "L'Afrique dans les mappemondes circulaires arabes médiévales: Typologie d"un représentation." In *Cartographier l'Afrique: Construction, transmission et circulation des savoirs géographiques du Moyen Âge au XIX siècle; Cartes et Géomatique; Revue du comité français de cartographie,* edited by Robin Seignobos and Vincent Hiribarren, Décembre, 19–36.

Durak, Koray. 2010. "Who Are the Romans? The Defintion of Bilād al-Rūm (Land of the Romans) in Medieval Islamic Geographies." *Journal of Intercultural Studies* 31, no. 3: 285–98.

Dzhafri, Raziia. 1985. *Geograficheskaia karta mira al-Khorezmi: Po knige "Surat al- arz," vvedenie i interpretatsiia karty S. Raziii Dzhafri, vvodnoe issledovanie I.U. S Maltseva, predislovie i redaktsiia K. S. Aini.* Dushanbe: Izd-vo "Donish."

Edson, Evelyn, and Emilie Savage-Smith. 2000. "An Astrologer's Map: A Relic of Late Antiquity." *Imago Mundi* 52: 7–29

———. 2004. *Medieval Views of the Cosmos.* Oxford: Bodleian Library.

Elverskog, Johan. 2010. *Buddhism and Islam on the Silk Road.* Philadelphia: University of Pennsylvania Press.

Elwell-Sutton, L. P. 1977. *The Horoscope of Asadullāh Mirza.* Nisaba 6. Leiden: Brill.

Encyclopaedia Iranica. 1985–2014 (current). Edited by Ehsan Yarshater. 15+ vols. London: Routledge & Kegan Paul; Costa Mesa, CA: Mazda. Available online at http://www.iranicaonline.org/.

The Encyclopaedia of Islam, Second Edition (EI2). 1960–2002. Edited by P. Bearman, Th. Bianquis, C. E. Bosworth, E. van Donzel, W. P. Heinrichs. 11 vols. Leiden: Brill. Available online at http://referenceworks.brillonline.com/encyclopaedia-of-islam-2.

The Encyclopaedia of Islam, Third Edition (EI Three). Edited by K. Fleet, G. Krämer, D. Matringe, J. Nawas, and E. Rowson. Available online at http://referenceworks.brillonline.com/encyclopaedia-of-islam-3.

Encyclopaedia of the History of Science, Technology, and Medicine in Non-Western Cultures. Edited by Helaine Selin. Dordrecht, Netherlands: Kluwer Academic.

Encyclopaedia of the Qur'ān. 2001–2006. General Editor: Jane Dammen McAuliffe. 6 vols. Leiden: Brill. Available online at http://referenceworks.brillonline.com/encyclopaedia-of-the-quran/.

Evans, James. 1998. *The History and Practice of Ancient Astronomy.* New York: Oxford University Press.

Fahd, Toufic. 1966. *La divination arabe: Etudes religieuses, sociològiques et folkloriques sur le milieu natif de l'Islam.* Leiden: Brill.

———. 1978. "Le merverilleux dans la faune, la flore et les minéraux." In *L'Étrange et le merveilleux dans le islam medieval.* Paris: Editions J. A.

Fahmy, A. M. 1950. *Muslim Sea-Power in the Eastern Mediterranean, from the Seventh to the Tenth century A.D: Studies in Naval Organisation.* London: Tip. Don Bosco.

Farag, W. 1990. "The Aleppo Question: a Byzantine-Fatimid Conflict of Interests in Northern Syria in the Later 10th Century." *Byzantine and Modern Greek Studies* 14: 44–60.

al-Farghānī. 1998. *Kitāb jawāmī' 'ilm al-nujūm wa-uṣūl al-ḥarakāt al-samāwīyah.* Arabic ed. and Turkish trans. by Yavuz Unat. Cambridge, MA: Harvard University Press.

Felix, W. 1981. *Byzanz und die islamische welt in früheren 11. Jahrhundert: Geschichte der politische Beziehungen von 1101 bis 1055.* Vienna: Verlag der Österreichischen Akademie der Wissenschaften.

Fierro, Maribel. 1996. "Bāṭinism in al-Andalus. Maslama b. Qāsim al-Qurṭubī (d. 353/964), author of the Rutbat al-Ḥakīm and the Ghāyat al-Ḥakīm (Picatrix)." *Studia Islamica* 84: 87–112.

Flood, Finbarr Barry. 2009. *Objects of Translation: Material Culture and Medieval "Hindu-Muslim" Encounter.* Princeton, NJ: Princeton University Press.

Fodor, A. 1974. "Malhamat Daniyal." In *The Muslim East: Studies in Honour of Julius Germanus.* Edited by G. Káldy-Nagy. Budapest, 84–133 + 26 pp. Arabic.

Fontaine, R. 2000. "Medieval Jewish Authors on the Inhabited and Unihabited Parts of the Earth." *Arabic Science and Philosophy* 10, no. 1: 101–37.

Forbes, Lesley. 2014. "Intellectual Gold? Oxford's Book of Curiosities and Its Importance for Research on the Middle East and Islamic World." In *Books and Bibliophiles: Studies in Honour of Paul Auchterlonie on the Bio-Bibliography of the Muslim World,* 79–88. Exeter: Gibb Memorial Trust.

Forcada, Miquel. 1998. "Books of anwā' in al-Andalus." In *The Formation of al-Andalus: Part 2; Language, Religion, Culture and the Sciences,* edited by Maribel Fierro and Julio

Samsó, 305–28. The Formation of the Classical Islamic World 47. Aldershot, UK: Ashgate.

———. 2000. "Astrology and Folk Astronomy: The Mukhtaṣar min al-Anwāʾ of Aḥmad b. Fāris." *Suhayl* 1: 107–205.

Forêt, Philippe, and Andreas Kaplony. 2008. *The Journey of Maps and Images on the Silk Road*. Leiden: Brill.

Fortleben im Abendland: Historische Darstellung, Teil I. 2000. Geschichte der arabischen Schrifttums 10. Frankfurt-am-Main: Institut für Geschichte der Arabisch-Islamischen Wissenschaften an der Johann Wolfgang Goethe-Universität.

Foss, Clive. 1988. "Strobilos and Related Sites." *Anatolian Studies* 38: 147–77; reprinted in C. Foss, *History and Archaeology of Byzantine Asia Minor*. Aldershot, UK: Ashgate, Variorum, 1990.

———. 1994. "The Lycian Coast in the Byzantine Age." *Dumbarton Oaks Papers* 48: 1–52; reprinted in C. Foss, *Cities, Fortresses and Villages of Byzantine Asia Minor*. Aldershot, UK: Ashgate Variorum, 1996.

Freeman-Grenville, G. S. P. 1962. *The East African Coast Select Documents from the First to the Earlier Nineteenth Century (compiled with maps)*. Oxford: Clarendon.

———. 1981. *The Book of the Wonders of India: Buzurg ibn Shariyar of Ramhormuz*. London: East-West.

Garcin, J-C. 1983. "Ibn Hawqal, l'Orient et le Maghreb." *Revue de l'Occident Musulman et de la Méditerranée* 35: 77–91.

Gari, Lutfallah. 2008. "About al-Shayzarī and Ibn Bassām: Who Preceded the Other?" *Studies in Islam and the Middle East* 5, no. 1.

Gascoigne, Alison L. 2007. "The Water Supply of Tinnīs: Public Amenities and Private Investment." In *Cities in the Pre-Modern Islamic World: The Urban Impact of Religion, State and Society*, edited by Amira Bennison and Alison L. Gascoigne, 161–76. London: Routledge.

Gaspar, Joaquim Alves. 2008. "Dead Reckoning and Magnetic Declination: Unveiling the Mystery of Portolan Charts." *e-Perimetron* 3, no. 4: 191–203.

Gautier-Dalché, Patrick. 1995. *Carte marine et portulan au xiie siecle: Le liber de existencia riveriarum et forma maris nostri Mediterranei (Pise, circa 1200)*. Rome: École française de Rome.

———. 2002. "Portulans and the Byzantine World." In *Travel in the Byzantine World: Papers from the Thirty-Fourth Spring Symposium of Byzantine Studies, Birmingham, April 2000*, edited by Ruth Macrides. Aldershot, UK: Ashgate Variorum.

———. 2005. *Du Yorkshire à l'Inde: Une "géographie" urbaine et maritime de la fin du XIIe siècle (Roger de Howden?)*. Geneva: Droz.

———. 2009. *La géographie de Ptolémée en Occident (IVe-XVIe siècle)*. Turnhout, Belgium: Brepols.

———. 2011. "Les cartes marines: Origins, caractèrea, usages; Á propos de deux ouvrages récents." *Geographia Antiqua* 20–21: 217–20.

Gertwagen, R. 1996. "Geniza letters: Maritime Difficulties along the Alexandria-Palermo Route." In *Communication in the Jewish Diaspora: The Pre-Modern World*, edited by Sophia Menache, 73–92. Leiden: Brill.

Gimaret, Daniel. 1971. *Le livre de Bilawhar et Būdāsf selon la version arabe ismaélienne*. Hautes études islamiques et orientales d'histoire comparée 3. Geneva: Droz.

Goitein, S. D., and Mordechai Akiva Friedman. 2009. *India Traders of the Middle Ages: Documents from the Cairo Geniza; India Book*. Leiden: Brill.

Goldberg, Jessica. 2012. *Trade and Institutions in the Medieval Mediterranean: The Geniza Merchants and Their Business World.* Cambridge: Cambridge University Press.

Goldstein, Bernard R. 1967. *The Arabic Version of Ptolemy's Planetary Hypotheses.* Transactions of the American Philosophical Society 57, pt. 4. Philadelphia: American Philosophical Society.

———. 1971. *Al-Biṭrūjī: On the Principles of Astronomy: Edition, Translation, and Commentary.* 2 vols. New Haven, CT: Yale University Press.

Goldstein, Bernard R., and Noel Swerdlow. 1970. "Planetary Distances and Sizes in an Anonymous Arabic Treatise Preserved in Bodleian Ms. Marsh 621." *Centaurus* 15: 135–70.

Gole, Susan. 1983. *Indian Maps and Plans from Earliest Times to the Advent of European Surveys.* New Delhi: Manohar.

Golvin, Lucien. 1979. "Mahdiya à l'époque fatimide." *Revue de l'Occident musulman et de la Mediterranée* 27: 75–97.

Gunderson, Lloyd L. 1980. *Alexander's Letter to Aristotle about India.* Meisenheim am Glam: Hain.

Habib, I. 1982. *An Atlas of the Mughal Empire: Political and Economic Maps with Detailed Notes, Bibliography, and Index.* Aligarh: Centre of Advanced Study in History, Aligarh Muslim University; Delhi: Oxford University Press.

Halm, Heinz. 1984 "Der Mann auf dem Esel. Der Aufstand des Abū Yazīd gegen die Fatimiden nach einem Augenzeugenbericht." *Die Welt des Orients* 15: 144–204.

———. 1996. *The Empire of the Mahdi: The Rise of the Fatimids.* Translated from the German by Michael Bonner. Leiden: Brill.

———. 1997. *The Fatimids and Their Traditions of Learning.* London: I. B. Tauris.

Hansen, Valerie. 2012. *The Silk Road: A New History.* Oxford: Oxford University Press.

Harley, J. B., and David Woodward, eds. 1987. *The History of Cartography.* Vol. 1, *Cartography in Prehistoric, Ancient, and Medieval Europe and the Mediterranean.* Chicago: University of Chicago Press.

———, eds. 1992. *History of Cartography.* Vol. 2, *Book 1: Cartography in the Traditional Islamic and South Asian Societies.* Chicago: University of Chicago Press.

———, eds. 1994. *The History of Cartography.* Vol. 2, *Book Two: Cartography in the Traditional East and Southeast Asian Societies.* Chicago: University of Chicago Press.

Hartner, Willy. 1938. "The Pseudo-Planetary Nodes of the Moon's Orbit in Hindu and Islamic Iconography." *Ars Islamica* 5: 121–59.

Hārūn, ʿAbd al-Salām Muḥammad. 1951. *Nawādir al-Makhṭūṭat.* 5 vols. in 2. Cairo: Maṭbaʿat Lajnat al-Taʾlīf wa-al-Tarjamah wa-al-Nashr; reprinted Beirut: Dār al-Jīl, 1991.

al-Hāshimī, ʿAlī ibn Sulaymān. 1981. *The Book of Reasons behind Astronomical Tables (Kitāb fī ʿilāl al-zījāt): A Facsimile Reproduction of the Unique Arabic Text Contained in the Bodleian MS Arch. Seld. A.11.* Translated by F. I. Haddad and E. S. Kennedy, comm. by David Pingree and E. S. Kennedy. Delmar, NY: Scholars' Facsimiles & Reprints.

Heck, P. 2002. *The Construction of Knowledge in Islamic Civilization: Qudāma ibn Jaʿfar and His Kitāb al-Kharāj wa Ṣināʿat al-Kitāba.* Leiden: Brill.

Heinen, Anton M. 1982. *Islamic Cosmology: A Study of as-Suyūṭī's al-Hayʾa as-sanīya fī al-hayʾa as-sunnīya with Critical Edition, Translation, and Commentary.* Beiruter Texte und Studien 27. Beirut: Franz Steiner Verlag.

———. 1987. "An Unknown Treatise by Sanad ibn ʿAlī on the Relative Magnitudes of the Sun, Earth, and Moon." In *From Deferent to Equant: A Volume of Studies in the History of Science in the Ancient and Medieval Near East in Honor of E. S. Kennedy, edited by D. A. King and G. Saliba*, 167–74. Annals of the New York Academy of Sciences 500. New York: New York Academy of Sciences.

al-Hijji, Y., and V. Christides, eds. 2002. *Aspects of Arab Seafaring: An Attempt to Fill in the Gaps of Maritime History.* Athens: Institute for Graeco-Oriental and African Studies and Kuwait Foundation for the Advancement of Science.

Hild, Friedrich. 2000. "Die lykischen Bistümer Kaunos, Panormos und Markiane." In *Λιθοστρωτον: Studien zur byzantinischen Kunst und Geschichte. Festschrift für Marcell Restle*, edited by Birgitt Borkopp and Thomas Steppan, 107–16. Stuttgart: Anton Hiersemann.

Hill, Donald R. 1993. *Islamic Science and Engineering. Islamic Surveys.* Edinburgh: Edinburgh University Press.

Hine, Harry M. 2002. "Seismology and Vulcanology in Antiquity?" In *Science and Mathematics in Ancient Greek Culture*, edited by C. J. Tuplin and T. E. Rihll, 56–75. Oxford: Oxford University Press.

Hippocrates. 1923. *Hippocrates: Volume I.* With an English Translation by W. H. S. Jones. Cambridge, MA: Harvard University Press.

———. 1969. *Kitāb Buqrāṭ fi'l-amrāḍ al-bilādiyya. Hippocrates: On Endemic Diseases (Airs, Waters and Places).* Edited and translated by J. N. Mattock and M. C. Lyons. Arabic Tehnical and Scientific Texts 6. Cambridge: W. Heffer.

Hirschler, Konrad. 2011. *The Written Word in Medieval Arabic Lands: A Social and Cultural History of Reading Practices.* Edinburgh: Edinburgh University Press.

Hoffmann, Eva R. 2000. *Muqarnas* 17: 37–52.

Hogendijk, Jan P. 2007. "A New Look at the Barber's Astrolabe in the Arabian Nights." In *O ye Gentlemen: Arabic Studies on Science and Literary Culture in Honour of Remke Kruk*, 65–76. Leiden: Brill.

Hopkins, J. F. P. 1990. "Geographical and Navigational Literature." In *Religion, Learning and Science in the ʿAbbasid Period*, edited by M. J. L. Young, J. D. Latham, and R. B. Serjeant, 301–27. The Cambridge History of Arabic Literature. Cambridge: Cambridge University Press.

Hopkins, J. F. P., and N. Levtzion. 1981. *Corpus of Early Arabic Sources for West African History.* Fontes Historiae Africanae Series Arabica 4. Cambridge: Cambridge University Press.

Horden, Peregrine, and Nicholas Purcell. 2000. *The Corrupting Sea: A Study of Mediterranean History.* Oxford: Blackwell.

———. 2006. "AHR Forum: The Mediterranean and 'the New Thalassology.'" *American Historical Review* 111: 722–40.

Horton, Mark. 1987. "The Swahili Corridor." Scientific American 257, no. 3: 86–93.

———. 2007. "Artisans, Communities, and Commodities: Medieval Exchanges between North-western India and East Africa." *Ars Orientalis* 34: 62–80.

———. 2015. *Zanzibar and Pemba, the Archaeology of an Indian Ocean Archipelago.* London: British Institute in Eastern Africa.

Horton, Mark, Nicole Boivin, Alison Crowther, Ben Gaskell, Chantal Radimilahy, and Henry Wright. 2017. "East Africa as a Source for Fatimid Rock Crystal: Workshops from Kenya to Madagascar." In *Gemstones in the First Millennium AD: Mines, Trade, Workshops, and Symbolism.* International conference, October 20th–22nd, 2015, eds. Alexandra Hilgner, Susanne Greiff, Dieter Quast, 1–15. Mainz, Germany: Verlag des Römisch-Germanischen Zentalmuseums.

Horton, Mark, and John Middleton. 2000. *The Swahili: The Social Landscape of a Mercantile Society.* Oxford: Blackwell.

Hubner, Wolfgang. 2000. "The Ptolemaic View of the Universe." *Greek, Roman and Byzantine Studies* 41: 59–93.

Ḥudūd al-ʿālam. 1970. *"The Regions of the World": A Persian Geography 372 ah–982ad.* Translated by V. Minorsky. E. J. W. Gibb Memorial Series, n.s. 11. Cambridge: Trustees of the E. J. W. Gibb Memorial. 2nd ed. with preface by V. V. Barthold, edited by C. E. Bosworth. London: Luzac.

Huxley, G. 1976. "A Porphyrogenitan Portulan." *Greek, Roman and Byzantine Studies* 17, no. 3: 295–300.

Ibn Abī Uṣaybiʿah. 1882–84. *ʿUyūn al-anbāʾ fī ṭabaqāt al-aṭibbāʾ*. Edited by A. Müller. 2 vols. Cairo: al-Maṭbaʿah al-Wahbiyya; Königsberg: Selbstverlag.

Ibn al-Faqīh al-Hamadhānī, Aḥmad ibn Muḥammad. 1885. *[Mukhtaṣar Kitāb al-Buldān], Compendium libri kitâb al-buldân,* edited by Michael Jan de Goeje. Bibliotheca Geographicorum Arabicorum 5. Leiden: Brill; reprinted 1967.

Ibn al-Qifṭī. 1903. *Taʾrīkh al-ḥukamāʾ.* Edited by Julius Lippert. Leipzig: Dieterich'sche Verlagsbuchhandlung.

Ibn al-Ukhuwwah. 1938. *Maʿālim al-qurba fī aḥkām al-ḥisba of Ḍiyāʾ al-Dīn Muḥammad ibn Muḥammad al-Qurashī al-Shāfiʿī Known as ibn as-Ukhuwwa.* Translated by Reuben Levy. Gibb Memorial Series, n.s. xii. London: Luzac.

Ibn Bassām, Muḥammad ibn Aḥmad. 1967. *Kitāb Anīs al-Jalīs fī Akhbār Tinnīs.* Edited by Jamāl al-Dīn al-Shayyāl. In *Majallat al-Majmāʿ al-ʿIlmī al-ʿIrāqī* 14: 151–189; reprinted as a booklet, *Anīs al-jalīs fī akhbār Tinnīs.* Cairo: Maktabat al-Thaqāfah al-Dīnīyah, 2000.

Ibn Fātik, Mubashshir Abū al-Wafāʾ. 1958. *Mukhtār al-Ḥikam wa-Maḥāsin al-Kalim.* Edited by ʿAbd al-Raḥmān Badawī. Madrid: al-Maʿhad al-Miṣrī lil-Dirāsāt al-Islāmīyah.

Ibn Ḥawqal. 1873. *Kitāb Ṣūrat al-arḍ. Edited by M. J. de Goeje.* Bibliotheca Geographorum Arabicorum 2. Leiden: Brill.

———. 1938–39. *Kitāb Ṣūrat al-arḍ, Opus geographorum auctore Ibn Ḥauḳal (Kitāb Ṣūrat al-arḍ).* Edited by J. H. Kramers. Bibliotheca Geographorum Arabicorum 2, 2nd ed. 2 vols. Leiden: Brill.

———. 1964. *Ibn Hauqal: Configuration de la terre (Kitāb ṣūrat al-arḍ).* Translated by J. H. Kramers and edited by G. Wiet. 2 vols. Collection UNESCO d'oeuvres représentatives, série arabe. Paris: G.-P. Maisonneuve & Larose.

Ibn Hibintā. 1987. *The Complete Book on Astrology: Al-Mughnī fī aḥkām al-nujūm.* Facsimile ed. Edited by Fuat Sezgin, M. Amawi, A. Iokhosha, and E. Neubauer. 2 vols. Frankfurt am Main: Institut für Geschichte der Arabisch-Islamischen Wissenschaften.

Ibn Kathīr. 1987. *Tafsīr al-Qurʾān al-āẓim.* 4 vols. Beirut: Dār al-Maʿrifah.

Ibn Khaldūn. 1958. *The Muqaddimah: An Introduction to History.* Translated by Franz Rosenthal. 3 vols. Princeton, NJ: Bollingen Foundation.

Ibn Khurradādhbih. 1889. *Kitāb al-Masālik wa-al-mamālik (Kitâb al-Masâlik wa'l-mamâlik) (Liber viarum et regnorum).* Edited by Michael Jan de Goeje. Bibliotheca Geographicorum Arabicorum 6. Leiden: Brill; reprinted 1967.

Ibn al-Mujāwir, Yūsuf ibn Yaʿqūb (d. 1291 or 1292). 1951–54. *Ṣifat bilād al-Yaman wa-Makkah wa-baʿḍ al-Ḥijāz al-musammā Tārīkh al-mustabṣir.* Edited by Oscar Löfgren. 2 vols. Leiden: Brill.

Ibn Qutaybah. 1956. *Kitāb al-anwāʾ.* 2 vols. Hyderabad: Dāʾirat al-Maʿārif.

Ibn Shaddād, Muḥammad ibn ʿAlī. 1962. *Al-Aʿlāq al-khaṭīrah fī dhikr umarāʾ al-Shām wa-al-Jazīrah: Tārīkh Lubnān wa-al-Urdunn wa-Filasṭīn*. Edited by Sāmī al-Dahhān. Damascus: al-Maʿhad al-Faransī lil-Dirāsāt al-ʿArabīyah.

Ibn Waṣīf Shāh, Ibrāhīm. 1898. *L'abrégé des merveilles: Traduit de l'arabe d'aprés les manuscrits de la Bibliothèque nationale de Paris, par Le Bon Carra De Vaux*. Paris: C. Klincksieck.

Ibn Sīdah, ʿAlī ibn Ismāʿīl. 1898–1903. *Kitāb Mukhaṣṣaṣ*. 17 vols. Cairo/Būlāq: al-Maṭbaʿah al-Kubrá al-Amīrīyah.

Idris, Hady Roger. 1962. *La Berbérie orientale sous les Zırıdes, Xe–XIIe siècles*. 2 vols. Paris: Adrien-Maisonneuve.

Idrīs. 1973. *'Uyūn al-akhbār wa-funūn al-āthār fī faḍā'il a'immat al-āthār*. Edited by Muṣṭafā Ghālib. 5 vols. Beirut: Dar al-Andalus.

———. 1985. *Ta'rīkh al-khulafā' al-Fāṭimiyīn bil-Maghrib: Al-qism al-khāṣṣ min kitāb 'Uyūn al-akhbār*. Edited by Muḥammad al-Ya'lāwī. Beirut: Dar al-Gharb al-Islāmī.

al-Idrīsī. 1970–76. *Nuzhat al-mushtāq fī ikhtirāq al-āfāq, Opus geographicum, sive "Liber ad eorum delectationem qui terras peragrare studeant" (Kitāb nuzhat al-mushtaq)*. Edited by Alessio Bombaci, Umberto Rizzitano, Roberto Rubinacci, and Laura Veccia Vaglieri. 9 parts. Naples: Istituto universitario orientale di Napoli and Istituto italiano per il medio ed estremo oriente.

Ikhwān al-Ṣafā'. 1928. *Rasā'il Ikhwān al-Ṣafā' wa-khullān al-wafā'*. Edited by Khayr al-Dīn Zirkilī. Cairo: al-Maktabah al-Tijārīyah al-Kubrá.

———. 2008. *Rasā'il Ikhwān al-Ṣafā'*. 4 vols. Beirut: Dar Sader.

———. 2011. *Epistles of the Brethren of Purity: On Magic I; An Arabic Critical Edition and English Translation of Epistle 52a*. Edited and translated by Godefroid de Callataÿ and Bruno Halflants. Oxford: Oxford University Press.

———. 2013. *Epistles of the Brethren of Purity: On the Natural Sciences; An Arabic Critical Edition and English Translation of Epistles 15–21*. Edited and translated by Carmela Baffioni. Oxford: Oxford University Press.

———. 2015a. *Epistles of the Brethren of Purity: On Astronomia; An Arabic Critical Edition and English Translation of Epistle 3*. Edited and translated by F. Jamil Ragep and Taro Mimura. Oxford: Oxford University Press.

———. 2015b. *Epistles of the Brethren of Purity: Sciences of the Soul and Intellect, Part I; An Arabic Critical Edition and English Translation of Epistles 32–36*. Edited and translated by Paul E. Walker, Ismail K. Poonawala, David Simonowitz, and Godefroid de Callataÿ. Oxford: Oxford University Press.

Isaksen, Leif. 2012. "Ptolemy's Geography and the Birth of GIS." Paper presented at the annual international conference of the Alliance of Digital Humanities Organizations (ADHO), Hamburg. http://www.dh2012.uni-hamburg.de/conference/programme/abstracts/ptolemys-geography-and-the-birth-of-gis.1.html. (Accessed December 12, 2014.)

al-Iṣṭakhrī. 1870. *[Kitāb al-Masālik wa-al-mamālik] Viae regnorum descriptio ditionis moslemicae*. Edited by Michael Jan de Goeje. Bibliotheca Geographorum Arabicorum 1. Leiden: Brill; reprinted 1927, 1967.

Jābir ibn Ḥayyān. 1928. *The Arabic Works of Jâbir ibn Ḥayyân*. Edited with translations into English and critical notes by Eric John Holmyard. Paris: P. Geuthner.

Jackson, Peter. 1999. *The Delhi Sultanate: A Political and Military History*. Cambridge: Cambridge University Press.

Jacoby, D. 1997. "Byzantine Crete in the Navigation and Trade Networks of Venice and Genoa." In *Oriente e occidente tra Medioevo ed éta moderna: Studi in onore di Geo Pista-*

rino, edited by Laura Balletto. Genova: G. Brigati, 517–40; reprinted in idem., *Byzantium*, Latin Romania and the Mediterranean. Aldershot, UK: Ashgate, 2001.

———. 1998. "What Do We Learn about Byzantine Asia Minor from the Documents of the Cairo Genizah." In *Hē Vyzantinē Mikra Asia* 6.-12. ai. [Byzantine Asia Minor, 6th-12th cent.], edited by Stelios Lamprakēs, 83–95. Athens: Ethniko Hidryma Ereunōn, Instituto Vyzantinōn Ereunōn.

———. 2000a. "Byzantine Trade with Egypt from the Mid-Tenth Century to the Fourth Crusade." *Thesaurismata* 30: 102–32; reprinted in idem., *Commercial Exchange across the Mediterranean: Byzantium, the Crusader Levant, Egypt and Italy*. Aldershot, UK: Ashgate Variorum, 2005.

———. 2000b. "Diplomacy, Trade, Shipping and Espionage between Byzantium and Egypt in the Twelfth Century." In *Polypleuros nous: Miscellanea für Peter Schreiner zu seinem 60. Geburtstag*, edited by Cordula Scholz und Georgios Makris. Munich: Saur.

Janos, Damien. 2011. "Moving the Orbs: Astronomy, Physics, and Metaphysics, and the Problem of Celestial Motion according to Ibn Sīnā." *Arabic Science and Philosophy* 21: 165–214.

Jansen, P. Jansen, A. Nef, and C. Picard, et al. 2000. *La Méditerranée entre pays d'Islam et monde latin: Milieu Xe-milieu XIIIe siècle*. Paris: SEDES.

Jenks, Stuart. 1983. "Astrometeorology in the Middle Ages." *ISIS* 74: 185–210.

Johns, Jeremey. 2002. *Arabic Administration in Norman Sicily: The Royal Dīwān. Cambridge Studies in Islamic Civilization*. Cambridge: Cambridge University Press.

———. 2004. "Una nuova fonte per la geografia e la storia della Sicilia nell'XIo secolo: Il Kitāb Gharā'ib al-funūn wa-mulaḥ al-'uyūn." In *Mélanges de l'École française de Rome (Moyen Âge)* 116: 409–49.

Johns, Jeremy, and Emilie Savage-Smith. 2003. "The Book of Curiosities: A Newly Discovered Series of Islamic Maps." *Imago Mundi* 55: 7–24 and plates 1–7.

Juste, David. 2007. *Les Alchandreana Primitifs: Études sur les plus anciens traités astrologiques latins d'origine arabe (Xe siècle)*. Brill's Studies in Intellectual History, 152. Leiden: Brill.

Kahlaoui, T. 2008. "The Depiction of the Mediterranean in Islamic Cartography (11th–16th Centuries): The Images of the Mediterranean from the Bureaucrats to the Sea Captains." PhD diss., University of Pennsylvania.

Kamal, Youssouf. 1926–52. *Monumenta cartographica Africae et Aegypti*. 5 vols. in 16 fascs. Privately printed and distributed; reprinted (much reduced in size) in 6 vols. Series D-3, 1–6. Frankfurt am Main: Institut für Geschichte der Arabisch-Islamischen Wissenschaften, 1987.

Kaplony, A. 2008. "Ist Europe eine Insel? Europa auf der rechteckigen Weltkarte des arabischen 'Book of Curiosities' (Kitāb Ġarā'ib al-funūn)." In *Europa im Weltbild des Mittelalters*, edited by Baumgärtner, Ingrid und Kugler, Hartmu (Hgg.), 143–56. Orbis mediaevalis. Vorstellungswelten des Mittelalters 10. Berlin: Akademie Verlag.

Kawatoko, Mutsuo. 1993a. "On the Tombstones Found at the Bādiᶜ Site, the Rīḥ Island." *Kush* 16: 186–202.

———. 1993b. "Preliminary survey of ᶜAydhāb and Bādiᶜ sites." *Kush* 16: 203–24.

Kemp, Peter. 1976. *The Oxford Companion to Ships and the Sea*. Oxford: Oxford University Press.

Kennedy, E. S. 1957. "Comets in Islamic Astronomy and Astrology." *Journal of Near Eastern Studies* 16: 44–51.

———. 1980. "Astronomical Events from a Persian Astrological Manuscript." *Centaurus* 24: 162–77.

———. 1987. "Suhrāb and the World-map of Ma'mūn." In *From Ancient Omens to Statistical Mechanics: Essays on the Exact Sciences Presented to Asger Aaboe*, edited by J. L. Berggren and B. R. Goldstein, 113–19. *Acta Historica Scientiarum Naturalium et Medicinalium* 39. Copenhagen: University Library.

Kennedy, E. S., and M. H. Kennedy. 1987. *Geographical Coordinates of Localities from Islamic Sources.* Frankfurt-am-Main: Institut für Geschichte der Arabisch-Islamischen Wissenschaften an der Johann Wolfgang Goethe-Universität.

Khalilieh, H. 2006. *Admiralty and Maritime Laws in the Mediterranean Sea (ca. 800–1050): The Kitāb Akriyat al-Sufun vis-á-vis the Nomos Rhodion Nautikos.* Leiden: Brill.

al-Khwārazmī. 1926. *[Kitāb Ṣūrat al-arḍ], Das Kitāb Ṣūrat al-arḍ des Abū Ğa'far Muḥammad ibn Mūsā al-Khuwārizmī.* Edited by Hans von Mžik. Bibliothek arabischer Historiker und Geographen 3. Leipzig: Otto Harrassowitz.

King, David A. 1986. *A Survey of the Scientific Manuscripts in the Egyptian National Library.* American Research Center in Egypt, Catalogs 5. Winona Lake, IN: Eisenbrauns.

———. 1993. "Folk Astronomy in the Service of Religion: The Case of Islam." In *Astronomies and Cultures*, edited by Clive L. N. Ruggles and Nicholas J. Saunders, 124–38. Niwot: University Press of Colorado.

———. 1999. *World-Maps for Finding the Direction and Distance to Mecca: Innovation and Tradition in Islamic Science.* Islamic Philosophy, Theology and Science: Texts and Studies 36. Leiden: Brill; London: al-Furqān Foundation.

———. 2004a. *In Synchrony with the Heavens: Studies in Astronomical Timekeeping and Instrumentation in Medieval Islamic Civilization.* Volume One: *The Call of the Muezzin.* Islamic Philosophy, Theology and Science, Texts and Studies 55. Leiden: Brill.

———. 2004b. "A Hellenistic Astrological Table Deemed Worthy of Being Penned in Gold Ink: The Arabic Tradition of Vettius Valens." In *Studies in the History of the Exact Sciences in Honour of David Pingree*, edited by Charles Burnett, Jan P. Hogendijk, Kim Plofker, and Michio Yano, 666–714. Leiden: Brill.

King, David A., and Julio Samsó. 1984. *Astronomy for Landlubbers and Navigators: The Case of the Islamic Middle Ages.* Série Separatas, Centro de Estudos de História e Cartografia Antiga, Instituto de Investigação Científica Tropical, 164. Lisbon: Instituto de Inventigação Científica Tropical.

———. 1994. *Die Wissenschaft under den ägyptischen Fatimiden.* Hildesheim: Georg Olms Verlag.

———. 2000. "Too Many Cooks . . . A New Account of the Earliest Muslim Geodetic Measurements." *Suhayl* 1: 207–41.

———. 2001. "Astronomical Handbooks and Tables from the Islamic World (750–1900): An Interim Report." *Suhayl* 2: 9–105.

Kline, Naomi Reed. 2001. *Maps of Medieval Thought: The Hereford Paradigm.* Woodbridge, UK: Boydell.

Köhler, Bärbel, 1994. *Die Wissenschaft unter den ägptischen Fatimiden.* Hildesheim, Germany: Georg Olms Verlag.

Komorowska, J. 2004. *Vettius Valens of Antioch: An Intellectual Monography.* Krakow: Ksiegarnia Akademicka.

Kopf, L. 1956. "The Zoological Chapter of the Kitāb al-Imtāʿ wa-ʾl-Muʾānasa of Abū Ḥayyān al-Tauḥīdī (10th Century): Translated from the Arabic and Annotated." *Osiris* 12: 390–466.

Koutelakis, Haris. 2008. *Αιγαίο και Χάρτες με ανατρέπτικη ματία (Αναμοχλεύοντας την Ιστορία του Αιγαίου από την Προϊστορία μέχρι σήμερα)* [*The Aegean and Its Maps from a*

Radical Viewpoint (Unearthing and Reconstructing its History)—Mistakes, Transcription Errors, Considerations, Transformations and Displacements of Locations], with Synopsis in English by Michael Boussios. Athens: Editions Erinni.

Koutrakou, N. 1995. "Diplomacy and Espionage: Their Role in Byzantine Foreign Relations, 8th–10th Centuries." *Graeco-Arabica* 6.

Kramers, J. H. Kramers. 1932. "Le question Balkhī-Iṣṭakhrī-Ibn Ḥawḳal et l'Atlas de l'Islam." *Acta Orientalia* 10: 9–30.

Kruk, Remke. 1999. "On Animals: Excerpts of Aristotle and Ibn Sīnā in Marwazī's Ṭabāʾiʿ al-ḥayawān." In *Aristotle's Animals in the Middle Ages and Renaissance.* Edited by Carlos Steel, Guy Guldemond, and Pieter Beullens, 96–126. Leuven, Belgium: Leuven University Press.

———. 2001a. "Of Rukhs and Rooks, Camels and Castles." *Oriens* 36: 288–98.

———. 2001b. "Timotheus of Gaza's On Animals in the Arabic Tradition." *Le Muséon* 114: 389–421.

———. 2001c. "Ibn abi-l-Ashʿath's Kitāb al-ḥayawān: A Scientific Approach to Anthropology, Dietetics and Zoological Systematics." *ZGAIW* 14: 119–68.

Kulke, Hermann, and Dietmar Rothermund. 1998. *A History of India.* 3rd ed. London: Routledge.

Kunitzsch, Paul. 1959. *Arabische Sternnamen in Europa.* Wiesbaden: Otto Harrassowitz.

———. 1961. *Untersuchungen zur Sternnomenklatur der Araber.* Wiesbaden: Otto Harrassowitz.

———. 1974. *Der Almagest: Die Syntaxis Mathematica des Claudius Ptolemäus in arabisch-latinischer Überlieferung.* Wiesbaden: Harrassowitz.

———. 1981. "Stelle beibenie—al-kawākib al-biyābāniya: Ein Nachtrag." *Zeitschrift der Deutschen Morgenländischen Gesellschaft* 131: 263–67; reprinted P. Kunitzsch, *The Arabs and the Stars.* Variorum Reprints, CS 307, no. xiv. Northampton, UK: Ashgate, 1989.

———. 1983. *Über eine anwāʾ-Traditions mit bischer unbekannten Sternnamen.* Beiträge zur Lexikographie des klassischen Arabisch, 4; Bayerische Akademie der Wissenschaften, phil.-hist. Klasse, Sitzungsberichte, 1983, Heft 5. Munich: Verlag der Bayerischen Akademie der Wissenschaften, in Kommission bei der C. H. Beck'schen Verlagsbuchhandlung.

———. 1993. "The Chapter on the Fixed Stars in Zarādusht's Kitāb al-Mawālīd." *ZGAIW* 8: 241–49.

———. 1994. "ʿAbd al-Malik b. Ḥabīb's Book on the Stars." *ZGAIW* 9: 161–94; reprinted in The Formation of al-Andalus: Part 2; Language, Religion, Culture and the Sciences, edited by Maribel Fierro and Julio Samsó, 161–86. The Formation of the Classical Islamic World 47. Aldershot, UK: Ashgate, 1998.

———. 1997. "ʿAbd al-Malik b. Ḥabīb's Book on the Stars (Conclusion)." *ZGAIW* 11: 179–88.

———. 2001. "Liber de stellis beibeniis: Textus Arabicus et Translatio Latina." In *Hermetis Trismegisti Astrologica et Divinatoria,* edited by G. Bos, C. Burnett, T. Charmasson, P. Kunitzsch, F. Lelli, and P. Lucentini, 9–99. Corpus Christianorvm, Continuatio Mediaeualis 144. Turnhout, Belgium: Brepols.

Kunitzsch, Paul, and Manfred Ullmann. 1992. *Die Plejaden in den Vergleichen der arabischen Dichtung.* Bayerische Akademie der Wissenschaften, phil.-hist. Kl. 4. Munich: Verlag der Bayerischen Akademie der Wissenschaften, in Kommission bei der C. H. Beck'schen Verlagsbuchhandlung.

Kūshyār ibn Labbān. 1997. *Introduction to Astrology.* Edited and translated by Michio Yano. Studia Culturae Islamicae 62. Tokyo: Institute for the Study of Languages and Cultures of Asia and Africa.

Lane, Edward William. 1863–93. *An Arabic-English Lexicon.* 8 pts. London: Williams and Norgate.

Langermann, Tzvi. 1985. "The Book of Bodies and Distances of Ḥabash al-Ḥāsib." *Centaurus* 28: 108–28.

———. 2017. *Ibn al-Haytham's On the Configuration of the World.* London: Routledge. Available in electronic online version

Lanman, Jonathon T. 1987. *On the Origin of Portolan Charts.* Chicago: Newberry Library.

Lelli, Fabrizio. 2001. "Sefer Hermes: Textus Hebraicus et Translatio Anglica." In *Hermetis Trismegisti Astrologica et Divinatoria,* edited by G. Bos, C. Burnett, T. Charmasson, P. Kunitzsch, F. Lelli, and P. Lucentini, 109–37. Corpus Christianorvm, Continuatio Mediaeualis 144. Turnhout, Belgium: Brepols.

Lemay, Richard. 1997. "Astrology in Islam." In *Encyclopaedia of the History of Science, Technology, and Medicine in Non-Western Cultures,* edited by Helaine Selin, 81–83. Dordrecht, Netherlands: Kluwer Academic.

Leslie, D. D. 1982. "Chinese Cities in Arabic and Persian Sources." *Papers for Far Eastern History* 25: 1–38.

Lettinck, Paul. 1999. *Aristotle's Meteorology and Its Reception in the Arab World, With an Edition and Translation of Ibn Suwār's Treatise on Meteorological Phenomena and Ibn Bajja's Commentary on the Meteorology.* Aristoteles Semitico-Latinus 10. Leiden: Brill.

Lev, Yaacov. 1984. "The Fāṭimid Navy, Byzantium and the Mediterranean Sea 909–1036 C.E./297–427 A.H." *Byzantion* 54.

———. 1991. *State and Society in Fatimid Egypt.* Leiden: Brill, 1991.

———. 1995. "The Fatimids and Byzantium, 10th–12th Centuries." *Graeco-Arabica* 6: 190–208.

———. 1999. "Tinnīs: An Industrial Medieval Town." In *L'Égypte fatimide: Son art et son histoire; Actes du colloque organisé à Paris les 28, 29 et 30 mai,* edited by Marianne Barrucand, 83–96. Paris: Presses de l'Université de Paris-Sorbonne.

———. 2012. "A Mediterranean Encounter: The Fatimids and Europe, Tenth to Twelfth Centuries." In *Shipping, Trade and Crusade in the Medieval Mediterranean: Studies in Honour of John Pryor,* edited by Ruthy Gertwagen and Elizabeth Jeffreys. Farnham, UK: Ashgate.

Levey, M. 1966. *The Medical Formulary or Aqrābādhīn of al-Kindī.* Madison: University of Wisconsin Press.

Levtzion, N. 2000. "Arab Geographers, the Nile, and the History of Bilad al-Sudan." In *The Nile: Histories, Culture, Myths,* edited by H. Ehrlich and I. Gershoni, 71–76. Boulder, CO: Lynne Rienner.

Lewicki, T. 1977. "Les voies maritimes de la Méditerranée dans le haut Moyen Age d'après les sources arabes." *Navigazione Mediterranea nell'alto Medioevo* 2: 439–69. Spoleto, Italy: APR.

Lézine, Alexandre. 1965. *Mahdiya. Recherches d'archéologie islamique. Archéogie méditerranéenne.* Paris: Librairie C. Klincksiek for the Centre National de la Recherche Scientifique.

Liber Lunae or Book of the Moon, Being British Library Sloane MS 3826, fols. 84–96v with Supplementary Material from fols. 57–83v, Transcribed, Annotated, and Introduced with a

Contemporary English Version by Don Karr, together with a Facsimile of A. W. Greenup's Edition of Sepher ha-Levanah from Oriental MS 6360 with and English Translation by Calanit Nachshon. 2011. Singapore: Golden Hoard Press.

Liddell, Henry George, and Robert Scott. 1940. *A Greek-English Lexicon: A New Edition, Revised and Augmented Throughout,* edited by Henry Stuart Jones and Roderick McKenzie. Oxford: Clarendon; reprinted 1958.

Lippincott, Kristen, and David Pingree. 1987. "Ibn al-Ḥātim on the Talismans of the Lunar Mansions." *Journal of the Warburg and Courtauld Institute* 50: 57–81.

Lowick, N. M. 1983. *Numismatic Digest* 7: 62–69.

MacLean, Derryl N. 1989. *Religion and Society in Arab Sind.* Leiden: Brill.

Madan, P. L. 1997. *Indian Cartography: A Historical Perspective.* New Delhi: Manohar.

Maddison, F., and E. Savage-Smith. 1997. *Science, Tools, and Magic.* 2 vols. London: Azimuth; Oxford: Oxford University Press.

Mahdi, M. 1984–94. *The Thousand and One Nights (Alf layla wa-layla) from the Earliest Known Sources.* Leiden: Brill.

Majumdar, R. C. Majumdar, ed. 1955. *The History and Culture of the Indian People.* Vol. 4, The Age of Imperial Kanauj. Bombay: Bhartiya Vidya Bhavan.

Malamut, E. 2004. "The Region of Serçe Limanı in Byzantine Times." In *Serçe Limani: An Eleventh-Century Shipwreck,* edited by George Bass et al. The Nautical Archaeology Series, vol. 1, no. 4, *The Ship and Its Anchorage, Crew and Passengers.* College Station: Texas A&M University Press.

Maqbul Ahmad, S. 1960. *India and the Neighbouring Territories in the Kitāb nuzhat al-mushtāq fī ikhtirāq al-āfāq of al-Sharīf al-Idrīsī.* Leiden: Brill.

———. 1992. "Cartography of al-Sharīf al-Idrīsī." In Harley and Woodward 1992, 156–74.

———. 1995. *A History of Arab-Islamic Heography: 9th–16th Century A.D.* With a foreword by Muḥammad ʿAdnān al-Bakhit. Mafraq, Jordan: Al al-Bayt University.

al-Maqrīzī, Taqī al-Dīn Aḥmad ibn ʿAlī ibn ʿAbd al-Qādir. 1967–71. *Ittiʿāẓ al-ḥunafāʾ bi-akhbār al-aʾimmah al-Faṭimīyīn al-khulafāʾ.* Edited by Jamāl al-Dīn al-Shayyāl and Muḥammad Ḥilmī Muḥammad Aḥmad. Cairo: Lajnat Iḥyāʾ al-Turāth al-Islāmīyah.

———. 2002. *al-Mawāʿiẓ wa-al-iʿtibār fī dhikr al-khiṭaṭ wa-al-āthār.* Edited by Ayman Fu'ad Sayyid. 4 vols. (in 5). London: Mu'assasat al-Furqān lil-Turāth al-Islāmī.

Marçais, Georges. 1954. *L'architecture musulmane d'Occident.* Tunisie, Algérie, Moroc, Espagne et Sicile. Paris: Arts et mètiers graphiques.

Marcotte, D. 2012. "Le Périple dans son genre et sa tradition textuelle." In Boussac et al. 2012, 7–25.

Margariti, Roxani. 2007. *Aden and the Indian Ocean Trade: 150 Years in the Life of a Medieval Arabian Port.* Chapel Hill: University of North Carolina Press,.

———. 2009. "Thieves or Sultans? Dahlak and the Rulers and Merchants of Indian Ocean Port Cities, 11th to 13th centuries AD." In Blue et al. 2009, 155–63.

———. 2012. "An Ocean of Islands: Islands, Insularity, and the Historiography of the Indian Ocean." In *The Sea: Thalassography and Historiography,* edited by Peter Miller. Ann Arbor: University of Michigan Press.

Maróth, M. 1980. "Ptolemaic Elements and Geographical Actuality in al-Huwārizmī's Description of Central Asia." *Acta Antiqua Academiae Scientiarum Hungaricae* 28: 317–52; reprinted in From Hecataeus to al-Huwārizmī. Bactrian, Pahlavi, Sogdian, Persian, Sanskrit, Syriac, Arabic, Chinese, Greek and Latin Sources for the History of Pre-Islamic Central Asia, edited by J. Harmata. Budapest: Akadémiai Kiadó, 1984.

al-Marzūqī, Aḥmad ibn Muḥammad. 1914. *Kitāb al-azminah wa-al-amkinah*. 2 vols. Hyderabad: Maṭbaʿat Majlis Dāʾirat al-Maʿārif.

al-Masʿūdī, ʿAlī ibn al-Ḥusayn. 1894. *Kitāb Al-tanbīh wa-al-ishrāf.* Edited Michael Jan de Goeje. Leiden: Brill.

———. 1938. *Akhbār al-zamān*. Edited by ʿAbd Allāh al-Ṣāwī. Cairo: ʿAbd al-Ḥamid Aḥmad al-Ḥanafi.

———. 1962–97. *Murūj al-dhahab wa-maʿādin al-jawhar.* French: *Les prairies d'or; traduction française de Barbier de Meynard et Pavet de Courteille; revue et corrigée par Charles Pellat*. 5 vols. Paris: Société Asiatique.

———. 1965–79. *[Kitāb Murūj al-dhahab wa-maʿādim al-jawhar] Les prairies d'or.* Edited by C. Barbier de Meynard and Pavet de Courteill. Rev. ed. by Charles Pellat. Qism al-Dirāsāt al-Taʾrīkhīyah 10. 7 vols. Beirut: Manshūrāt al-Jāmiʿah al-Lubnānīyah.

McGeer, E. 1995. *Sowing the Dragon's Teeth: Byzantine Warfare in the Tenth Century.* Washington, DC: Dumbarton Oaks Research Library and Collection.

Mercier, Raymond P. 1992. "Geodesy." In Harley and Woodward 1992, 175–88.

———. 1996. "Accession and Recession: Reconstruction of the Parameters." In *From Baghdad to Barcelona: Studies in the Islamic Exact Sciences in Honour of Prof. Juan Vernet*, edited by Josep Casulleras and Julio Samsó, 1:299–348. Anuari de filologia (Universitat de Barcelona) 19 B2. 2 vols. Barcelona: Instituto "Millás Vallicrosa" de Historia de la Ciencia Arabe.

Metcalfe, Alex. 2009. *The Muslims of Medieval Italy.* Edinburgh: Edinburgh University Press.

Micheau, Françoise. 2008. "Baghdad in the Abbasid Sea: A Compolitan and Multi-Confessional Capital." In *The City in the Islamic World*, edited by Salma K. Jayyusi, Renata Holod, et al., 221–45. Brill: Boston.

Michot, Yahya J. 2000a. "Ibn Taymiyya on Astrology: Annotated Translation of Three Fatwas." *Journal of Islamic Studies* 11: 147–208; reprinted with Addenda et corrigenda in Savage-Smith 2004, 277–340.

———. 2000b. "Variétés intellectuelles . . . L'impasse des rationalismes selon le Rejet de la Contradiction d'Ibn Taymiyyah." In *Religion versus Science in Islam: A Medieval and Modern Debate*, edited by Carmela Baffioni. Oriente Moderno 19, no. 3.

Miller, Konrad. 1928. *Weltkarte des Arabers Idrisi vom Jahre 1154*. Stuttgart: Brockhaus/Antiquarium.

Miller, Naomi. 2003. *Mapping the City: The Language and Culture of Cartography in the Renaissance.* London: Continuum.

Miquel, André. 1967–80. *Le Géographie humaine du monde musulman jusqu' au milieu de 11e siècle.* Centre de Recherches Historiques. Civilisations et Sociétés 7. 3 vols. Paris: Moutan & Co.

Mishra, S. M. 1977. *Yasovarman of Kanauj.* New Delhi: Abhinav.

al-Muqaddasī. 1877. *[Aḥsan al-taqāsīm fī maʿrifat al-aqālīm], Descripto imperii moslemici.* Edited by Michael Jan de Goeje. Bibliotheca Geographorum Arabicorum 3. Leiden: Brill; reprinted 1906.

———. 1994. *The Best Divisions for Knowledge of the Regions: A Translation of Ahsan al-Taqasim fi Maʿrifat al-Aqalim.* Translated by Basil Anthony Collins, reviewed by Muhammad Hamid al-Tai. Reading, UK: Garnet Publishing and The Centre for Muslim Contribution to Civilisation.

Montgomery, James. 2006. "Spectral Armies, Snakes, and a Giant from Gog and Magog:

Ibn Faḍlān as Eyewitness Among the Volga Bulghārs." *Medieval History Journal* 9: 63–87.

al-Munajjid, Ṣalāḥ al-Dīn. 1958. "Qiṭʿah min kitāb mafqūd: al-masālik wa'l-mamālik li'l-Muhallabī." *Majallat Maʿhad al-Makhṭūṭāt al-ʿArabiyya* 4: 43–72.

Muradyan, Gohar. 2005. *Physiologus: The Greek and Armenian Versions with a Study of Translation Technique.* Leuven, Belgium: Peeters.

al-Musabbiḥī, al-Amīr al-Mukhtār ʿIzz al-Mulk Muḥammad ibn ʿUbayd Allāh ibn Aḥmad. 1978–84. *al-Juzʾ al-arbaʿūn min Akhbār Miṣr.* Edited by Ayman Fuʾād Sayyid and Thiery Bianquis. 2 vols. Cairo: al-Maʿhad al-ʿIlmī al-Faransī lil-Āthār al-Sharqīyah.

Mžik, Hans von. 1916. *Africa nach der arabischen Bearbeitung der Γεωγραφικὴ ʼυφήγησις des Claudius Ptolemaeus von Muḥammad ibn Mūsā al-Ĥuwārizmi.* Kaiserliche Akademie der Wissenschaften in Wien, phil.-hist. Klasse 59, no. 4. Vienna: Oesterreichische Akademie der Wissenschaften.

Nallino, C. A. 1939–48. "Al-Huwârismî e il suo rifacimento della Geografia di Tolomeo." In *Raccolta di scritti editi e inediti*, edited by Maria Nallino, 5:458–532. 6 vols. Rome: Istituto per l'Oriente.

Nāser-e Khosraw. 1986. *Nāser-e Khosraw's Book of Travels (Safarnāma).* Translated from Persian, with introduction and annotation by W. M. Thackston, Jr. Albany, NY: Bibliotheca Persica.

Nash, Harriet, and Dionisius A. Agius. 2011. "The Use of Stars in Agriculture in Oman." *Journal of Semitic Studies* 56: 167–87.

Naysābūrī, Aḥmad ibn Ibrāhīm. 2011. *A Code of Conduct: A Treatise On the Etiquette of the Fatimid Ismaili Mission: A Critical Edition of the Arabic Text and English Translation of Aḥmad B. Ibrāhīm Al-Naysābūrī's Al-Risāla Al-mūjaza Al-kāfiya Fī Ādāb Al-du'āt.* Edited by Verena Klemm, Paul Ernest Walker, and Susanne Karam. London ; New York: I. B. Tauris.

Nef, Annliese, ed. 2013. *A Companion to Medieval Palermo: The History of a Mediterranean City from 600 to 1500.* Leiden: Brill.

Netton, Ian Richard. 1989. *Allāh Transcendent: Studies the Structure and Semiotics of Islamic Philosophy, Theology and Cosmology.* London: Routledge.

Neugebauer, Otto. 1962. "Thābit ben Qurra 'On the Solar Year' and 'On the Motion of the Eighth Sphere.'" *Proceedings of the American Philosophical Society* 106, no. 3: 290–99.

Neuhäuser, R., and P. Kunitzsch. 2014. "A Transient Event in AD 775 Reported by al-Ṭabarī: A Bolide—Not a Nova, Supernova, or Kilonova." *Astronomische Nachrichten* 88: 789–800. Published online.

New Dictionary of Scientific Biography (New DSB). 2008 +. Detroit: Charles Scribner's Sons. [Ongoing publication that includes supplemental material and postscripts updating material in the original edition as well as entirely new entries. Available online through Gale Virtual Reference Library.]

Nicolai, R. 2015. "The Premedieval Origin of Portolan Charts: New Geodetic Evidence." *Isis* 106, no. 3: 517–43.

Noble, Joseph V., and Derek J. de Solla Price. 1968. "The Water Clock in the Tower of the Winds." *American Journal of Archaeology* 72: 345–55.

Nordenskiöld, A. E. 1897. *Periplus: An Essay on the Early History of Charts and Sailing-Directions.* Translated from the Swedish by Francis A. Bather. Stockholm: P. A. Norstedt & Söner.

Norris, Harry Thirlwall. 1972. *Saharan Myth and Saga.* Oxford: Clarendon.

North, J. D. 1986. *Horoscopes and History.* Warburg Institute Surveys and Texts 13. London: Warburg Institute.

al-Nuʿmān ibn Muḥammad, Abū Ḥanīfah al-Qāḍī. 1969–72. *Taʾwīl al-Daʿāʾim.* 3 vols. Cairo: Dār al-Maʿārif.

———. 2006. *Founding the Fatimid State: The Rise of an Early Islamic Empire; An Annotated English Translation of al-Qāḍī al-Nuʿmān's Iftitāḥ al-Daʿwah.* Translated by Hamid Haji. London: I. B. Tauris in association with the Institute of Ismaili Studies.

Obrist, Barbara. 1997. "Wind Diagrams and Medieval Cosmology." *Speculum* 72: 33–84.

Oesterle, Jenny. 2013. "Missionaries as Cultural Brokers at the Fatimid Court in Cairo." In *Cultural Brokers at Mediterranean Courts in the Middle Ages,* edited by Marc von der Höh, Nikolas Jaspert, and Jenny Rahel Oesterle. Munich: Wilhelm Fink.

Papaconstantinou, Arietta Papaconstantinou. 2007. "They Shall Speak the Arabic Language and Take Pride in It: Reconsidering the Fate of Coptic after the Arab Conquest." *Le Muséon* 120: 273–99.

Park, Hyunhee. 2012. *Mapping the Chinese and Islamic Worlds: Cross-Cultural Exchange in Pre-Modern Asia.* Cambridge: Cambridge University Press.

Parker, Kenneth S. 2013. "Coptic Language and Identity in Ayyūbid Egypt." *Al-Masāq* 23: 222–39.

Pasachoff, Jay M. 2000. *A Field Guide to the Stars and Planets.* 4th edition. The Peterson Field Guide Series. Boston: Houghton Mifflin.

Paulys Realencyclopädie der classischen Altertumswissenschaft, Neue Bearbeitung, Bd. 8A,2 (P. Vergilius Maro bis Windeleia). 1958. Started by Georg Wissowa, continued by W. Kroll and K. Mittelhaus. Stuttgart: Alfred Druckenmüller Verlag.

Pedersen, Olaf. 2011. *A Survey of the Almagest.* Rev. ed. with annotation and new commentary by Alexander Jones. New York: Springer.

Pellat, Charles. 1955–56. "Dictons rimés, *anwāʾ* et mansions lunaires chez les arabes." *Arabica* 2: 17–41.

———. 1986. *Cinq calendriers égyptiens.* Cairo: Institut Français d'Archéologie Orientale du Caire.

Pezzini, Elena. 2013. "Palermo in the 12th Century: Transformation in *forma urbis.*" In *Nef* 2013, 195–232.

Physiologus. 2009. Translated by Michael J. Curley. Chicago: University of Chicago Press.

Picard, Christophe. 1997. *La mer et les musulmans d'occident au Moyen Âge, VIIIe-XIIIe siècle.* Paris: Presses Universitaires de France.

———. 2009. "Le port 'construit' sur let littoraux du monde musulman méditerranéen et atlantique (VIIIe—Xve siècles) d'après les sources arabes." In *Les ports et la navigation en Méditerranée au moyen âge: Actes du colloque de Lattes, 12, 13, 14 novembre 2004, musée archéologique Henri Prades,* edited by Ghislaine Fabre, Daniel Le Blévec, and Denis Menjot, 217–28. Paris: Manuscrit.

———. 2011. "Espaces maritimes et polycentrisme dans l'Islam abbasside." *Annales Islamologiques* 45: 23–46.

Picatrix: Ghayat al Hakim—The Goal of the Wise, Vol. 1. 2000. Translated from the Arabic by Hashem Atallah, edited by William Kiesel. Seattle: Ouroboros Press.

Picatrix: Ghayat al Hakim—The Goal of the Wise, Vol. 2. 2008. Translated from the Arabic by Hashem Atallah and Geylan Holmquest, edited by William Kiesel. Seattle: Ouroboros Press.

Pingree, David. 1963. "Astronomy and Astrology in India and Iran." *Isis* 54: 229–46.

———. 1970. "The Fragments of the Works of al-Fazārī." *Journal of Near Eastern Studies* 29: 103–23.

———. 1972. "Precession and Trepidation in Indian Astronomy before AD 1200." *Journal for the History of Astronomy* 3: 27–35.

———. 1976. *Dorothei Sidonii Carmen Astrologicum.* Leipzig: Teubner.

———. 1980. "Some of the Sources of the Ghayat al-hakim." *Journal of the Warburg and Courtauld Institute* 43: 1–15.

———. 1990. "Astrology." In *Religion, Learning and Science in the Abbasid Period,* edited by M. J. L. Young, J. D. Latham, and R. B. Serjeant, 290–300. Cambridge: Cambridge University Press.

———. 1997. *From Astral Omens to Astrology: From Babylon to Bīkānēr.* Rome: Istituto Italiano per l'Africa et l'Oriente.

Pinna, Margherita. 1996. *Il Mediterraneo e la Sardegna nella Cartografia Musulmana (dall'VIII al XVI secolo).* 2 vols. Nuoro, Italy: Regione Autonoma della Sardegna, Istituto Superiore Regionale Etnografico.

Pinto, Karen. 2016. *Medieval Islamic Maps: An Exploration.* Chicago: University of Chicago Press.

Pormann, P. E. Pormann, ed. 2010. *Islamic Medical and Scientific Tradition.* 4 vols. London: Taylor & Francis.

Power, Tim. 2008. "The Origin and Development of the Sudanese Ports (ʿAydhāb, Bādiʿ, Sawākin) in the Early Islamic Period." *Arabian Humanities* 15: 92–110.

———. 2009. "The Expansion of Muslim Commerce in the Red Sea Basin, c. AD 833—969." In Blue et al. 2009, 111–18.

Pradines, S. 2004. *Fortifications et urbanisation en Afrique orientale.* Cambridge Monographs in African Archaeology 58. Oxford: Archaeopress.

———. 2009. "L'île de Sanjé ya Kati (Kilwa, Tanzanie): Un mythe Shirâzi bien réel." *Azania: Archaeological Research in Africa* 44, no. 1: 49–73.

———. 2013. "The Rock Crystal of Dembeni, Mayotte Mission Report 2013." *Nyame Akuma* 80: 59–72.

Pradines, S., and P. Blanchard. 2005. "Kilwa al-Mulûk: Premier bilan des travaux de conservation-restauration et des fouilles archéologiques dans la baie de Kilwa, Tanzanie." *Annales islamologiques* 39: 25–80.

Pryor, John H. 1988. *Geography, Technology and War: Studies in the Maritime History of the Mediterranean,* 649–1571. Cambridge: Cambridge University Press.

———. 1994. "The Voyages of Saewulf." In *Peregrinationes tres: Saewulf, Johannes Wirziburgensis, Theodericus,* edited by R. B. C. Huygens, 35–57. Turnhout, Belgium: Brepols.

———. 2003. "Byzantium and the Sea (900–1025)." In *War at Sea in the Middle Ages and Renaissance,* edited by John B. Hattendorf and Richard W. Unger, 83–104. Woodbridge, CT: Boydell.

Pryor, John H., and Elizabeth M. Jeffreys. 2006. *The Age of the Dromon: The Byzantine Navy ca. 500–1204.* With an appendix translated from the Arabic of Muḥammad Ibn Mankali by Ahmad Shboul. Leiden: Brill.

Ptolemy. 1940. *Tetrabiblos.* Edited and translated by F. E. Robbins. Loeb Classical Library. Cambridge, MA: Harvard University Press; reprinted 1980.

———. 1984. Ptolemy's "Almagest." Translated by G. J. Toomer. London: Duckworth.

Pujades, Ramon J. I Battaller. 2007. *Les cartes portolanes: La representació medieval d'una mar solcada.* With an English version of the text entitled "Portolan charts: The medieval representation of a ploughed sea," pp. 401–526. Barcelona: Institut Cartogràfic de Catalunya; Institut d'Estudis Catalans; Institut Europeu de la Mediterrània; Lunwerg.

al-Qabīṣī, Abū al-Ṣaqr ʿAbd al-Azīz ibn ʿUthmān. 2004. *Al-Qabīṣī (Alcabitius): The Introduction to Astrology; Editions of the Arabic and Latin texts and an English translation.* Edited by Charles Burnett, Keiji Yamamoto, and Michio Yano. Warburg Institute Studies and Texts 2. London: Warburg Institute.

Qaddūri (Kaddouri), Samīr. 2005. "Aḥmad ibn Muḥammad al-Yaḥṣabī al-Qurṭubī (al-qarn 12/6) wa-kitābuhu al-tabyīn fī maʿrifat dukhūl al-shuhūr wa-al-sinīn." *Suhayl* 5 (Arabic section): 69–104.

al-Qalqashandī. 1913–22. *Ṣubḥ al-aʿshā.* 14 vols. Cairo: Dār al-Kutub.

al-Qazwīnī, Zakariyā ibn Muḥammad. 1960. *Āthār al-bilād wa-akhbār al-ʿibād.* Beirut: Dār Ṣādir.

———. 1977. *ʿAjā'ib al-makhlūqāt wa-gharā'ib al-mawjūdāt.* Edited by Fārūq Sa'd. Beirut: Dār al-Āfāq al-Jadīdah.

Qudāmah (Qudāma ibn Jaʿfar al-Baghdādī). 1889. *[Kitāb al-Kharāj] Kitâb al-Kharâdj.* Edited by Michael Jan de Goeje. Bibliotheca Geographorum Arabicorum 6. Leiden: Brill; reprinted 1967.

———. 1981. *Al-Kharāj wa-ṣināʿat al-kitāba.* Edited by Muḥammad Ḥusayn al-Zubaydī. Baghdad: Wizārat al-Thaqāfa wa-al-Iʿlām; Dār al-Rashīd.

al-Qummī, Abū Naṣr Ḥasan ibn ʿAlī. 1997. *Tarjame-ye al-madhal elā ʿelm-e ahkām-al-nojum.* Edited by Jalil Azavan Zanjānī. Iran [n.p.]: Elmi va Farhangi Publishing Co.

Qutbuddin, Tahera. 2005. *Al-Mu'ayyad al Shirazi and Fatimid Da'wa Poetry: A Case of Commitment in Classical Arabic Literature.* 412 pp., vol. 57 in the series Islamic History and Civilization, edited by Wadad Qadi and Rotraud Wielandt. Leiden: Brill.

———. 2011. "Fatimid Aspirations of Conquest and Doctrinal Underpinnings in the Poetry of al-Qā'im bi-Amr Allāh, Ibn Hānī' al-Andalusī, Amīr Tamīm b. al-Mu'izz, and al-Mu'ayyad al-Shīrazī." In *Poetry and History: The Value of Poetry in Reconstructing Arab History,* edited by Ramzi Baalbaki, Saleh Said Agha, and Tarif Khalidi, 195–246. Beirut: American University of Beirut Press.

———. 2013. *A Treasury of Virtues: Sayings, Sermons, and Teachings of ʿAlī.* Facing-page edition and translation of the *Dustur ma'alim al-hikam* compiled by al-Qāḍī al-Quḍāʿī, with the *Mi'at kalimah (One Hundred Proverbs)* attributed to the compilation of al-Jāḥiẓ. New York: New York University Press, in the series Library of Arabic Literature.

Rabbat, Nasser O. 1995. *The Citadel of Cairo: A New Interpretation of Royal Mamluk Architecture.* Leiden: Brill.

Rada, W. S. 1999–2000. "A Catalogue of Medieval Arabic and Islamic Observations of Comets during the Period AD 700–1600." *ZGAIW* 13: 72–91.

Ragep, F. Jamil. 1996. "Al-Battānī, Cosmology, and the Early History of Trepidation in Islam." In *From Baghdad to Barcelona: Studies in the Islamic Exact Sciences in Honour of Prof. Juan Vernet,* edited by Josep Casulleras and Julio Samsó. Anuari de filologia (Universitat de Barcelona) 19 B2, 1:267–98. 2 vols. Barcelona: Instituto "Millás Vallicrosa" de Historia de la Ciencia Arabe.

Ramaḍān, ʿAbd al-ʿAzīz. 2009. "The Treatment of Arab Prisoners of War in Byzantium, 9th- 10th centuries." *Annales islamologiques* 43: 155–94.

Rapoport, Y. 2008. "The Book of Curiosities: A Medieval Islamic View of the East." In *The Journey of Maps and Images on the Silk Road,* edited by Philippe Forêt and Andreas Kaplony. Brill's Inner Asian Library 21, pp. 151–71. Leiden: Brill.

———. 2010. "The View from the South: The Maps of the Book of Curiosities and the Commercial Revolution of the Eleventh Century." In *Histories of the Middle East: Studies in Middle Eastern Society, Economy, and Law in Honor of A.L. Udovitch*, edited by R. Margariti, A. Sabra, and P. Sijpesteijn. Leiden: Brill.

———. 2012. "Reflections of Fatimid Power in the Maps of Island Cities in the 'Book of Curiosities.'" In *Herrschaft verorten: Politische Kartographie des Mittelalters und der frühen Neuzeit*, edited by Martina Stercken and Ingrid Baumgärtner (Medienwandel—Medienwechsel—Medienwissen), 183–210. Zürich: Chronos.

Rapoport, Y., and E. Savage-Smith. 2004. "Medieval Islamic View of the Cosmos: The Newly Discovered Book of Curiosities." *Cartographic Journal* 41: 253–59.

———. 2008. "The Book of Curiosities and a Unique Map of the World." In *Cartography in Antiquity and the Middle Ages: Fresh Perspectives, New Methods*, edited by Richard Talbert and Richard Unger, 121–38 and plates 4–6. Technology and Change in History 10. Leiden: Brill.

———. 2014. *An Eleventh-Century Egyptian Guide to the Universe: The "Book of Curiosities."* Edited with an annotated translation by Yossef Rapoport and Emilie Savage-Smith. Islamic Philosophy, Theology and Science, Texts and Studies 87. Leiden: Brill.

Rashed, Roshdi Rashed. 2007a. "The Celestial Kinematics of Ibn al-Haytham." *Arabic Sciences and Philosophy* 17: 7–55.

———2007b. "The Configuration of the Universe: A Book by al-Ḥasan ibn al-Haytham?" *Revue d'histoire des sciences* 60: 47–63.

Rawadieh, al-Mahdī ʿĪd. 2011. *Kitāb gharāʾib al-funūn wa-mulaḥ al-ʿuyūn li-muʾallif majhūl. Dirāsah wa-taḥqīq al-Mahdī ʿĪd al-Rawāḍieh.* Beirut: Dār Ṣādir.

Red Sea Pilot. 1967. *Red Sea and Gulf of Aden Pilot, Comprising the Suez Canal, the Gulfs of Suez and Aqaba, the Red Sea, the Gulf of Aden, the South-eastern Coast of Arabia from Ras Baghashwa to Ras al Hadd, the Coast of Africa from Ras Asir to Ras Hafun, Socotra and Its Adjacent Islands.* 11th ed. London: Hydrographer of the Navy.

Reinert, Stephen W. 1998. "The Muslim Presence in Constantinople, 9th–15th Centuries: Some Preliminary Observations." In *Studies on the Internal Diaspora of the Byzantine Empire*, edited by Hélène Ahrweiler and Angeliki E. Laiou. Washington, DC: Dumbarton Oaks Research Library and Collection.

Rex, Smith. 2008. *A Traveller in Thirteenth-Century Arabia: Ibn al-Mujāwir's Tārīkh al-mustabṣir.* Translated from Oscar Löfgren's Arabic text and edited with revisions and annotations by G. Rex Smith. Aldershot, UK: Ashgate.

Ridpath, Ian. 2011. *Collins Stars and Planets.* London: Harper Collins.

Rochberg-Halton, F. 1988. "Elements of the Babylonian Contribution to Hellenistic Astrology." *J. Amer. Or. Soc.* 108: 51–62.

Romm, James. 1992. *The Edges of the Earth in Ancient Thought: Geography, Exploration and Fiction.* Princeton, NJ: Princeton University Press.

———. 2010. "Continents, Climates, and Cultures: Greek Theories of Global Structure." In *Geography and Ethnography: Perceptions of the World in Pre-Modern Societies*, edited by Kurt A. Raaflaub and Richard J. A. Talbert, 215–35. Oxford: Wiley-Blackwell.

Rosenthal, F. 1961. "Al-Mubashshir ibn Fatik: Prolegomena to an Abortive Edition." *Oriens* 13–14: 132–58.

Russel, Jeffrey Burton. 1991. *Inventing the Flat Earth: Columbus and Modern Historians.* New York: Praeger.

Sabra, A. I. 1998. "One Ibn al-Haytham or Two: An Exercise in Reading the Bio-Bibiliographical Sources." *ZGAIW* 12: 1–40.

Saliba, George. 1992. "The Role of the Astrologer in Medieval Islamic Society." *Bulletin d'études orientales* 44: 45–67; reprinted in reprinted in Savage-Smith 2004, pp. 341–70.

Samarrai, Alauddin. 1993. "Beyond Belief and Reverence: Medieval Mythological Ethnography in the Near East and Europe." *Journal of Medieval and Renaissance Studies* 23, no. 1: 19–42.

Samsó, Julio. 2008. "Lunar Mansions and Timekeeping in Western Islam." *Suhayl* 8: 121–61.

Samsó, Julio Samsó, and Hamid Berrani. 1999. "World Astrology in Eleventh-Century al-Andalus: The Epistle on Tasyīr and the Projection of Rays by al-Istijjī." *Journal of Islamic Studies* 10: 293–312.

Sauvaget, Jean, ed. *Akhbār al-Ṣīn wa'l-Hind. Relation de la Chine et de l'Inde, rédigée en 851.* 1948. Traduit et commenté par Jean Sauvaget. Paris: Belles Lettres.

Savage-Smith, Emilie. 1985. *Islamicate Celestial Globes: Their History, Construction, and Use. Smithsonian Studies in History and Technology* 46. Washington, DC: Smithsonian Institution Press.

———. 1992. "Celestial Mapping." In Harley and Woodward 1992, 12–70 and plates 1–2.

———. 2003. "Memory and Maps." In *Culture and Memory in Medieval Islam. Essays in Honour of Wilferd Madelung,* edited by Farhad Daftary and Josef W. Meri, 109–27 and figs. 1–4. London: I. B. Tauris.

———, ed. 2004. *Magic and Divination in Early Islam.* The Formation of the Classical Islamic World 42. Aldershot, UK: Ashgate.

———. 2009. "Maps and Trade." In *Byzantine Trade (4th–12th centuries): The Archaeology of Local, Regional and International Exchange,* edited by M. Mango, 15–29. Farnham, UK: Ashgate.

———. 2011. "Tradition des étoiles et pratique de l'astrologie dans le Livre des curiosités." In *Image et Magie, Picatrix entre Orient et Occident,* edited by Jean-Patrice Boudet, Anna Caiozzo, and Nicolas Weill-Parot. Sciences, Techniques et Civilisations du Moyen Âge à l'Aube des Lumières, 233–51 and figs. 9–11. Collection dirigée par Danielle Jacquart et Claude Thomasset 13. Paris: Honoré Champion Éditeur.

———. 2013. "The Stars in the Bright Sky: The Most Authoritative Copy of ʿAbd al-Raḥmān al-Ṣūfī's Tenth-Century Guide to the Constellations." In *God Is Beautiful: He Loves Beauty; The Object in Islamic Art and Culture,* edited by Sheila Blair and Jonathan Bloom, 122–55. New Haven, CT: Yale University Press, in association with the Qatar Foundation, Virginia Commonwealth University, and Virginia Commonwealth University School of the Arts in Qatar.

———. 2014a. "The Universality and Neutrality of Science." In *Universality in Islamic Thought: Rationalism, Science and Religious Belief,* edited by Michael G. Morony, 153–88. London: I. B. Tauris.

———. 2014b. "Cartography." In *A Companion to Mediterranean History,* edited Peregrine Horden and Sharon Kinoshita, 184–99. Chichester, UK: Wiley-Blackwell.

Savage-Smith, Emilie, and Marion B. Smith 2004. "Islamic Geomancy and a Thirteenth-Century Divinatory Device: Another Look." In *Magic and Divination in Early Islam,* edited by E. Savage-Smith, 211–76. The Formation of the Classical Islamic World 42. Aldershot, UK: Ashgate.

Savage-Smith, Emilie, and Anna Dorothee von den Brincken. 2005. *Der mittelalterliche Kosmos: Karten der christlichen und islamischen Welt.* Translated by Thomas Ganschow. Darmstadt, Germany: Primus Verlag.

Sayılı, Aydin. 1960. *The Observatory in Islam and its Place in the General History of the Ob-*

servatory, vii, 38. Publications of the Turkish Historical Society. Ankara: Türk Tarih-Kuatumu Basimevi.

Scafi, Alessandro. 2006. *Mapping Paradise: A History of Heaven on Earth*. London: British Library.

Schmidl, Petra G. 2006. "On Timekeeping by the Lunar Mansions in Medieval Egypt." In *Proceedings of the Conference "Time and Astronomy in Past Cultures," Toruń, March 30–April 1, 2005*, edited by A. Sołtysiak, 75–87. (Warsar/Toruń).

———. 2007. *Volkstümliche Astronomie im islamischen Mittelalter: Zur Bestimmung der Gebetszeiten und der Qigla bei al-Aṣbaḥī, Ibn Raḥīq und al-Fārisī*. 2 vols. Islamic Philosophy, Theology, and Science: Texts and Studies 68. Leiden: Brill.

Schwartzberg, Joseph E. 1992a. "Part Two: South Asian Cartography." In Harley and Woodward.

———. 1992b. *A Historical Atlas of South Asia*. With the collaboration of Shiva G. Bajpai et al. 2nd impression with additional material. New York: Oxford University Press.

Seaver, Kirsten A. 2004. *Maps, Myths and Men: The Story of the Vinland Map*. Redwood City, CA: Stanford University Press.

Seed, Patricia. 2014. *The Oxford Map Companion: One Hundred Sources in World History*. New York: Oxford University Press.

Seignobos, Robin. 2017. "L'origine occidentale du Nil dans le géographie latine at arabe avant le XIVe siècle." In *Orbis disciplinae. Hommages en l'honneur de Patrick Gautier Dalché*. Textes réunis par Nathalie Bouloux, Anca Dan, et Georges Tolias, 371–94. Turnhout: Brepols.

———. 2011. "L'Île de Bilāq dans le Kitāb Nuzhat al- muštāq d'al-Idrīsī (XIIe siècle)." *Afriques: Débats, méthods et terrains d'histoire* (online). Posted February 24, 2001; accessed February 29, 2012. http://afriques.revues.org/807.

Sen, Tansen. 2003. *Buddhism, Diplomay and Trade: The Realignment of Sino-Indian relations, 600–1400*. Honolulu: University of Hawai'i Press.

Serikoff, Nikolai. 1996. "Rūmī and Yūnānī: Towards the understanding of the Greek Language in the Medieval Muslim World." In *East and West in the Crusader States: Context—Contacts—Confrontations; Acta of the Congress Held at Hernen Castle in May 1993*, edited by Krijnie Ciggaar, Adelbert Davids, and Herman Teule, 169–94. Leuven, Belgium: Uitgeverij Peters.

Seymore, Jennifer Ann. 2001. "The Life of Ibn Riḍwān and His Commentary on Ptolemy's Tetrabiblos." PhD diss., Columbia University.

Sezgin, Fuat. 1978. *Astronomie bis ca 430 H.* Geschichte des arabischen Schrifttums 6. (Sezgin GAS VI.) Leiden: Brill.

———. 1979. *Astrologie—Meteorologie und Verwandtes bis ca 430 H.* Geschichte des arabischen Schrifttums 7. (Sezgin GAS VII.) Leiden: Brill.

———. 1987. *The Contribution of the Arabic-Islamic Geographers to the Formation of the World Map*. Franfurt-am-Main: Institut für Geschichte der Arabisch-Islamischen Wissenschaften.

———. 2000a. *Mathematische Geographie und Kartographie im Islam und ihr Fortleben im Abendland: Historische Darstellung, Teil 2*. Geschichte der arabischen Schrifttums 11. (Sezgin GAS XI.) Frankfurt-am-Main: Institut für Geschichte der Arabisch-Islamischen Wissenschaften an der Johann Wolfgang Goethe-Universität.

———. 2000b. *Mathematische Geographie und Kartographie im Islam und ihr Fortleben im Abendland: Kartenband*. Geschichte der arabischen Schrifttums 12. (Sezgin GAS XII.)

Frankfurt-am-Main: Institut für Geschichte der Arabisch-Islamischen Wissenschaften an der Johann Wolfgang Goethe-Universität.

———. 2000–2007. *Mathematical Geography and Cartography in Islam and Their Continuation in the Occident.* 3 vols. English version of (vols. 10–12) of Geschichte des Arabischen Schrifttums. Frankfurt am Main: Institute for the History of Arabic-Islamic Science.

Sezgin, Ursula. 2001. "Pharaonische wunderweke bei Ibn Waṣīf Shāh und al-Masʿūdī, part III." *ZGAIW* 14.

Shafiq, Suhanna. 2013. *Seafarers of the Seven Seas: The Maritime Culture in the Kitāb ʿAjāʾib al-Hind [The Book of the Marvels of India]*, by Buzurg ibn Shahriyār (d. 399/1009). Berlin: Klaus Schwarz Verlag.

Shefer-Mossensohn, M., and K. Abou Hershkovitz. 2013. "Early Muslim Medcine and the Indian Context: A Reinterpretation." *Medieval Encounters* 19: 274–99.

Silverstein, Adam J. 2007a. *Postal Systems in the Pre-Modern Islamic World.* Cambridge: Cambridge University Press.

———. 2007b. "From Markets to Marvels: Jews on the Maritime Rout to China c. 850–c. 950 CE." *Journal of Jewish Studies* 58: 91–104.

———. 2010. "The Medieval Islamic Worldview: Arabic Geography in Its Historical Context." In *Geography and Ethnography: Perceptions of the World in Pre-Modern Societies*, edited by Kurt A. Raaflaub and Richard J. A. Talbert, 273–90. Oxford: Wiley-Blackwell.

Simon, Jesse. 2013. "Chorography Reconsidered: An Alternative Approach to Ptolemaic Precision." In *Mapping Medieval Geographies: Geographical Encounters in the Latin West and Beyond, 300–1600*, edited by Keith Lilley, 23–44. Cambridge: Cambridge University Press.

Sircar, D.C. 1971. *Studies in the Geography of Ancient and Medieval India.* Delhi: Motilal Banarsidass.

Soudan, Frédérique. 1990. "Al-Mahdiyya et son histoire d'après le récit de voyage d'al-Tiǧānī." *Revue d'études islamiques* 58: 135–88.

Stadiasmus. 1855. *Anonymi Stadiasmus sive Maris Magni, in Geographi Graeci Minores.* Edited by C. Müller. 2 vols and 1 volume of maps, 1: 427–514 and maps 19–27. Paris: A. Firmin Didot.

Starr, J. 1939. *The Jews in the Byzantine Empire, 641–1204.* Athens: Verlag der "Byzantinisch-Neugriechischen Jahrbücher."

Stern, S. M. 1955. "Heterodox Ismaʿilism at the time of al-Mu'izz." *BSOAS* 17: 10–33.

Stowasser, Barbara Freyer. 2014. *The Day Begins at Sunset: Perceptions of Time in the Islamic World.* London: I. B. Tauris.

Stylianou, Andreas, and Judith A. Stylianou. 1980. *The History of the Cartography of Cyprus.* Nicosia: Cyprus Research Centre.

Suarez, T. 1999. *Early Mapping of Southeast Asia.* Hong Kong: Periplus.

al-Ṣūfī. ʿAbd al-Raḥmān. 1954. *Ṣuwaru'l-kawākib or Uranometry (Description of the 48 Constellations), with the urjūza of Ibn u'Ṣ-Ṣūfī, based on the Ulugh Bēg Royal Codex, Arabe 5036 of the Bibliothèque Nationale, Paris.* Hyderabad: Osmania Oriental Publications Bureau.

Suhrāb. 1930. *[Kitāb ʿAjāʾib al-aqālīm al-sabʿah], Das Kitāb ʿaǧāʾib al-aḳālim as-sabʿa des Suhrāb.* Edited by Hans von Mžik. Bibliothek arabischer Historiker und Geographen 5. Leipzig: Otto Harrassowitz.

Swerdlow, Noel. 1968. "Ptolemy's Theory of the Distances and Sizes of the Planets: A Study of the Scientific Foundations of Medieval Cosmology." PhD diss., Yale University.

al-Ṭabarī. 1969. *Jāmiʿ al-bayān ʿan taʾwīl al-Qurʾān*. Edited by Muḥammad Muḥammad Shākir. 2nd ed. 16 vols. Cairo: Dār al-Maʿārif.

al-Tadmurī, ʿUmar ʿAbd al-Salām. 1978. *Taʾrīkh Ṭarābulus al-siyāsī wa-al-ḥaḍārī ʿabra al-ʿuṣūr*. Tripoli: Maṭābiʿ Dār al-Bilād.

Talbi, Mohamed Talbi. 1966. *L'Émirat aghlabide, 184–296/800–909: Histoire politique*. Paris: Librairie d'Amérique et d'Orient.

Tannery, Paul. 1920–50. *Mémoires scientifiques*. Edited by J.-L. Heiberg. 17 vols. Toulousse: Édouard Privat; Gauthier-Villars.

Thorndike, Lynn. 1950. *Latin Treatises on Comets between 1238 and 1368 A.D.* Chicago: University of Chicago Press.

Tibbetts, G. R. 1971. *Arab Navigation in the Indian Ocean before the Coming of the Portuguese*. London: Royal Asiatic Society.

———. 1979. *A Study of the Arabic Texts Containing Material on Southeast Asia*. Leiden: Brill.

———. 1992a. "The Beginnings of a Cartographic Tradition." In Harley and Woodward 1992, 90–107.

———. 1992b. "The Balkhī School of Geographers." In Harley and Woodward 1992, 108–36.

———. 1992c. "Later Cartographic Developments." In Harley and Woodward 1992, 137–55.

———. 1992d. "The Role of Charts in Islamic Navigation in the Indian Ocean." In Harley and Woodward 1992, 256–62.

al-Tijānī, Abū Muḥammad ʿAbd Allāh ibn Muḥammad ibn. 1958. *Riḥlat al-Tijānī, qāma bihā fī al-bilād al-Tūnisīyah wa-al-quṭr al-Ṭarābulusī min sanat 706 ilá sanat 708 H*. Edited by Ḥasan Ḥusnī ʿAbd al-Wahhāb. Tunis: Kitābat al-Dawlah lil-Maʿārif.

Till, W. 1936. "Eine koptische Bauernpraktik." Deutsches Institut für aegyptische Altertumskunde in Kairo. *Mitteilungen* 6: 108–49; Nachtrag 175–76.

Timotheus. 1949. *On Animals [Peri zōōn]: Fragments of a Byzantine Paraphrase of an Animal-Book of the 5th Century A. D.* Translation, commentary and introduction by F. S. Bodenheimer and A. Rabinowitz. Paris, Académie internationale d'histoire des sciences.

Tolmacheva, M. 2006. "Geography." In *Medieval Islamic Civilization: An Encyclopedia*, edited by J. Meri (New York), 1:284–88.

Toral-Niehoff, Isabel. 2004. *Kitāb Ǧiranīs: Die arabische Übersetzung der ersten Kyranis des Hermes Trismegistos und die griechischen Parallelen*. Munich: Utz.

Tourkin, Sergei. 2004. "Medical Astrology in the Horoscope of Iskandar Sulṭān." In *Sciences, techniques et instruments dans le monde iranien*, edited by N. Poujarvady and Ž. Vesel, 105–9. Tehran: IFRI.

Trimingham, J.S. 1975. "The Arab Geographers and the East African Coast." In *East Africa and the Orient*, edited by H. N. Chittick and R. I. Rotberg, 115–46. New York: Africana.

Tripathi, R. S. 1959. *History of Kanauj to the Moslem Conquest*. Delhi: M. Banarsidass.

Troupeau, G. 1954. "La 'description de la Nubie' d'al-Uswānī." *Arabica* 1: 276–88.

Udovitch, A. L. 1987. "L'énigme d'Alexandrie: Sa position au Moyen Age d'après les documents de la Geniza du Caire." *Revue de l'occident Musulman et de la Méditerranée* 46: 71–79.

———. 1993. "11th c. Islamic Treatise on the Law of the Sea." *Annales Islamologiques* 27: 37–54.

Ullmann, Manfred. 1972. *Die Natur und Geheimwissenschaften im Islam.* Handbuch der Orientalistik, Abteilung 1, Ergänzungsband vi, Abschnitt 2. Leiden: Brill.

Vallet, Éric Vallet. 2010. *L'Arabie marchande: État et commerce sous les sultans Rasūlides du Yémen, 626–858/1229–1454.* Paris: Publications de la Sorbonne.

———. 2012. "Le Périple au miroir des sources arabes médiévales: Le cas des produits du commerce." In Boussac et al. 2012, 359–80.

van Bladel, Kevin. 2009. *The Arabic Hermes: From Pagan Sage to Prophet of Science.* Oxford: Oxford University Press.

———. 2011. "The Bactrian Background of the Barmakids." In *Islam and Tibet—Interactions along the Musk Routes,* edited by Anna Akasoy, Charles Burnett, Ronit Yoeli-Tlalim, 43–88. Farnham, UK: Ashgate.

van Donzel, Emeri, and Andrea Schmidt. 2010. *Gog and Magog in Early Eastern Christian and Islamic Sources: Salam's Quest for Alexander's Wall.* With a contribution by Claudia Ott. Brill's Inner Asian Library 22. Leiden: Brill.

Varisco, Daniel M. 1989. "The Anwāʾ Stars According to Abū Isḥāq al-Zajjāj." *Zeitschrift für Geschichte der Arabischen-Islamischen Wissenschaften* 5: 145–66.

———. 1991. "The Origin of the anwāʾ in Arab Tradition." *Studia Islamica* 74: 5–28. Reprinted in Daniel M. Varisco, *Medieval Folk Astronomy and Agriculture in Arabia and the Yemen,* 1997.

———. 1994. *Medieval Agriculture and Islamic Science: The Almanac of a Yemni Sultan.* Seattle: University of Washington Press.

———. 1995. "The Magical Significance of the Lunar Stations in the 13th-Century Yemeni Kitāb al-Tabṣira fī ʿilm al-nujūm of al-Malik al-Ashraf." *Quaderni di Studi Arabi* 13: 19–40.

———. 2000. "Islamic Folk Astronomy." In *Astronomy across Cultures: The History of Non-Western Astronomy,* edited by Helaine Serin, 615–50. Dordrecht, Netherlands: Kluwer.

von Hees, Syrinx. 2005. "The Astonishing: A Critique and Re-reading of ʿAǧāʾib literature." *Middle Eastern Literatures* 8 (July): 101–20.

Walker, Paul E. 1993. "The Ismaili Daʿwa in the Reign of the Fatimid Caliph al-Ḥakim." *Journal of American Research Center in Egypt (JARCE)* 30: 161–82.

———. 1997 [2008]. "Fatimid Institutions of Learning." *JARCE* 34: 179–200; reprinted in Walker, *Fatimid History and Isamili Doctrine* (Variorum, 2008), 1–41.

Wallis Budge, E. A. 1889. *The History of Alexander the Great.* Cambridge: The University Press.

———. 1896. *The Life and Exploits of Alexander the Great.* London: C. J. Clay and Sons.

Weinstock, Stefan. 1949. "Lunar Mansions and Early Calendars." *Journal of Hellenic Studies* 69: 48–69.

Westsrem, Scott. 2011. *The Hereford Map: A Transcription and Translation of the Legends with Commentary.* Turnhout, Belgium: Brepols.

Wink, André. 1990. *Al-Hind, the Making of the Indo-Islamic world.* Vol. 1, *Early Medieval India and the Expansion of Islam, 7th–11th Centuries.* Leiden: Brill.

Woodward, David, and G. Malcolm Lewis. 1998. *The History of Cartography.* Vol. 2, *Book Three: Cartography in the Traditional African, American, Arctic, Australian, and Pacific Societies.* Chicago: University of Chicago Press.

Yāqūt. 1866–73. *[Kitāb Muʿjām al-buldān], Jacut's geographisches Wörterbuch.* Edited by Ferdinand Wüstenfeld, 6 vols. Leipzig: F. A. Brockhaus.

Zadeh, Travis. 2011. *Mapping Frontiers across Medieval Islam: Geography, Translation, and the ʿAbbāsid Empire.* London: New York: I. B. Tauris.

Zeitschrift für Geschichte der Arabisch-Islamischen Wissenschaften (ZGAIW)

Zorić, Vladimir. 1998. "La catena portuale: Sulle difese passive dei porti prima e dopo l'adozione generalizzata delle bocche da fuoco." In P*alermo medieval: Testi del'VIII colloquio medievale, Palermo 27–27 aprile 1989*, edited by C. Roccaro, 75–108. Palermo: Officina di Studi Medievali.

al-Zuhrī, Muḥammad ibn Abī Bakr. 1968. [Kitāb al-Jaʿrāfīyah], "Kitāb al-Djaʿrāfiyya." Edited by Muḥammad Hadj-Sadok. *Bulletin d'Études Orientales* 21: 7–312.

Index